T0190100

Environmental Footprints and Eco-design of Products and Processes

Series Editor

Subramanian Senthilkannan Muthu, Head of Sustainability - SgT Group and API, Hong Kong, Kowloon, Hong Kong

Indexed by Scopus

This series aims to broadly cover all the aspects related to environmental assessment of products, development of environmental and ecological indicators and eco-design of various products and processes. Below are the areas fall under the aims and scope of this series, but not limited to: Environmental Life Cycle Assessment; Social Life Cycle Assessment; Organizational and Product Carbon Footprints; Ecological, Energy and Water Footprints; Life cycle costing; Environmental and sustainable indicators; Environmental impact assessment methods and tools; Eco-design (sustainable design) aspects and tools; Biodegradation studies; Recycling; Solid waste management; Environmental and social audits; Green Purchasing and tools; Product environmental footprints; Environmental management standards and regulations; Eco-labels; Green Claims and green washing; Assessment of sustainability aspects.

Aldo Alvarez-Risco ·
Subramanian Senthilkannan Muthu ·
Shyla Del-Aguila-Arcentales
Editors

Footprint and Entrepreneurship

Global Green Initiatives

Editors
Aldo Alvarez-Risco
Universidad de Lima
Lima, Peru

Subramanian Senthilkannan Muthu
SgT Group and API
Hong Kong, Kowloon, Hong Kong

Shyla Del-Aguila-Arcentales
Escuela de Posgrado
Universidad San Ignacio de Loyola
Lima, Peru

ISSN 2345-7651 ISSN 2345-766X (electronic)
Environmental Footprints and Eco-design of Products and Processes
ISBN 978-981-19-8897-4 ISBN 978-981-19-8895-0 (eBook)
https://doi.org/10.1007/978-981-19-8895-0

This Springer imprint is published by the registered company Springer Nature Singapore Pte Ltd.
The registered company address is: 152 Beach Road, #21-01/04 Gateway East, Singapore 189721, Singapore

Preface

Entrepreneurship has been seen as an alternative for social development in the world. Various literature reports various efforts in different economic realities and social dynamics. Thus, entrepreneurship has been seen as a survival mechanism for people with limited resources or few job opportunities, often linked to rural populations or migrants to the city. It has also been seen that entrepreneurship has been the mainstay for several families related to such entrepreneurship in a small town.

Also known are the efforts made by governments and NGOs to support the development of entrepreneurship, generating training and financing programs for entrepreneurs, generating the economic development of several families involved, and generating jobs. Also, some ventures have been guided to develop sales at national and even international levels so that global entrepreneurship has also been achieved. However, little has been explicitly reported about green entrepreneurship, which is increasing due to the pandemic COVID-19 has taken a more significant role. Because of this, this book seeks to cover different conceptual aspects, supported by the broad approach to sustainability, providing concrete examples of entrepreneurship, addressing the importance of the entrepreneurship mindset, and developing topics that should be part of future research to support the further growth of green entrepreneurship in the world.

This book shows the view of authors and editors in a highly relevant and timely global change movement: green entrepreneurship. Highlighted in the book are relevant literature and academic knowledge. We expect the book to guide students, governments, teachers, universities, schools, firms, and citizens. Also, we trust that this work supports creating more entrepreneurship initiatives and can be greener, collaborating with the UN Sustainable Development Goals.

Lima, Peru — Aldo Alvarez-Risco
Hong Kong — Subramanian Senthilkannan Muthu
Lima, Peru — Shyla Del-Aguila-Arcentales

About the Authors, Reviewers, Editors, and Their Contributions

This book would not have been possible without the incredible and dedicated efforts and contributions of its many authors, who wrote chapters and often assisted in reviewing the chapters of others.

Aldo Alvarez-Risco is coauthor of all the chapters. Also, he is one of the editors and reviewers of the book.
María de las Mercedes Anderson-Seminario is the author of the first to eleventh chapters.
Camila Almanza-Cruz is coauthor of the second chapter.
Marián Arias-Meza is coauthor of the fourth chapter.
Myreya De-La-Cruz-Diaz is coauthor of the fifth and ninth chapters.
Shyla Del-Aguila-Arcentales is coauthor of all the chapters. Also, she is one of the editors and reviewers of the book.
Marco Calle-Nole is coauthor of the eleventh chapter.
Nilda Campos-Dávalos is coauthor of the third chapter.
Sarahit Castillo-Benancio is coauthor of the second and sixth chapters.
Anguie Contreras-Taica is coauthor of the eleventh chapter.
Berdy Briggitte Cuya-Velásquez is coauthor of the fourth and seventh chapters.
Sharon Esquerre-Botton is coauthor of the second and sixth chapters.
Romina Gómez-Prado is coauthor of the fourth and tenth chapters.
Micaela Jaramillo-Arévalo is coauthor of the fifth and ninth chapters.
Luis Juarez-Rojas is coauthor of the third chapter.
Luigi Leclercq-Machado is coauthor of the second and eighth chapters.
Maria F. Lenti—Dulong is coauthor of the third chapter.
Gabriel-Mauricio Martinez-Toro is coauthor of the twelveth chapter.
Flavio Morales-Rios is coauthor of the sixth chapter.
Arianne Andre Ortiz-Guerra is coauthor of the fifth chapter.
Paula Viviana Robayo-Acuña is coauthor of the twelveth chapter.
Mercedes Rojas-Osorio is coauthor of the twelveth chapter.
Jorge Sánchez-Palomino is coauthor of the tenth chapter.
Melissa Velarde-Alzamora is coauthor of the fifth chapter.

Contents

Editors and Contributors

About the Editors

Dr. Aldo Alvarez-Risco is an associate professor at Universidad de Lima in Peru. He obtained his Ph.D. from the Universidad Autónoma de Nuevo León in Mexico. Also, he is Doctor in Pharmacy and Biochemistry, Master in Pharmacology and Pharmacist at the Universidad Nacional Mayor de San Marcos in Peru, and Master in Pharmaceutical Care at Universidad de Granada in Spain. He has six years of regulatory affairs experience in pharmaceutical companies and eight years of experience in the Ministry of Health in Peru. He has published more than 50 research publications, written some book chapters, and authored/edited books in the areas of pharmaceutical care, sustainability, entrepreneurship, and circular economy. He is a lecturer in postgraduate programs since 2004 and a speaker in academic events in 21 countries.

Dr. Subramanian Senthilkannan Muthu currently works for Green Story as Chief Sustaiinability Officer, Canada, and is based out of Hong Kong. He earned his Ph.D. from the Hong Kong Polytechnic University and is a renowned expert in the areas of environmental sustainability in textiles and clothing supply chain, product life cycle assessment (LCA), and product carbon footprint assessment (PCF) in various industrial sectors. He has five years of industrial experience in textile manufacturing, research and development and textile testing and over a decade's of experience in life cycle assessment (LCA), carbon and ecological footprints assessment of various consumer products. He has published more than 100 research publications, written numerous book chapters, and authored/edited over 100 books in the areas of carbon footprint, recycling, environmental assessment, and environmental sustainability.

Prof. Shyla Del-Aguila-Arcentales is a professor researcher in Escuela de Posgrado, Universidad San Ignacio de Loyola. She is also a master in Industrial Pharmacy at Universidad Nacional Mayor de San Marcos in Peru. Also, she is a pharmacist at the

Universidad Nacional de la Amazonia Peruana in Peru. She has 12 years of experience in pharmaceutical companies' processes in manufacturing, regulatory affairs, and control quality. She has published more than 45 research publications, written some book chapters, and authored/edited books in the areas of circular economy, pharmaceutical care, management, smart cities, sustainability, and entrepreneurship. Also, she is a lecturer in undergraduate and postgraduate programs since 2010.

Contributors

Camila Almanza-Cruz Universidad de Lima, Lima, Peru

Aldo Alvarez-Risco Universidad de Lima, Lima, Peru

Marián Arias-Meza Universidad de Lima, Lima, Peru

Marco Calle-Nole Universidad de Lima, Lima, Peru

Nilda Campos-Dávalos Universidad de Lima, Lima, Peru

Sarahit Castillo-Benancio Universidad de Lima, Lima, Peru

Anguie Contreras-Taica Universidad de Lima, Lima, Peru

Berdy Briggitte Cuya-Velásquez Universidad de Lima, Lima, Peru

Myreya De-La-Cruz-Diaz Universidad de Lima, Lima, Peru

María de las Mercedes Anderson-Seminario Universidad de Lima, Lima, Peru

Shyla Del-Aguila-Arcentales Escuela de Posgrado, Universidad San Ignacio de Loyola, Lima, Peru

Sharon Esquerre-Botton Universidad de Lima, Lima, Peru

Romina Gómez-Prado Universidad de Lima, Lima, Peru

Micaela Jaramillo-Arévalo Universidad de Lima, Lima, Peru

Luis Juarez-Rojas Universidad de Lima, Lima, Peru

Francis Julca-Zamalloa Universidad de Lima, Lima, Peru

Luigi Leclercq-Machado Universidad de Lima, Lima, Peru

Gabriel-Mauricio Martinez-Toro Universidad Autónoma de Bucaramanga, Bucaramanga, Colombia

Sabina Mlodzianowska Universidad de Lima, Lima, Peru

Flavio Morales-Rios Universidad de Lima, Lima, Peru

Paula Viviana Robayo-Acuña Fundación Universitaria Konrad Lorenz, Bogota, Colombia

Mercedes Rojas-Osorio Escuela de Posgrado, Universidad San Ignacio de Loyola, Lima, Peru

Jorge Sánchez-Palomino Universidad de Lima, Lima, Peru

Creating a Green Circular Entrepreneurship Mindset in Students

María de las Mercedes Anderson-Seminario and Aldo Alvarez-Risco

Abstract Today in the world, there are still entrepreneurs who prefer a model of linear growth and consumerism, which means that the objective is to increase the production and sales volume of products in the market and their profit, which also requires a greater consumption of resources. While other entrepreneurs consider incorporating new business models, where processes that allow resource efficiency, respect for the environment, and sustainability are applied. Likewise, the OECD considers that entrepreneurs must evolve and change their values and processes because of the risks that may arise in the future due to environmental and climate challenges. How can we develop entrepreneurs with a circular and green mentality? By generating awareness in the new generations, it is possible to improve the competitiveness, efficiency, and effectiveness of the different production and consumption processes, leading to economic growth.

Keywords Entrepreneurship · Green entrepreneurship · Circular entrepreneurship · Green education · Circular · Circularity · Circularity

1 Entrepreneurship and Its Evolution

Like many concepts, entrepreneurship has been evolving, having many changes and reorientations in classical theories, modern theories of economics, and business theory but always framed in the origin of this [86]. The definition of entrepreneurship changes from a risky person and not approved for being innovative in the thirteenth and fourth centuries to becoming a political entrepreneur or for the government in the sixteenth and seventeenth centuries. Then, in the eighteenth and nineteenth centuries, it became a person who carried out some business. The most significant change occurs when the entrepreneur from a political entrepreneur becomes a market entrepreneur belonging to the private sector, such as in the agricultural sector, manufacturing, and trade, and where there is uncertainty and risk of future prices in the market [31, 86].

M. de las Mercedes Anderson-Seminario · A. Alvarez-Risco (✉)
Universidad de Lima, Lima, Perú
e-mail: aralvarez@ulima.edu.pe

A. Alvarez-Risco et al. (eds.), *Footprint and Entrepreneurship*, Environmental Footprints and Eco-design of Products and Processes,
https://doi.org/10.1007/978-981-19-8895-0_1

In 1730, the concept of entrepreneurship was defined to people who worked for the government as private suppliers of goods and services or also called at the time political entrepreneur, obtaining predetermined income, but with uncertain future costs, which is how Cantillon applies the term explaining that the word entrepreneur existed and changed the orientation from a political entrepreneur to a market entrepreneur [12, 85, 86]. New contributions were later made by Say and Mill, such as the success of an entrepreneur and his essential contribution to society and himself, and he defined the entrepreneur as a leader, who foresees, assumes risks, evaluates projects, and manages the factors of production to satisfy the needs of consumers and generate high productivity [77]. Likewise, for Say, entrepreneurship includes many economic activities, including planning, organization, supervision, innovation, and capital financing.

While Mill considers that the development of entrepreneurship is an important activity in the growth and development of a country to manage a venture requires special skills, which Marshall reinforced in 1980, who recognizes that the entrepreneur is a leader and is willing to take risks, Mill also incorporated the word "entrepreneur" into the English language with the same meaning "entrepreneurship" [31, 57]. von Mises [91] examined the concept of entrepreneurship, focusing on the capabilities of entrepreneurs in the development and management of business being for Knight: ability, effort and luck and for von Mises: evaluator, entrepreneur, and withstanding uncertainty [31, 86].

Entrepreneurship and its evolution allowed the emergence of two currents, first the Environmental School where we can consider Schumpeter as the predecessor where he says that entrepreneurship is "Entrepreneurship is essentially a creative activity, that is, an innovative function" [34]. The soul of Schumpeterian entrepreneurship is innovation, which is "The introduction of something new: a new idea, method, or device" [79]. Innovation is the key to the economic development of any company, region of the country, or the country itself, and as technologies change, existing products scale down, and old or obsolete industries shrink. Inventions and innovators are the pillars of the future of any economic unit [56]. Therefore, the Schumpeterian entrepreneurial theory explains the outbreaks of entrepreneurship, which originated a series of entrepreneurial manifestations generating the so-called Entrepreneurial Era [10].

The second current is the Psychological School, whose representatives are [90] [1966] and [45], who focus on finding the personal traits of a successful entrepreneur, in search of a pattern of characteristics that can be part of the entrepreneur or that can be optimized through education, but if they are one of the characteristics of the entrepreneurial individual. Kirzner sees the entrepreneur as a creator because he finds a viable opportunity to develop. Likewise, Kirzner states that entrepreneurial discoveries constitute "steps through which markets tend to achieve coordination, gradually replacing earlier states of generalized mutual ignorance with successively better-coordinated states of society".

In summary, we can say that entrepreneurs drive economic growth by introducing innovative products, services, methods, production, technologies, and processes that improve the productivity of the company and increase competition in the market [7] and as stated by Say is not a simple manager but also has the vision to project himself in the market [77].

2 Does Entrepreneurship Incentivize the Growth of Economies?

Institutions define rules and norms as well as reflect the heterogeneity of how and why economic activities are organized within a given structure which varies, and it is individuals and organizations that tend to comply to maintain their positions and legitimacy [5, 6, 13].

In different regions, due to the regulations, entrepreneurship has different modes [33, 81], implying that human behavior is conditioned by the institutional environment [62]; therefore, the decision to develop a new venture is also determined by institutions, within the framework of production. The way institution's structure economic benefits influence nature, effort, and entrepreneurship activities [14, 27, 59]. Formal and informal institutions incentivize individuals' behavior, which influences the evolution and behavior of entrepreneurial activity in an economy [5, 82, 88].

Entrepreneurship is attracted to institutions that provide security, such as rules related to intellectual property protection, employment, contract enforcement, market regulation, education, and others [9, 39]. Informal institutions influence entrepreneurship and trust [38, 83]. For high-impact entrepreneurship, the most important thing is an institutional environment full of new opportunities created by the knowledge and capital necessary for their implementation and development [82].

The behavior of the institutional environment generates measures that support the development and growth of a country. The economic production of a country (GDP) and GDP per capita measure the level of income and standard of living of an economy determined by investment, consumption, trade, and public spending. These variables reflect the internal demand and business opportunities to be entrepreneurship opportunities [51]. According to the literature, it is indicated that there is a direct relationship between economic condition and entrepreneurship using GDP (growth) and GDP per capita indicators [15, 50, 51]. We can analyze how the GDP has developed, which means higher income and greater demand for goods and services if it is positive, and the opposite if the indicator presents a negative figure. If the figure is positive, we can say that there are opportunities for business, employment, and higher income, which is a start for entrepreneurship [93]. Therefore, it can be considered that any country with negative indicators does not stimulate entrepreneurship.

3 Entrepreneurship in Education

Universities are the institutions that develop the competencies and skills of students and turn them into future entrepreneurs [5, 84], and having entrepreneurship training generates new businesses during their studies at university or later [1, 68]. Hence, education and training in entrepreneurship is a concern and a goal to be fulfilled for many states [53].

Universities should direct training toward local or international entrepreneurship, which possibly depends on the nature of each program. It should be emphasized that in an international venture, a high level of innovation and proactivity is required, which allow a broad reach in the international arena, and the ability of a company to internationalize through a venture depends on its ability to grow and expand internationally [20, 60]. Many firms engage in international entrepreneurship to strengthen their competitiveness [60]. International entrepreneurship is the finding, evaluating, developing, and implementing new opportunities outside one's own country [66]. As an inherently entrepreneurial process, it involves creating value across borders that determine the benefits and risks of venturing into the world, which depends on knowledge and networking as essential factors of speed and success.

Whatever the training route with local or international entrepreneurship, it should be clear that there is a close relationship between research, curriculum, and teaching [75]. The different subjects can encourage the development of innovation and entrepreneurship education through the curriculum that allows students to improve and develop a competitive graduate profile and achieve their own goals [52]. Entrepreneurship at the undergraduate level offers a different context [8] because the student is in a training stage which is directed toward self-employment and entrepreneurial awareness [80] and in the efficiency and effectiveness of entrepreneurship [1, 65, 71]. The achievement and success of an educational system with entrepreneurship training should be reflected in the intention of students to become self-employed upon graduation [5].

Entrepreneurship education is considered a dynamic process because of the acquisition of new knowledge and the perfection of skills and competencies related to the creation of enterprises and new market opportunities, which are related to pre-existing knowledge that in the end, the combination of all these allows achieving functional learning outcomes [40].

Entrepreneurship education have three objectives: (1) creating awareness of entrepreneurship, (2) preparation through knowledge, and (3) applying knowledge through practices [47]. In addition to motivating students in entrepreneurship through knowledge and cognitive and functional skills, universities must provide the necessary resources (infrastructure, technology, counseling) for students to start new businesses [71, 92]. Likewise, other educational institutions apply to learn processes in entrepreneurship through a strategic collaboration between universities and companies [94].

Here, we have an essential question: Is the learning received by students sufficient to generate growth in a country's economy? In recent decades, we must consider that companies have been created and developed as part of the integration of economies. In a globalized world, they have been driven to more significant trade in the face of lower tariff barriers, in search of lower costs and better information and communication technologies. This advance has led to an increase in transaction volumes and the development of new business models impacting ecosystems and generating problems in the world's economic, social and ecological development [3]. It means a transition from a linear model to a circular model generated by environmental problems, resource depletion, economic challenges, and social problems leading to the non-sustainability of companies and economies [35]. Therefore, there is a shift and an adaptation in entrepreneurship education that universities should provide in search of the planet's prosperity. How do future entrepreneurs develop their personal experiences in this new knowledge of circular green entrepreneurship?

4 New Business Models: Circular and Green Entrepreneurship

Increased competition, climate change, resource scarcity, price volatility, and instability have developed a global concern. This situation has led companies to adapt to new production processes, moving from a linear process to a circular one, aiming at competitiveness and respect for the environment.

The first model, the circular economy, is an alternative to the linear model, presented on the need to generate more sustainable development and balance ecological, economic, social, and environmental conditions [43]. Likewise, there is a growing current that advocates for an alternative model of processes for production and consumption, allowing the sustainability of the economy; likewise, there is the opinion of researchers who consider that it should include cooperation between different actors and generate the incentive for the development of public inventions to promote sustainable production patterns [25, 26, 35, 67, 74].

The adoption of the circular economy needs companies to shift from an approach focused on their operability to an ecosystemic vision where different actors interact dynamically to create environmental and socio-economic value [70, 95] while generating business and economic opportunities for the benefit to society. [28] consider that understanding the circular economy model is not complicated, but what is complicated is putting it into practice due to the complexity and diversity of processes that make up a company or an area of it. According to [32], the process of changing from a linear to a circular model requires three levels: macro-level (agreements with the state to reduce the impact of climate change and promote innovations at the public and private level), meso-level (policies that promote change in companies), and micro-level (economic incentives for the generation of the adoption of renewable energy and recycling models). These different levels show us that the development change is complex due to the need for radical changes at the different levels mentioned [26].

The second model, green entrepreneurship, is defined by the United Nations [87] as an efficient model in the use of resources, oriented to reducing or eliminating environmental problems and socially inclusive. In a green economy, employment and income should be fostered by public and private investment in economic and infrastructure activities that reduce carbon emissions, make improvements in energy use, generate resource efficiency, and prevent the loss of biodiversity.

Social entrepreneurship enters as a new model where economic activities are used to provide resources to the organization to achieve its social mission based on values [17]. Social entrepreneurship is born with a purpose for change that has enabled the creation of organizations that focus on addressing social and environmental problems and enable the design, development, innovation, and promotion of sectoral regulations and assistance to facilitate the transition process toward green economies green growth. Social entrepreneurship is considered one of the main contributors to the green economy. Social enterprises that work and contribute to the improvement of the environment are referred to as green (social) enterprises practicing green (social) entrepreneurship [63].

Green entrepreneurship aims to create and implement solutions to environmental problems and promote social change not to harm the environment. It is also considered that green entrepreneurship could be a new paradigm of entrepreneurship rather than entrepreneurship, as green entrepreneurs have broader motivations than just launching environmentally friendly products and services for a niche market [44]. Green or environmental entrepreneurship can be explained based on theories on entrepreneurship and the economics of environment and well-being as a critical element for successful sustainable entrepreneurship [23]. Green entrepreneurship favors those sectors where lifestyles and consumer health and safety aspects are considered very important, such as green tourism, green foods, green construction, eco-friendly fashion retailing, and others [76].

The question is born, what is the best model? Any circular model should also include a green model [67]. The circular model sees by the different processes of product reuse, remanufacturing and refurbishment, lower resource, and energy use which allows products to be more economical than conventional recycling of materials which helps to reduce negative environmental impacts and stimulates new business opportunities [46]. While the green business model initiative is essential in recognizing new green business opportunities, it has a crucial role in transforming the existing business paradigm in a more sustainable direction. Environmental and social perspectives are considered and purely economic benefits [78].

What differences do we find between a traditional entrepreneur with circular and green entrepreneurship?

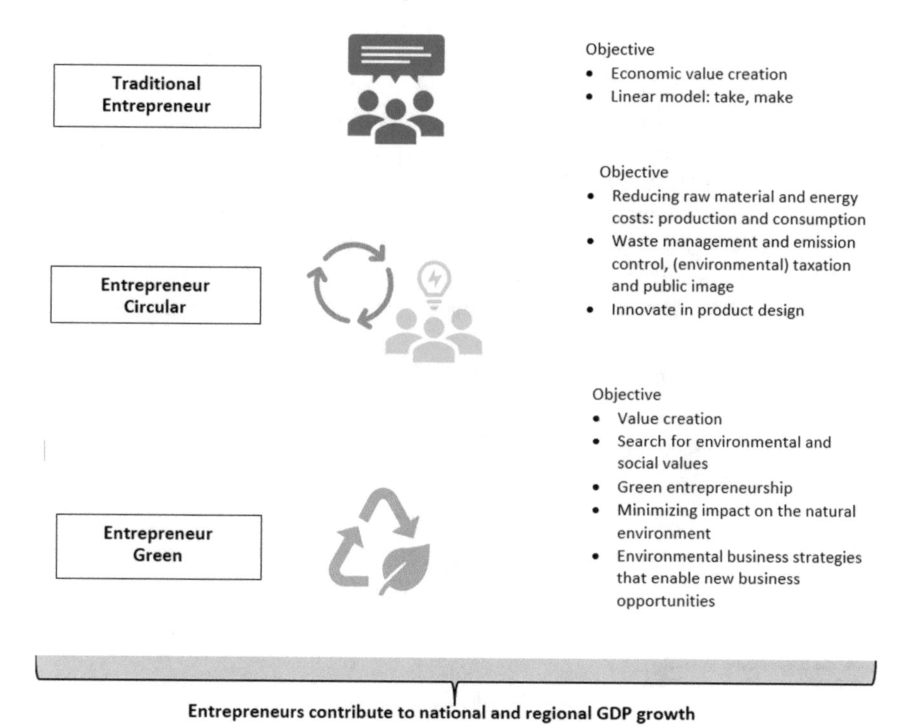

Fig. 1 Differences between a traditional entrepreneur with circular and green entrepreneurship

Green and circular entrepreneurs are important players in the business ecosystem that will help implement more sustainable production processes that will lead to sustainable consumption. Sustainable consumption and production aim to develop competitiveness based on sustainable principles that enable economic development while reducing environmental and social costs and providing workers with a better working environment and climate (Fig. 1).

5 Education and the Circular and Green Entrepreneurship Mindset

About the term "entrepreneurial mindset," [54] introduced the term as well as defined it, although the term mindset had been previously mentioned by [64]. The first definition of entrepreneurial management was "a way of thinking about business that captures the benefits of uncertainty" [54]. It is a behaviorally analyzed definition that seeks the best opportunities through disciplined, coordinated, and focused execution [69].

Likewise, [54] focused their research on successful entrepreneurs to determine how they think and behave. After 20 years, [49] took up McGrath and MacMillan's idea and added a third question: How do they feel? The three questions led [49] to define the proposed macro-vision of the entrepreneur's mindset, composed of the cognitive, behavioral, and emotional aspects. These three elements comprehensively define entrepreneurial management, differentiating it from other entrepreneurs [54]. Thus, [69] developed a conceptual model that develops it under the idea that the mindset is not born from nothing, but it is essential to consider the convictions or expectations of opportunity that trigger the cognitive responses of entrepreneurs.

Entrepreneurial mindset conceptual model

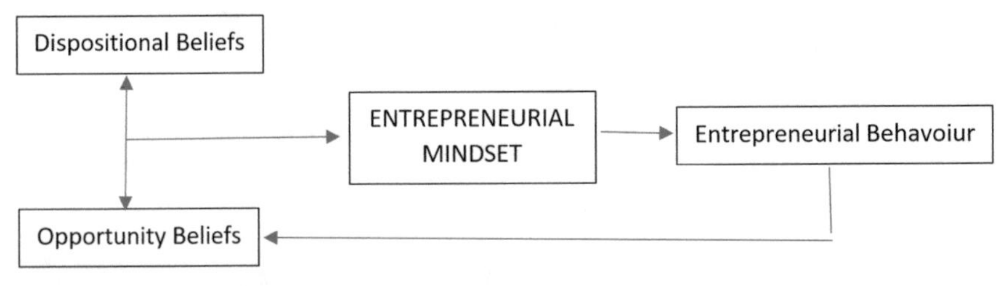

Source Adapted from [69]

There is a low level of research even though there is broad agreement that the entrepreneur's mindset is a vital mechanism in the entrepreneurial process that stimulates recognizing opportunities and creates new companies and opportunities [48]. But also, [61] makes known that despite the limitations in research and the lack of understanding of the concept itself, two currents have been developed which together allow understanding the mindset of the entrepreneur does because it is the foundation of what it is and the second current claims to be a cognitive ability, a way of thinking of the person. It can be concluded that people either have or do not have an entrepreneurial mindset, but it is also said that the entrepreneurial mindset is a strategic cognitive response generated by a perceived opportunity stimulus [37, 69] which is based on knowledge and experience.

The entrepreneurial mindset is based on the opportunity by goal-oriented entrepreneurial behavior produced by its conceptualization and when the value of autonomy, proactivity, innovativeness, competitiveness, and risk-taking are integrated with specific opportunity beliefs [69]. In some people, these traits are carried with them, and therefore the entrepreneurial mindset is easy to develop, while other individuals do not have these traits developed, and therefore, developing the entrepreneurial mindset requires more effort or is not achieved. According to [55], the entrepreneurial mindset recognizes an opportunity formed in response to the environment in which the individual develops in which he or she perceives, recognizes, and believes that an opportunity exists, representing a cognitive response to an environmental stimulus. Therefore, the entrepreneurial mindset can be developed or improved over time through information, knowledge, and experiences [69]. We can then say that all expectations and trends related to the opportunities found in a market and all strategic thinking in response to external factors can be developed through training.

The universities should generate knowledge, skills, and good practices, aiming to train professionals in production, innovation, consumption, and management strategies of sustainable models and generate green and circular entrepreneurs from them [11]. In the literature reviewed, some authors mention the importance of entrepreneurship as the requirement of entrepreneurs who develop sustainable business ideas and who can find and develop business opportunities, turning them into a product or service to be marketed in the national and international market. Incorporating strategies and sound environmental practices in the company promotes the development of new products, innovations, efficiency in using resources, and generates greater productivity [72].

Teaching methods are essential to make entrepreneurship studies more interactive, proactive, and practical. Thus, universities should develop or reform their curricula and incorporate subjects or academic areas related to entrepreneurship. The subjects in addition to entrepreneurship training courses should be complemented with the practice of entrepreneurship, entrepreneurship and the soft skills required by an entrepreneur [52]. Thus, the participation of students and teachers in green and circular business plans should be generated, as well as the development of practical training in start-ups and existing companies. Other methodologies that allow students to develop their skills and knowledge are case studies, business simulation games, and prototypes of green circular products. All these methodologies and others to be developed should be part of their subjects to promote the development and mindset of entrepreneurship [29, 52, 89].

Likewise, green and circular entrepreneurship should be a knowledge that should be transversal throughout a curriculum and remain in a specific subject and should be part of the subject matter of compulsory subjects such as specialization. The curriculum should inculcate through its subjects the importance of change toward sustainable development objectives and thus train them to create and manage future companies with a responsible attitude toward the environment. Just becoming aware of the necessary change is not easy; currently, entrepreneurs require adopting new sustainable practices or models. They lack experience, knowledge, and capabilities to incorporate green and circular models [42]. New companies entering the market should consider good environmental practices and sustainable models from the beginning and have professionals with the skills to develop and generate appropriate environmental strategies about production and communication strategies and consumption toward consumers [73].

Recent suggested lectures about green approach

- Waste reduction and carbon footprint [36]
- Material selection for circularity and footprints [58]
- 3D print, circularity, and footprints [21]
- Virtual tourism [21]

- Leadership for sustainability in crisis time [2]
- Virtual education and circularity [18]
- Circular economy for packaging [16]
- Students oriented to circular learning [22]
- Food and circular economy [19]
- Water footprint and food supply chain management [41]
- Waste footprint [4]
- Measuring circular economy [24]
- Carbon footprint [30]

6 Closing Remarks

Companies require an entrepreneurial mindset that enables changes, transformation, innovation, competitiveness, and management strategies that allow a circular green production and consumption. As time progresses, environmental values regain importance in society, and consumers in their different ages change their perception or learn of its importance and levels of need and satisfaction, which allow companies to change and assume new models and sustainable business practices to satisfy society. Universities should focus their study programs or reinforce them to generate a more significant development of the mentality of an entrepreneur and channel it to the development of new projects that generate growth and welfare of the economy of a country, such as managing companies in a responsible way about the environment.

References

1. Abd Razak NSN, Buang NA, Kosnin H (2018) The influence of entrepreneurship education towards the entrepreneurial intention in 21st century learning. J Soc Sc Res 502–507:506
2. Alvarez-Risco A, Del-Aguila-Arcentales S, Villalobos-Alvarez D, Diaz-Risco S (2022) Leadership for sustainability in crisis time. In: Alvarez-Risco A, Muthu SS, Del-Aguila-Arcentales S (eds) Circular economy: impact on carbon and water footprint. Springer Singapore, pp 41–64. https://doi.org/10.1007/978-981-19-0549-0_3
3. Anderson-Seminario M, Alvarez-Risco A (2020) Circular economy and pricing strategies [Economía circular y las estrategias de precios]. Retrieved 01/01/2022 from https://repositorio.ulima.edu.pe/handle/20.500.12724/13075
4. Arias-Meza M, Alvarez-Risco A, Cuya-Velásquez BB, de las Mercedes Anderson-Seminario M, Del-Aguila-Arcentales S (2022) Fashion and textile circularity and waste footprint. In: Alvarez-Risco A, Muthu SS, Del-Aguila-Arcentales S (eds) Circular economy: impact on carbon and water footprint. Springer Singapore, pp 181–204. https://doi.org/10.1007/978-981-19-0549-0_9
5. Ayob AH (2021) Institutions and student entrepreneurship: the effects of economic conditions, culture and education. Educ Stud 47(6):661–679. https://doi.org/10.1080/03055698.2020.1729094
6. Baumol WJ (1996) Entrepreneurship: productive, unproductive, and destructive. J Bus Ventur 11(1):3–22. https://doi.org/10.1016/0883-9026(94)00014-X
7. Bedarkar M, Deokar B (2021) Venture capital and entrepreneurship

8. Bergmann H, Hundt C, Sternberg R (2016) What makes student entrepreneurs? On the relevance (and irrelevance) of the university and the regional context for student start-ups. Small Bus Econ 47(1):53–76. https://doi.org/10.1007/s11187-016-9700-6
9. Bjørnskov C, Foss N (2013) How strategic entrepreneurship and the institutional context drive economic growth. Strateg Entrep J 7(1):50–69. https://doi.org/10.1002/sej.1148
10. Bodrožić Z, Adler PS (2018) The evolution of management models: a neo-Schumpeterian theory. Adm Sci Q 63(1):85–129. https://doi.org/10.1177/0001839217704811
11. Bonaccorsi A, Colombo MG, Guerini M, Rossi-Lamastra C (2013) University specialization and new firm creation across industries. Small Bus Econ 41(4):837–863. https://doi.org/10.1007/s11187-013-9509-5
12. Brown C, Thornton M (2013) How entrepreneurship theory created economics. Q J Austrian Econ 16(4)
13. Bruton GD, Ahlstrom D, Li HL (2010) Institutional theory and entrepreneurship: where are we now and where do we need to move in the future? Entrep Theory Pract 34(3):421–440. https://doi.org/10.1111/j.1540-6520.2010.00390.x
14. Calcagno PT, Sobel RS (2014) Regulatory costs on entrepreneurship and establishment employment size. Small Bus Econ 42(3):541–559. https://doi.org/10.1007/s11187-013-9493-9
15. Carree M, Van Stel A, Thurik R, Wennekers S (2002) Economic development and business ownership: an analysis using data of 23 OECD countries in the period 1976–1996. Small Bus Econ 19(3):271–290. https://doi.org/10.1023/A:1019604426387
16. Castillo-Benancio S, Alvarez-Risco A, Esquerre-Botton S, Leclercq-Machado L, Calle-Nole M, Morales-Ríos F, … Del-Aguila-Arcentales S (2022) Circular economy for packaging and carbon footprint. In: Alvarez-Risco A, Muthu SS, Del-Aguila-Arcentales S (eds) Circular economy: impact on carbon and water footprint. Springer Singapore, pp 115–138. https://doi.org/10.1007/978-981-19-0549-0_6
17. Chichevaliev S, Ortakovski T (2020) Social entrepreneurship-the much-needed accelerator of the modern "green" economy. AICEI Proceedings 15(1):29–38. https://doi.org/10.5281/zenodo.4393532
18. Contreras-Taica A, Alvarez-Risco A, Arias-Meza M, Campos-Dávalos N, Calle-Nole M, Almanza-Cruz C, … Del-Aguila-Arcentales S (2022) Virtual education: carbon footprint and circularity. In: Alvarez-Risco A, Muthu SS, Del-Aguila-Arcentales S (eds) Circular economy: impact on carbon and water footprint. Springer Singapore, pp 265–285. https://doi.org/10.1007/978-981-19-0549-0_13
19. Cuya-Velásquez BB, Alvarez-Risco A, Gomez-Prado R, Juarez-Rojas L, Contreras-Taica A, Ortiz-Guerra A, … Del-Aguila-Arcentales S (2022) Circular economy for food loss reduction and water footprint. In: Alvarez-Risco A, Muthu SS, Del-Aguila-Arcentales S (eds) Circular economy: impact on carbon and water footprint. Springer Singapore, pp 65–91, https://doi.org/10.1007/978-981-19-0549-0_4
20. Dai L, Maksimov V, Gilbert BA, Fernhaber SA (2014) Entrepreneurial orientation and international scope: the differential roles of innovativeness, proactiveness, and risk-taking. J Bus Ventur 29(4):511–524. https://doi.org/10.1016/j.jbusvent.2013.07.004
21. De-la-Cruz-Diaz M, Alvarez-Risco A, Jaramillo-Arévalo M, de las Mercedes Anderson-Seminario M, Del-Aguila-Arcentales S (2022) 3D print, circularity, and footprints. In: Alvarez-Risco A, Muthu SS, Del-Aguila-Arcentales S (eds) Circular economy: impact on carbon and water footprint. Springer Singapore, pp 93–112. https://doi.org/10.1007/978-981-19-0549-0_5
22. de las Mercedes Anderson-Seminario M, Alvarez-Risco A (2022) Better students, better companies, better life: circular learning. In: Alvarez-Risco A, Muthu SS, Del-Aguila-Arcentales S (eds) Circular economy: impact on carbon and water footprint. Springer Singapore, pp 19–40. https://doi.org/10.1007/978-981-19-0549-0_2
23. Dean TJ, McMullen JS (2007) Toward a theory of sustainable entrepreneurship: reducing environmental degradation through entrepreneurial action. J Bus Ventur 22(1):50–76. https://doi.org/10.1016/j.jbusvent.2005.09.003
24. Del-Aguila-Arcentales S, Alvarez-Risco A, Muthu SS (2022) Measuring circular economy. In: Alvarez-Risco A, Muthu SS, Del-Aguila-Arcentales S (eds) Circular economy: impact on

carbon and water footprint. Springer Singapore, pp 3–17. https://doi.org/10.1007/978-981-19-0549-0_1
25. del Mar Alonso-Almeida M, Perramon J, Bagur-Femenias L (2020) Shedding light on sharing ECONOMY and new materialist consumption: An empirical approach. J Retail Consum Serv 52:101900. https://doi.org/10.1016/j.jretconser.2019.101900
26. del Mar Alonso-Almeida M, Rodriguez-Anton JM, Bagur-Femenías L, Perramon J (2021) Institutional entrepreneurship enablers to promote circular economy in the European Union: impacts on transition towards a more circular economy. J Clean Prod 281:124841. https://doi.org/10.1016/j.jclepro.2020.124841
27. Dilli S, Elert N, Herrmann AM (2018) Varieties of entrepreneurship: exploring the institutional foundations of different entrepreneurship types through 'Varieties-of-Capitalism' arguments. Small Bus Econ 51(2):293–320. https://doi.org/10.1007/s11187-018-0002-z
28. Domenech T, Bahn-Walkowiak B (2019) Transition towards a resource efficient circular economy in Europe: policy lessons from the EU and the member states. Ecol Econ 155:7–19. https://doi.org/10.1016/j.ecolecon.2017.11.001
29. Donnellon A, Ollila S, Middleton KW (2014) Constructing entrepreneurial identity in entrepreneurship education. Int J Manage Educ 12(3):490–499. https://doi.org/10.1016/j.ijme.2014.05.004
30. Esquerre-Botton S, Alvarez-Risco A, Leclercq-Machado L, de las Mercedes Anderson-Seminario M, Del-Aguila-Arcentales S (2022) Food loss reduction and carbon footprint practices worldwide: a benchmarking approach of circular economy. In: Alvarez-Risco A, Muthu SS, Del-Aguila-Arcentales S (eds) Circular economy: impact on carbon and water footprint. Springer Singapore, pp 161–179. https://doi.org/10.1007/978-981-19-0549-0_8
31. Fernández-Salinero C, de la Riva B (2014) Entrepreneurial mentality and culture of entrepreneurship. Procedia Soc Behav Sci 139:137–143. https://doi.org/10.1016/j.sbspro.2014.08.044
32. Florido C, Jacob M, Payeras M (2019) How to carry out the transition towards a more circular tourist activity in the hotel sector. the role of innovation. Adm Sci 9(2). https://doi.org/10.3390/admsci9020047
33. Foster G, Shimizu C, Ciesinski S, Davila A, Hassan S, Jia N, Morris R (2013) Entrepreneurial ecosystems around the globe and company growth dynamics. World Economic Forum
34. Fritsch M (2017) The theory of economic development–an inquiry into profits, capital, credit, interest, and the business cycle. Reg Stud 51(4):654–655. https://doi.org/10.1080/00343404.2017.1278975
35. Geissdoerfer M, Savaget P, Bocken NM, Hultink EJ (2017) The circular economy–a new sustainability paradigm? J Clean Prod 143:757–768. https://doi.org/10.1016/j.jclepro.2016.12.048
36. Gómez-Prado R, Alvarez-Risco A, Sánchez-Palomino J, de las Mercedes Anderson-Seminario M, Del-Aguila-Arcentales S (2022) Circular economy for waste reduction and carbon footprint. In: Alvarez-Risco A, Muthu SS, Del-Aguila-Arcentales S (eds) Circular economy: impact on carbon and water footprint. Springer Singapore, pp 139–159. https://doi.org/10.1007/978-981-19-0549-0_7
37. Haynie JM, Shepherd D, Mosakowski E, Earley PC (2010) A situated metacognitive model of the entrepreneurial mindset. J Bus Ventur 25(2):217–229. https://doi.org/10.1016/j.jbusvent.2008.10.001
38. Hechavarria DM, Reynolds PD (2009) Cultural norms & business start-ups: the impact of national values on opportunity and necessity entrepreneurs. Int Entrepreneurship Manage J 5(4):417–437. https://doi.org/10.1007/s11365-009-0115-6
39. Henrekson M, Johansson D (2008) Competencies and institutions fostering high-growth firms. Now Publishers Inc.
40. Holcomb TR, Ireland RD, Holmes RM Jr, Hitt MA (2009) Architecture of entrepreneurial learning: exploring the link among heuristics, knowledge, and action. Entrep Theory Pract 33(1):167–192. https://doi.org/10.1111/j.1540-6520.2008.00285.x

41. Juarez-Rojas L, Alvarez-Risco A, Campos-Dávalos N, de las Mercedes Anderson-Seminario M, Del-Aguila-Arcentales S (2022) Water footprint in the textile and food supply chain management: trends to become circular and sustainable. In: Alvarez-Risco A, Muthu SS, Del-Aguila-Arcentales S (eds) Circular economy: impact on carbon and water footprint. Springer Singapore, pp 225–243. https://doi.org/10.1007/978-981-19-0549-0_11
42. Kirchherr J, Piscicelli L, Bour R, Kostense-Smit E, Muller J, Huibrechtse-Truijens A, Hekkert M (2018) Barriers to the circular economy: evidence from the European Union (EU). Ecol Econ 150:264–272. https://doi.org/10.1016/j.ecolecon.2018.04.028
43. Kirchherr J, Reike D, Hekkert M (2017) Conceptualizing the circular economy: an analysis of 114 definitions. Resour Conserv Recycl 127:221–232. https://doi.org/10.1016/j.resconrec.2017.09.005
44. Kirkwood J, Walton S (2010) What motivates ecopreneurs to start businesses? Int J Entrep Behav Res. https://doi.org/10.1108/13552551011042799
45. Kirzner I (1973) Competition and entrepreneurship. Chicago. Univ. In: Of Chicago Press
46. Korhonen J, Honkasalo A, Seppälä J (2018) Circular economy: the concept and its limitations. Ecol Econ 143:37–46. https://doi.org/10.1016/j.ecolecon.2017.06.041
47. Kourilsky ML, Carlson S (2000) Entrepreneurship education for youth: a curricular perspective. Entrepreneurship:193–213
48. Krueger NF Jr (2007) What lies beneath? The experiential essence of entrepreneurial thinking. Entrep Theory Pract 31(1):123–138. https://doi.org/10.1111/j.1540-6520.2007.00166.x
49. Kuratko DF, Fisher G, Audretsch DB (2021) Unraveling the entrepreneurial mindset. Small Bus Econ 57(4):1681–1691. https://doi.org/10.1007/s11187-020-00372-6
50. Lee SM, Peterson SJ (2000) Culture, entrepreneurial orientation, and global competitiveness. J World Bus 35(4):401–416. https://doi.org/10.1016/S1090-9516(00)00045-6
51. Liñán F, Fernandez-Serrano J (2014) National culture, entrepreneurship and economic development: different patterns across the European Union. Small Bus Econ 42(4):685–701. https://doi.org/10.1007/s11187-013-9520-x
52. Liu R, Huo Y, He J, Zuo D, Qiu Z, Zhao J (2021) The effects of institution-driven entrepreneurial education in Chinese universities: a qualitative comparative analysis approach. Front Psychol 12:719476–719476. https://doi.org/10.3389/fpsyg.2021.719476
53. Martin BC, McNally JJ, Kay MJ (2013) Examining the formation of human capital in entrepreneurship: a meta-analysis of entrepreneurship education outcomes. J Bus Ventur 28(2):211–224. https://doi.org/10.1016/j.jbusvent.2012.03.002
54. McGrath RG, MacMillan IC (2000) The entrepreneurial mindset: strategies for continuously creating opportunity in an age of uncertainty, vol 284. Harvard Business Press
55. McMullen JS, Kier AS (2016) Trapped by the entrepreneurial mindset: opportunity seeking and escalation of commitment in the Mount Everest disaster. J Bus Ventur 31(6):663–686. https://doi.org/10.1016/j.jbusvent.2016.09.003
56. Mehmood T, Alzoubi HM, Alshurideh M, Al-Gasaymeh A, Ahmed G (2019) Schumpeterian entrepreneurship theory: evolution and relevance. Acad Entrepreneurship J 25(4):1–10
57. Mill JS, Ashley SWJ, Ortíz T (1951) Principles of political economy: with some of its applications to social philosophy [Principios de economía política: con algunas de sus aplicaciones a la filosofía social]
58. Morales-Ríos F, Alvarez-Risco A, Castillo-Benancio S, de las Mercedes Anderson-Seminario M, Del-Aguila-Arcentales S (2022) Material selection for circularity and footprints. In: Alvarez-Risco A, Muthu SS, Del-Aguila-Arcentales S (eds) Circular economy: impact on carbon and water footprint . Springer Singapore, pp 205–221. https://doi.org/10.1007/978-981-19-0549-0_10
59. Murphy KM, Shleifer A, Vishny RW (1991) The allocation of talent: Implications for growth. Q J Econ 106(2):503–530. https://doi.org/10.2307/2937945
60. Narula R, Hagedoorn J (1999) Innovating through strategic alliances: moving towards international partnerships and contractual agreements. Technovation 19(5):283–294. https://doi.org/10.1016/S0166-4972(98)00127-8

61. Naumann C (2017) Entrepreneurial mindset: a synthetic literature review. Entrep Bus Econ Rev 5(3):149–172
62. North DC (2018) Institutional change: a framework of analysis. In: Social Rules. Routledge, pp 189–201
63. O'Neill K, Gibbs D (2016) Rethinking green entrepreneurship–Fluid narratives of the green economy. Environ Plann A: Econ Space 48(9):1727–1749. https://doi.org/10.1177/0308518X16650453
64. Osborne RL (1995) The essence of entrepreneurial success. Manag Decis. https://doi.org/10.1108/00251749510090520
65. Othman E, Yusoff MS, Aziz HA, Adlan MN, Bashir MJ, Hung Y-T (2010) The effectiveness of silica sand in semi-aerobic stabilized landfill leachate treatment. Water 2(4):904–915. https://doi.org/10.3390/w2040904
66. Oviatt BM, McDougall PP (2005) Defining international entrepreneurship and modeling the speed of internationalization. Entrep Theory Pract 29(5):537–553. https://doi.org/10.1111/j.1540-6520.2005.00097.x
67. Pattanaro G, Gente V (2017) Circular economy and new ways of doing business in the tourism sector. Eur J Serv Manage 21:45–50
68. Pfeifer S, Šarlija N, Zekić Sušac M (2016) Shaping the entrepreneurial mindset: entrepreneurial intentions of business students in Croatia. J Small Bus Manage 54(1):102–117
69. Pidduck RJ, Clark DR, Lumpkin G (2021) Entrepreneurial mindset: Dispositional beliefs, opportunity beliefs, and entrepreneurial behavior. J Small Bus Manage1–35. https://doi.org/10.1080/00472778.2021.190758
70. Pieroni MP, McAloone TC, Pigosso DC (2019) Business model innovation for circular economy and sustainability: a review of approaches. J Clean Prod 215:198–216. https://doi.org/10.1016/j.jclepro.2019.01.036
71. Politis D (2005) The process of entrepreneurial learning: a conceptual framework. Entrep Theory Pract 29(4):399–424. https://doi.org/10.1111/j.1540-6520.2005.00091.x
72. Porter M, Van der Linde C (1995) . The Dynamics of the eco-efficient economy: environmental regulation and competitive advantage. Green Competitive: Ending Stalemate:33
73. Qomariah A, Prabawani B (2020) The effects of environmental knowledge, environmental concern, and green brand image on green purchase intention with perceived product price and quality as the moderating variable. In: IOP conference series: earth and environmental science, vol 448. IOP Publishing, pp 012115. https://doi.org/10.1088/1755-1315/448/1/012115
74. Reike D, Vermeulen WJ, Witjes S (2018) The circular economy: new or refurbished as CE 3.0?—exploring controversies in the conceptualization of the circular economy through a focus on history and resource value retention options. Resour Conserv Recycl 135:246–264. https://doi.org/10.1016/j.resconrec.2017.08.027
75. Ronstadt R (1985) The educated entrepreneurs: A new era of entrepreneurial education is beginning. Am J Small Bus 10(1):7–23. https://doi.org/10.1177/104225878501000102
76. Saari U, Joensuu-Salo S (2019) Green entrepreneurship. Responsible consumption and production. https://doi.org/10.1007/978-3-319-71062-4_6-1
77. Say J (2001) Treatise on Political Economy [Tratado de Economía Política]. Fondo de Cultura Económica
78. Schaper M (2016) Understanding the green entrepreneur. In: Making ecopreneurs . Routledge, pp 27–40
79. Schumpeter JA (2013) Economic theory and entrepreneurial history. Harvard University Press
80. Shirokova G, Osiyevskyy O, Bogatyreva K (2016) Exploring the intention–behavior link in student entrepreneurship: moderating effects of individual and environmental characteristics. Eur Manag J 34(4):386–399. https://doi.org/10.1016/j.emj.2015.12.007
81. Stam E (2014) The Dutch entrepreneurial ecosystem. Available at SSRN 2473475
82. Stenholm P, Acs ZJ, Wuebker R (2013) Exploring country-level institutional arrangements on the rate and type of entrepreneurial activity. J Bus Ventur 28(1):176–193. https://doi.org/10.1016/j.jbusvent.2011.11.002

83. Taylor MZ, Wilson S (2012) Does culture still matter?: The effects of individualism on national innovation rates. J Bus Ventur 27(2):234–247. https://doi.org/10.1016/j.jbusvent.2010.10.001
84. Thai MTT, Turkina E (2014) Macro-level determinants of formal entrepreneurship versus informal entrepreneurship. J Bus Ventur 29(4):490–510. https://doi.org/10.1016/j.jbusvent.2013.07.005
85. Thornton M (2007) Cantillon, Hume, and the rise of antimercantilism. Hist Polit Econ 39(3):453–480. https://doi.org/10.1215/00182702-2007-018
86. Thornton M (2020) Turning the word upside down: how cantillon redefined the entrepreneur. Q J Austrian Econ 23(3–4), 265–280. https://doi.org/10.35297/qjae.010071
87. United Nations (2012) Future We Want. Retrieved 01/01/2022 from https://sustainabledevelopment.un.org/content/documents/733FutureWeWant.pdf
88. Urbano D, Alvarez C (2014) Institutional dimensions and entrepreneurial activity: an international study. Small Bus Econ 42(4):703–716. https://doi.org/10.1007/s11187-013-9523-7
89. Verzat C, O'Shea N, Jore M (2017) Teaching proactivity in the entrepreneurial classroom. Entrep Reg Dev 29(9–10):975–1013. https://doi.org/10.1080/08985626.2017.1376515
90. von Mises L (1949 [1966]) Human action: a treatise on economics
91. von Mises R (1951) Positivism: Study in human understanding. In: Harvard University Press
92. Walter SG, Parboteeah KP, Walter A (2013) University departments and self–employment intentions of business students: A cross–level analysis. Entrep Theory Pract 37(2):175–200. https://doi.org/10.1111/j.1540-6520.2011.00460.x
93. Wennekers S, Van Wennekers A, Thurik R, Reynolds P (2005) Nascent entrepreneurship and the level of economic development. Small Bus Econ 24(3):293–309. https://doi.org/10.1007/s11187-005-1994-8
94. Whitley R (1999) Divergent capitalisms: The social structuring and change of business systems. OUP Oxford
95. Zucchella A, Previtali P (2019) Circular business models for sustainable development: a "waste is food" restorative ecosystem. Bus Strateg Environ 28(2):274–285. https://doi.org/10.1002/bse.2216

Green Entrepreneurship—Added Value as a Strategic Orientation Business Model

Sarahit Castillo-Benancio, Aldo Alvarez-Risco, Camila Almanza-Cruz, Luigi Leclercq-Machado, Sharon Esquerre-Botton, María de las Mercedes Anderson-Seminario, and Shyla Del-Aguila-Arcentales

Abstract Traditional entrepreneurship focuses on a take-make-waste strategy to deliver goods on the market. Nonetheless, environmental and social concerns foment the necessity to adopt a new perspective of producing and selling products and services. It is worth mentioning that current businesses are constantly competing for differentiation methods to attract clients. This chapter aims to outline the importance of incorporating green strategies into the supply chain of entrepreneurship to deliver value. Also, we discuss the potential classification of green entrepreneurship according to technological opportunities and their business model. Finally, green innovation among the different subsystems of the value chain of entrepreneurship is crucial for the effective implementation of sustainable development. The findings show the current practices of businesses putting several green initiatives through technology and innovation. Nonetheless, there is a gap between green entrepreneurship and public and government awareness of these venture capitals. It is recommended that governments promote green business practices and ensure a suitable environment for entrepreneurs' stability.

Keywords Green · Ecology · Sustainable · Circular · SDG · Entrepreneur · Entrepreneurship · Business · International business

1 Introduction

As a consequence of climate problems growing continuously day by day, the necessity of creating and implementing new green technologies is increasing, too [123]. In the near past, innovations and perspectives have been focused on sustainable development, seeking the improvement of technologies and practices adopted by society [67].

S. Castillo-Benancio · A. Alvarez-Risco · C. Almanza-Cruz · L. Leclercq-Machado · S. Esquerre-Botton · M. de las Mercedes Anderson-Seminario
Universidad de Lima, Lima, Peru

S. Del-Aguila-Arcentales (✉)
Escuela de Posgrado, Universidad San Ignacio de Loyola, Lima, Peru
e-mail: sdelaguila@usil.edu.pe

A. Alvarez-Risco et al. (eds.), *Footprint and Entrepreneurship*, Environmental Footprints and Eco-design of Products and Processes,
https://doi.org/10.1007/978-981-19-8895-0_2

Nowadays, companies from diverse industries have to employ up-to-date methodologies, considering "environmental, social, and economic issues" for enterprise R&D [75]. Green technology models are created mainly to optimize energy use in all processes involved in creating, generating, and delivering a product or service. Likewise, when talking about sustainable and innovative products, they are offered to reduce the environmental impact, but instead of the operations and management, it is referred to their consumption [75, 108].

Economic, social, and, above all, environmental sustainability is of profound importance to people [4, 6, 8, 16, 28, 38–41, 50, 57, 81, 117, 138, 154, 191, 223, 228, 237] institutions [17, 21, 33, 34, 65, 69, 70, 145, 150, 180, 197, 205, 217, 228, 231, 232] and firms [7, 11, 27, 55, 60, 92, 101, 153, 181, 185, 208] and activities as tourism [5, 12, 76–78, 168, 178, 187], education[8, 17, 26, 30, 34, 36, 80, 186] circular economy approach [1, 19, 35, 59, 104, 139, 148, 152], prices [49, 52, 122, 126, 146, 165, 202], hospitality [13, 112, 119, 132, 155, 177, 227], intellectual property [2, 20, 25, 48, 58, 111], health [3, 9, 10, 18, 22–24, 29, 31, 42–44, 71, 84, 93, 104, 110, 159, 184, 192, 229, 230, 238–240], and research [87, 88, 140, 173, 182, 188].

In addition, green technology implemented by different brands has stimulated strategies that lead to "competitive advantages and environmentally sustainable value chains" [216]. This chapter analyzes green entrepreneurship business models, where they come from, how they work, how many are, and their value to traditional models, operational processes, management aspects, and logistic chains.

2 Entrepreneurship Traditional Approach

Entrepreneurship as a concept is an activity in which new companies are created, and existing ones are repowered and are expected to expand in the long term in the global market [109]. The policies currently being developed aim to increase the development of the internationalization of companies in which the public and private sectors have an essential role to play [46]. There is a relationship between the global productive functioning of nations, the economy, and the formation of an entrepreneurial culture that is important for the country's development [109].

Entrepreneurs recognize an opportunity being translated into a business and assume the risks that may occur within the life cycle of the enterprise [109]. Two fundamental functions fall to these people mentioned above; the first is to identify untapped profit opportunities, where the traditional model of having a purely economic outlook and deficiency in the technological field, the second is to be consistent in innovating with an emphasis on a production where improvement is needed because it is essential to provide a different value proposition to that offered by the market [163]. That is why companies with this model still in force have a linear process where they make high demands on natural resources, which are governed by extracting, manufacturing, consuming, and disposing of them, thus generating pollution that only harms humanity [156].

According to [210], in 2020, 61% of start-ups offered B2B solutions, a process of selling products or services to other companies, and 39% had a form of a business transaction directly with the end consumer. New companies worldwide are created, and others fail, many of them originating from China or the United States, but the development of the strategic plan and the business model differentiates them for success [174].

The different realities of countries can trigger the formation of a limiting factor for entrepreneurship with a traditional model, which can bear economic, social, or technological development [109]. The adaptation of new market scenarios that the economy demands is one of the challenges that must be faced. In this area, the issue of innovation must be addressed because it is necessary for an organization, especially when the trend is toward the employability of new technology seeking quality and efficiency.

3 Technology and Green Entrepreneurship Business Models

3.1 Sustainable Innovation and Green Technological Opportunities

For green technology, sustainability has been developed as a basis, emerging from the growing awareness and responsibility in economic, social, and environmental terms [103]. According to [108], sustainability helps to improve different characteristics that not only focus on the value proposition or the production process but also on the search for sustainable solutions for packaging development, the resolution of product standards, the reduction of waste-eco-efficiency, and the increase of upcycling. Tunn and Dekoninck [219] specify three ways in which sustainability functions as a contributing factor to the innovation of companies that choose to develop it, as follows:

- **Improving packaging solutions**: Companies seeking to apply sustainability to their product packaging must first identify which aspects they would like to change and the possible solutions. The most often appreciated solutions are the development of more biodegradable packaging that contributes to the circular economy; for this, they recycle old packaging or decide to change the distributor.
- **Reducing waste—Eco-efficiency**: To contribute to this aspect, manufacturers apply different ways, involving both production and after-sales processes. On the one hand, they seek to reduce energy and water consumption, besides limiting waste and transportation, in addition to the maximum use of resources, which derives from the improvement of design and the use of surplus materials as resources to be used in various projects. Also, to avoid product waste by consumers, companies can choose to improve the quality and durability of their

products, offer durable designs, and provide free maintenance, all to extend the life cycle of the products.
- **Other sustainable solutions for standard products**: When it comes to offering sustainable solutions, they are not always focused on impressing the consumer but on taking care of the environment and the health of society as much as possible. It can use more natural raw materials that can be recycled and allow a lower impact of carbon footprint, fair-trade ingredients, certifications of sustainable aspects, inputs that can be renewable, comply with government regulations and create standards within each company.
- **Upcycling**: This term refers to the reuse of materials, which are not necessarily part of the products marketed by companies but can be various items that are recycled to generate new products of more excellent value. For example, the creation of a bag from seat belts.

It has been shown that the transition of sustainable processes integrates constant innovation in a long-term period due to the change of structures, being these economic, industrial, technological, and cultural. However, it also integrates short-term results such as reforms at the business and governmental levels [62]. Sustainable innovation is distinctive from traditional innovation in three different ways: (1) It focuses on the reduction of environmental impact. (2) It constantly improves aspects at the business level, such as production, marketing, and administration. (3) It has a direct effect in external terms, such as environmental costs.

Schaltegger et al. [198] highlight the direct relationship between sustainable models and business drivers. It is essential to mention that to generate sustainable strategies, and business models need to be modified according to their new objectives. Similarly, if a company aims to achieve a specific business model, the strategies must include sustainable solutions. Just as innovation works in tandem with sustainability for the generation of new eco-solutions, there are times when sustainability hinders innovation. Tunn and Dekoninck [219] make mention of 3 ways:

- **Suppliers of sustainable materials**: Difficulty finding distributors that share the sustainable proposals that manufacturers suggest, especially for small companies that, when looking for distributors with such characteristics, only find a small number that is not certified, have low quality, or do not share the same standards.
- **Issues with consumers**: There are times when, in generating a new sustainable design, the esthetics of the previous design are lost, which may affect the preferences of new or less loyal consumers.
- **Cost**: The development of new sustainable solutions involves costs that can be perceived as high compared to conventional ones, which integrates the whole process of developing and creating sustainable strategies, the materials, and finally, the consumer's perception. Profit margins should be calculated based on the company's objectives; however, too high a price may decrease consumer demand.

Innovation is also a driver for new sustainable products, which, to be developed, also need support measures to achieve the desired results. In this way, "Innovation and investment are seen as drivers for the green technology development" [123, 193].

"The transition to green technologies may result in the loss of income and jobs in the brown industries" [123].

4 Classification of GreenTech Business Models

This section aims to identify those sustainable activities that contribute to sustainable business models. By taking into account case studies, sustainability rankings, and Webpages of organizations that promote industrial sustainability [64], literature has classified green business models in the following dimensions:

1. Maximize material and energy efficiency,
2. Close resource loops,
3. Substitute with renewables and natural processes [64, 118, 190, 216].

Regarding the first dimension, it aims at optimizing processes by reducing the use of material and energy resources to mitigate the environmental impact and contribute to the reduction of waste and emissions [64, 118, 190, 216]. For instance, lean manufacturing is a production method that seeks to reduce waste in the whole production process [64, 137, 158, 161]. We can consider Toyota Production System (TPS) as an example of this method [64, 190], as it eliminates waste at every stage of its processes by its seven wastes approach, which includes waste of overproducing, transport, waiting, inventory, processing, motion, and defects [86, 98].

When it comes to the second dimension, closing resource loops refers to not allowing anything to be wasted or discarded through reusing, recycling, and remanufacturing materials [64, 118, 190]. For example, the carpet manufacturer interface introduced a business model based on a cradle-to-cradle design, which means that a product can be remade or recycled at the end of its life cycle. Nevertheless, this business model failed because of an efficient recycling system. It was not until 2006 that interface became a pioneer of the first commercial recycle and reuse system, introducing green products for a closed-loop business model [216].

The last dimension to be mentioned is substituting with renewables and natural processes. This dimension seeks to make industrial processes more benign and contribute to reducing waste and pollution and the use of finite resource supplies. It aims to reduce emissions related to the use of non-renewable resources, such as fossil fuels, by introducing renewable sources [118, 189]. Some examples of renewable energy solutions include solar photovoltaic panels in developing markets [64], B2U storage [189], and onsite windmills for manufacturing processes [64].

Considering the previously mentioned dimensions plus technological opportunities (new enhanced and existing technology), [216] propose the following greentech business model archetypes (Table 1).

Table 1 Greentech business model archetypes

	Maximize energy efficiency	Maximize material efficiency	Close resource loops	Substitute with renewables and natural processes
New technology	I. Energy efficinnovator	IV. Material efficinnovator	VII. Recyclinnovator	X. Greenew substituter
Enhanced technology	II. Efficiency energizer	V. Efficiency materialenhancer	VIII. Recyclenhancer	XI. Greenhanced substituter
Existing technology	III. Energy efficreator	VI. Material efficreator	IX. Green technolooper	XII. Greentech substituter

Source Trapp and Kanbach [216]

5 Business Model Archetypes

5.1 Energy Efficinnovator

This archetype aims at promoting better use of energy resources, reducing emissions and pollution through the development of new technologies [216]. One example is the introduction of blockchain-based technologies, allowing customers to purchase power from local producers or microgrids [73, 212, 216]. Nevertheless, it cannot be decentralized due to the high costs associated with its management [73, 162, 226].

5.2 Efficiency Energizer

This archetype promotes better energy use by improving existing technologies [216]. A business case that can exemplify this approach is the one proposed by [194], a production company that shifted its production of traditional boilers to high-quality, sustainable ones that reduce the environmental impact by producing green energy. To achieve this, the company focused on technology specialization for modulating energy [194].

5.3 Energy Efficreator

The main objective of this archetype is to make better use of energy through existing technologies [216]. Some examples of this archetype have been executed in the hotel industry [107]. For instance, hotel companies in Italy have adopted technologies like the insulation of buildings to make conscientious use of energy resources [72]. The importance of hotels implementing this kind of technology is that this industry comprises more than 20% of the tourism sector's footprint so that this change makes a huge difference [160].

5.4 *Material Efficinnovator*

The Material Efficinnovator seeks to reduce waste by developing new technologies and a more accurate production [216]. To illustrate this, we can consider a Finnish company that developed a new production technology that enables to use of only a small amount of renewable materials, such as forest-based renewable resources, to produce wood-based packaging that is more durable and offer the same functionality as similar products [95].

5.5 *Efficiency Material Enhancer*

This archetype and the previous one aim to reduce the use of material-intensive products but enhance existing technologies [216]. For instance, [127] proposed the case of suppliers that provide material-efficiency services to improve existing processes. These enhanced processes can reduce costs and thus, become more competitive and attractive to customers [127].

5.6 *Material Efficreator*

The Material Efficreator archetype uses existing technologies to reduce the number of raw materials and, consequently, reduce waste [216]. One example that illustrates this archetype is the banking sector, which has become more environmentally sustainable during the past years [203]. In banks, significant amounts of paper, water, and energy are used as it is known. Some material-efficient processes have been implemented to solve this issue, such as digitalizing documents, e-learning, and electronic notepads for transactions [234].

5.7 *Recyclinnovator*

This archetype aims at using new technologies for turning low-value products into valuable ones for a new application. For example, according to [106], it is possible to develop low or high-tech innovations to manage residue streams and promote upcycling agricultural by-products in the agricultural sector.

5.8 Recyclenhancer

The Recyclenhancer seeks to eliminate waste and support the life cycle of products by turning them into valuable inputs through enhancing current technologies [216]. An illustrative example is a case of a Brazilian start-up called Preza that transforms waste into fashion accessories. For instance, they produce sunglasses made of wood from furniture companies [214].

5.9 Green Technolooper

The Green Technolooper archetype aims to adopt existing technologies that turn low-value products into high-value inputs [216]. For example, we can consider a biogas plant business model, which transforms agricultural by-products into heat, biogas, and digestor, working as an efficient way of managing organic waste [106, 195].

5.10 Greenew Substituter

The present archetype offers products that can substitute non-renewable resources with renewable ones through new technologies [216]. An illustrative case is the development of genetically modified algae biofuel in the aviation sector. The airline industry can implement this new technology instead of using petroleum, as it causes significantly fewer carbon emissions and is way less polluting than petroleum [169].

5.11 Greenhanced Substituter

This archetype and the previous one offer products or services that substitute non-renewable processes with natural ones by improving existing technologies [216]. A business case that illustrates the Greenhanced Substituter archetype is the Finnish firm Oulun Energia launched Farm Power. It offers users the opportunity to purchase power generated by natural and renewable processes (biogas, solar, wind, wood gasification, or hydropower) [135].

5.12 Greentech Substituter

The last archetype is the Greentech Substituter that aims at replacing non-renewable resources and processes with renewable ones using existing technologies [216]. Some examples of this archetype can be found in green transportation practices. It reduces greenhouse gas emissions by replacing fossil fuels with renewable ones like electric transportation powered by wind or solar power [224].

6 The 4 Gs and Green Entrepreneurship

The current environmental concern developed by consumers has led to the emergence of the green market [45, 149]. Moreover, this led to the consumption of eco-friendly goods, so they contribute in a specific way to reducing the environmental footprint [37, 149]. Nowadays, companies are looking for new solutions to address this problem [141]. For instance, environmentally-friendly and ecological products became one of the scopes that the private sector analyzed to create value [149]. As a result, companies with a social responsibility approach must develop "clean products" as an opportunity to achieve sustainable development [149]. The green entrepreneurship concept emerged as a solution to environmental problems [149]. It is a business model compatible with environmental needs and requires specific technology and design thinking to commercialize green products [149]. To understand how entrepreneurs can develop green strategies, [204] extended the green market concept into 4 dimensions: product, design, supply chain, and production.

6.1 Green Product

The development of a product is crucial for a business such as entrepreneurship since it represents the potential competitive advantage that this last may have compared to others [85]. When we refer to a green product, we must conceptualize it as a good design so that the environmental impact of its production is minimized [83, 113, 167]. Its development aims to use natural resources efficiently, avoid toxic chemicals that may injure the ecosystem, and reduce waste and pollution [63, 113]. The products are based on market opportunities where a gap must be filled, such as environmental concerns [142]. Also, it is essential to mention that the long-term benefits provided by green products are the added value being looked at by consumers [100]. The life cycle of green products, as a result, must be subjected to technological and scientific research [200].

6.2 Green Design

As it is known, product design is being evaluated from consumers' behavior purchasing this last [143]. This concept evokes the formalization of a solution into tangible products [164]. Generally, the design's scope was about manufacturing and product quality. Nonetheless, the increasing environmental concern resulted in implementing the environmental design [54, 201]. Designing environmental characteristics into a product is a must [220]. Indeed, implementing a sustainable perspective in product design ensures that the product fulfills consumers' requirements and achieves sustainable development [68, 105]. To analyze the importance of green perspective during product design, we should see the phases of its development, which consists of the followings:

- **Step 1**: Selection of raw materials to be used
- **Step 2**: Manufacturing process
- **Step 3**: Packaging and physical distribution
- **Step 4**: Usage
- **Step 5**: Disposal or end of Life [235]

As we can see, the traditional product design phases are a take-make-waste model and, as a result, an approach where environmental aspects may not be considered [209]. Nonetheless, realizing an audit of the product's life cycle ensures understanding the critical control points where improvement based on a sustainable strategy is necessary [82]. To sum up, green design promotes a relationship between humans and nature [124]. Indeed, this list is based on several principles. The first one is about maximizing the full use of natural resources, which means analyzing properties and renewability of raw materials to design an efficient manufacturing system that allows recycling, reusage, among others [54, 124, 225]. The second focuses on energy conservation, referring to the possibility of designing processes that do not use fossil fuels but solar power and wind energy [124]. Zero-pollution, the third principle, is about minimizing pollution and environmental footprint. The fourth and last one is about using the latest technology and focusing on sustainable development benefits, respectively [61, 124].

6.3 Green Supply Chain

Green Supply Chain, according to scholars, is an effective method to ensure environmental performance [218]. Based on what was stated by [128], this concept refers to the application of a sustainable perspective in all the supply chain activities. It is about green initiatives in all the processes from design to delivery and recovery of the output to ensure recycling, reuse, rethink, refurbish, among others, of this last for more outstanding environmental contribution [136, 157, 213]. Based on [125], we can consider 3 essential dimensions of GSC:

- **Internal Environmental Management**: Refers to the GSC initiatives incorporated in the overall process based on top-management decisions [125]. According to [222], managers and entrepreneurs should take action and decide whether to implement a green policy in a specific part of the supply chain to improve environmental performance.

- **Eco-Design**: Consists of developing eco-friendly products which minimize environmental impact throughout the product life cycle analysis mentioned above [15]. Eco-design is incorporated into the supply chain management through product design since the product idea is reflected from selecting raw materials to the disposal of the output [125, 221, 222].
- **Cooperation with Customers**: Consumers' support is a crucial driver to green supply chain implementation since their needs and expectations influence businesses, resulting in process improvement, eco-design, waste management, among others [125, 211].

Entrepreneurs can adopt green practices to satisfy current and future desires from customers [171]. To create value, they must deliver innovation such as green products with, for instance, eco-design and sustainable practices [125], which result in better economic, environmental, and social performance and, as a consequence, sustainable development [115].

6.4 Green Production

As part of the supply chain, production also requires a green perspective. Green production refers to implementing green practices in the product life cycle to ensure sustainability [183]. This last is about creating green products, which means a good with a lower environmental impact than conventional ones [134]. This concept focuses on reducing the environmental damage of a specific product through innovations such as renewable resources [236]. Nonetheless, it is essential to mention that external stakeholders such as consumers can affect this process and accelerate the transition into a green production system [63, 131]. Governments, NGOs, among others, influence and push businesses such as entrepreneurship to adopt a sustainable approach to their production process [233]. After discussing these 4 elements, [149] proposed model focuses on how entrepreneurship can take a green approach and ensure sustainable development. They surveyed the Science and Technology Park of Tehran University in Iran, obtaining 85 valid responses from green entrepreneurs. The findings highlighted the importance of green products, green supply chain, green design, and green production in the green entrepreneurship business model to achieve sustainable development [149] (Fig. 1).

Another study was done by Zhou et al. [241] in China. They surveyed the Yixing Environmental and Cleaner Technology Park (YECP), where 138 greentech ventures from this location participated. One of the most important findings is that information, communication, and technology (ICT) and a green economy provide several opportunities for business growth while efficiently using natural resources [241]. Moreover, technology orientation is a driver of greentech ventures. For instance, green product development helps green entrepreneurship grow [96, 149]. Nonetheless, it is commented that institutions promoting ICT for green entrepreneurship remain weak in several emerging countries [241].

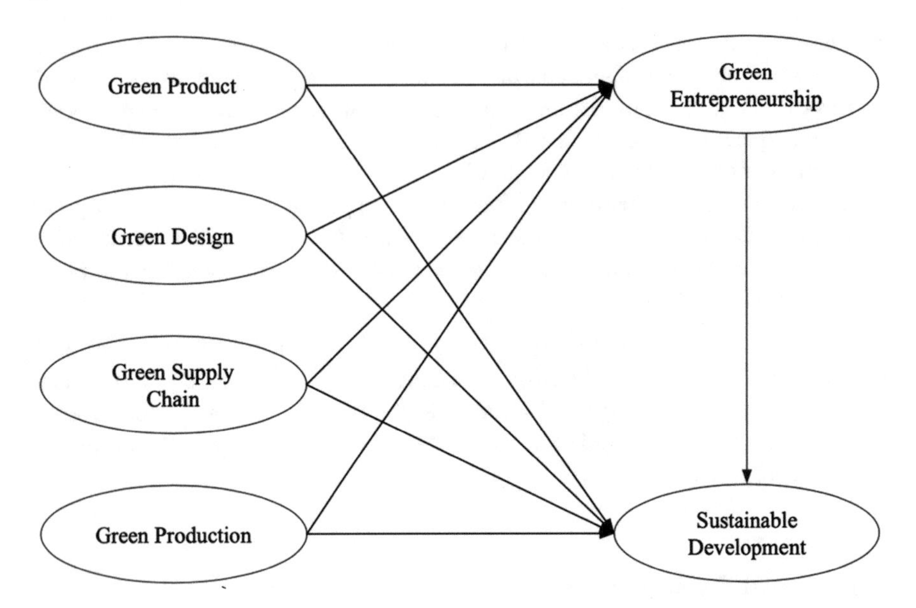

Fig. 1 Green entrepreneurship approach for sustainable development. *Source* Lotfi et al. [149]

7 Green Value-Added Process for Entrepreneurship

The last model shows how entrepreneurship can adopt several "green" approaches to achieve sustainable development. The Green Value-Added (GVA) process extends this model and focuses on how entrepreneurship can become sustainable by introducing environmental and social issues and, as a result, deliver value [56] (Fig. 2). The emergence of this framework is due to the importance that green entrepreneurship took in the race for sustainable development and green economy policy [14]. Nonetheless, environmental problems may not be resolved only through technological factors, since they should incorporate cultural, behavioral, and institutional variables that also affect entrepreneurship's performance [130, 170]. This extension of Porter's value chain is divided into 3 blocks: primary, internal support, and external support activities.

7.1 Primary Activities

Environmentally friendly practices can take several aspects such as product development, new design, advanced technology, green marketing, among others [172, 199]. New businesses can decide whether they innovate in any of these lasts related to green innovation [82]. For instance, [215] designed an algorithm to implement sustainable business models into primary activities in the food industry. These studies concluded that eco-design contributes to the environmental performance of SMEs

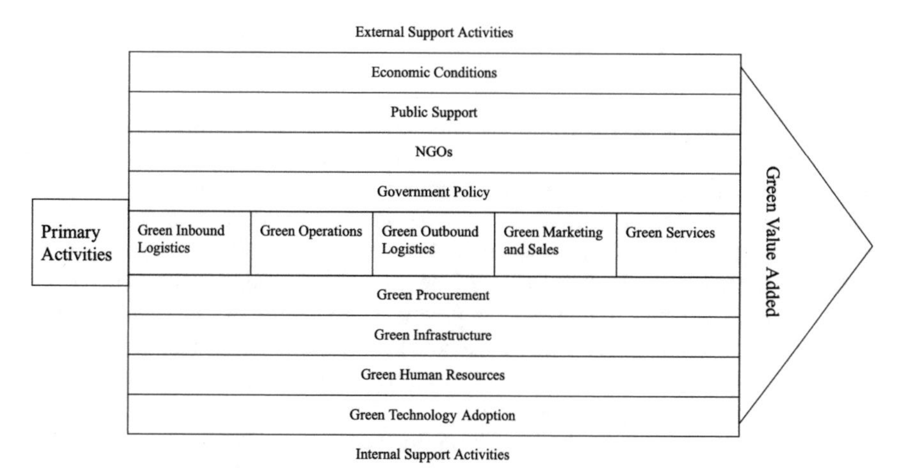

Fig. 2 Green value-added (GVA) process

and new entrepreneurship, corroborating what other researchers stated [15, 125]. Table 2 shows some companies that use green primary activities:

Globalization and climate change affect businesses opportunities since they face resource limitations and growing costs due to environmental costs [179], which influences how entrepreneurs design their strategies and implement a new core value: sustainability [196]. One way to share with the public the innovations is through green marketing. According to [144], green marketing can take several dimensions: partnerships, environmental commitment, green organizational culture, benchmarking, and compliance. This last can influence consumers to perceive specific entrepreneurship as green [207].

Table 2 Green entrepreneurship examples

Business	Value proposition	Country	Sources
Fuergy	Provide energy optimization among supply chain and for end-consumers	Slovakia	Fuergy [116]
Ecosaf	Uses waste materials to design new innovative products for the chemical sector	United Kingdom	ClimateLaunchpad [90]
Sages	Uses food waste materials to develop eco-friendly dyes for the fashion industry and end-consumers	United Kingdom	ClimateLaunchpad [90]

7.2 Internal Support Activities

It is important to understand how entrepreneurs can develop an organizational culture where environmental values are considered and considered when deciding a specific strategy related to green infrastructure [170]. This last concept is somehow related to green human resources that aim to enhance and develop a green culture among employees [170]. Indeed, researchers stated that ecological awareness among workers leads to a better understanding of the necessity to achieve sustainable development and, as a result, to increase the practical application of green policies [66, 129, 147]. Green Human Resource Management, connected to green infrastructure, is becoming necessary for businesses such as entrepreneurship to promote sustainable assignment and usage of resources [151].

It is about purchasing products with minimal environmental impact [176]. This concept is crucial for effective relationship development between suppliers and clients [170]. Since it is essential to increase green supply chain management practices, green procurement promotes the establishment of social and environmental goals among all the actors of the supply chain, including entrepreneurship itself [121]. Therefore, environmental performance increases since several methods, such as green purchasing management and procurement components, are implemented [47].

7.3 External Support Activities

Focusing on external support activities is about factors that the business cannot control [175]. In this category, we can find, according to [56], economic conditions, NGOs, government policy, and public support. Stakeholders such as the government can increase entrepreneurship business performance [206]. Nonetheless, the perception of green entrepreneurship among them is not visible [206]. It is crucial to promote and create awareness among this group about the current green entrepreneurship businesses on the market [206]. Indeed, several countries are under the average, indicating that governments must provide and promote adequate conditions to ensure the development of entrepreneurship, which can affect this group since green entrepreneurs are financially limited. The disadvantageous business environment makes their operations difficult [53].

Table 3 shows some of the strategies and policies adopted by governments to promote green entrepreneurship.

These government policies promote a sustainable approach from green entrepreneurs, which means that they should focus on energy efficiency and provide long-term environmental benefits [89]. Green entrepreneurs are essential for countries' development since they cooperate to achieve sustainable goals through innovations, initiatives, green production, among others [53]. In the case of public support, 45% of consumers worldwide are looking in 2020 for environmentally responsible

Table 3 Strategies adopted by other countries worldwide regarding green entrepreneurship

Country	Propositions for carbon footprint	Sources
Bulgaria	Bulgarian strategy 2020 Energy efficiency and green economy	Burzyńska [74]
Norway	Green and genius program	Burzyńska [74]
Romania	Green entrepreneurship program National sustainable development strategy Romania 2020–2030	Burzyńska [74]
Latvia	Sustainable development strategy of Latvia until 2030	Burzyńska [74]
Spain	The green entrepreneurs support network	Chygryn [89]

brands [210]. Both external factors affect entrepreneurship establishments. In the USA, for instance, 804 thousand businesses survived less than one year in 2020, compared to 569 thousand in 1994 [210]. Entrepreneurship must be encouraged to develop green initiatives and reduce waste [204].

Recent suggested lectures about green approach

- Waste reduction and carbon footprint [120]
- Material selection for circularity and footprints [166]
- 3D print, circularity, and footprints [97]
- Virtual tourism [97]
- Leadership for sustainability in crisis time [32]
- Virtual education and circularity [91]
- Circular economy for packaging [79]
- Students oriented to circular learning [99]
- Food and circular economy [94]
- Water footprint and food supply chain management [133]
- Waste footprint [51]
- Measuring circular economy [102]
- Carbon footprint [114]

8 Closing Remarks

Entrepreneurship is in constant innovation. For this reason, differentiation among the competitors is crucial. Entrepreneurs can adopt a green approach as an added value that follows global trends according to consumers' expectations. Indeed, they are following eco-friendly practices in all the supply chain processes, from selecting materials to the final delivery of the goods. In the long term, benefits are mostly related to reducing their environmental footprint, fomenting a green consumption habit, and creating differential value. Despite all these possibilities, government support is crucial for adopting green practices among entrepreneurship. That is due to several countries' lack of green entrepreneurship programs, an unstable economic condition, and low public support. Added value is part of the DNA of the business, and it can be developed through specific strategies like the ones we mentioned above. Sustainable innovation can be implemented in any part of the business model but ignoring this trend be fatal for the company's long-term growth.

References

1. Abad-Segura E, Batlles-delaFuente A, González-Zamar M-D, Belmonte-Ureña LJ (2021) Implications for sustainability of the joint application of bioeconomy and circular economy: a worldwide trend study. Sustainability 13(13):7182. https://doi.org/10.3390/su13137182
2. Abdaljaleel M, Singer EJ, Yong WH (2019) Sustainability in biobanking. In: Yong WH (ed) Biobanking: methods and protocols. Springer New York, pp. 1–6. https://doi.org/10.1007/978-1-4939-8935-5_1
3. Aberilla JM, Gallego-Schmid A, Stamford L, Azapagic A (2020) Environmental sustainability of cooking fuels in remote communities: life cycle and local impacts. Sci Total Environ 713:136445. https://doi.org/10.1016/j.scitotenv.2019.136445
4. Adam A (2015) Challenges of public finance sustainability in the European union. Procedia Econ Finan 23:298–302. https://doi.org/10.1016/S2212-5671(15)00507-9
5. Adam I, Agyeiwaah E, Dayour F (2021) Understanding the social identity, motivations, and sustainable behaviour among backpackers: a clustering approach. J Travel Tour Mark 38(2):139–154. https://doi.org/10.1080/10548408.2021.1887053
6. Adam JN, Adams T, Gerber J-D, Haller T (2021) Decentralization for increased sustainability in natural resource management? Two cautionary cases from Ghana. Sustainability 13(12). https://doi.org/10.3390/su13126885
7. Adam M (2018) The role of human resource management (HRM) for the implementation of sustainable product-service systems (PSS)—an analysis of fashion retailers. Sustainability 10(7). https://doi.org/10.3390/su10072518
8. Adams R, Martin S, Boom K (2018) University culture and sustainability: designing and implementing an enabling framework. J Clean Prod 171:434–445. https://doi.org/10.1016/j.jclepro.2017.10.032
9. Adesogan AT, Havelaar AH, McKune SL, Eilittä M, Dahl GE (2020) Animal source foods: sustainability problem or malnutrition and sustainability solution? Perspective matters. Glob Food Secur 25:100325. https://doi.org/10.1016/j.gfs.2019.100325
10. Adomah-Afari A, Chandler JA (2018) The role of government and community in the scaling up and sustainability of mutual health organisations: an exploratory study in Ghana. Soc Sci Med 207:25–37. https://doi.org/10.1016/j.socscimed.2018.04.044
11. Adomako S, Amankwah-Amoah J, Danso A, Konadu R, Owusu-Agyei S (2019) Environmental sustainability orientation and performance of family and nonfamily firms. Bus Strateg Environ 28(6):1250–1259. https://doi.org/10.1002/bse.2314
12. Adongo CA, Taale F, Adam I (2018) Tourists' values and empathic attitude toward sustainable development in tourism. Ecol Econ 150:251–263. https://doi.org/10.1016/j.ecolecon.2018.04.013
13. Agyeiwaah E (2019) Exploring the relevance of sustainability to micro tourism and hospitality accommodation enterprises (MTHAEs): evidence from home-stay owners. J Clean Prod 226:159–171. https://doi.org/10.1016/j.jclepro.2019.04.089
14. Ahmad NH, Halim HA, Ramayah T, Rahman SA (2015) Green entrepreneurship inclination among Generation Y: the road towards a green economy. Prob Perspect Manage 13(2):211–218
15. Al-Ghwayeen WS, Abdallah AB (2018) Green supply chain management and export performance. J Manuf Technol Manag 29(7):1233–1252. https://doi.org/10.1108/JMTM-03-2018-0079

16. Al-Naqbi AK, Alshannag Q (2018) The status of education for sustainable development and sustainability knowledge, attitudes, and behaviors of UAE University students. Int J Sustain High Educ 19(3):566–588. https://doi.org/10.1108/IJSHE-06-2017-0091
17. Aleixo AM, Leal S, Azeiteiro UM (2018) Conceptualization of sustainable higher education institutions, roles, barriers, and challenges for sustainability: an exploratory study in Portugal. J Clean Prod 172:1664–1673. https://doi.org/10.1016/j.jclepro.2016.11.010
18. Ali M, de Azevedo ARG, Marvila MT, Khan MI, Memon AM, Masood F, … Haq IU (2021) The influence of COVID-19-induced daily activities on health parameters—a case study in Malaysia. Sustainability 13(13):7465. https://doi.org/10.3390/su13137465
19. Alkhuzaim L, Zhu Q, Sarkis J (2021) Evaluating emergy analysis at the nexus of circular economy and sustainable supply chain management. Sustain Prod Consumption 25:413–424. https://doi.org/10.1016/j.spc.2020.11.022
20. Alonso-Fradejas A (2021) The resource property question in climate stewardship and sustainability transitions. Land Use Policy 108:105529. https://doi.org/10.1016/j.landusepol.2021.105529
21. Alshuwaikhat HM, Adenle YA, Saghir B (2016) Sustainability assessment of higher education institutions in Saudi Arabia. Sustainability 8(8). https://doi.org/10.3390/su8080750
22. Álvarez-Risco A, Arellano EZ, Valerio EM, Acosta NM, Tarazona ZS (2013) Pharmaceutical care campaign as a strategy for implementation of pharmaceutical services: experience Peru. Pharm Care Espana 15(1):35–37
23. Alvarez-Risco A, Dawson J, Johnson W, Conteh-Barrat M, Aslani P, Del-Aguila-Arcentales S, Diaz-Risco S (2020) Ebola virus disease outbreak: a global approach for health systems. Revista Cubana de Farmacia 53(4):1–13, Article e491.
24. Alvarez-Risco A, Del-Aguila-Arcentales S (2015) Prescription errors as a barrier to pharmaceutical care in public health facilities: experience Peru. Pharm Care Espana 17(6):725–731
25. Alvarez-Risco A, Del-Aguila-Arcentales S (2021) A note on changing regulation in international business: the world intellectual property organization (wipo) and artificial intelligence. In: Progress in international business research, vol 15, pp 363–371. https://doi.org/10.1108/S1745-886220210000015020
26. Alvarez-Risco A, Del-Aguila-Arcentales S (2021) Public policies and private efforts to increase women entrepreneurship based on STEM background. In: Contributions to management science, pp 75–87. https://doi.org/10.1007/978-3-030-83792-1_5
27. Alvarez-Risco A, Del-Aguila-Arcentales S (2022) Sustainable Initiatives in International Markets. In: Contributions to management science, pp 181–191. https://doi.org/10.1007/978-3-030-85950-3_10
28. Alvarez-Risco A, Del-Aguila-Arcentales S, Delgado-Zegarra J, Yáñez JA, Diaz-Risco S (2019) Doping in sports: findings of the analytical test and its interpretation by the public. Sport Sci Health 15(1):255–257. https://doi.org/10.1007/s11332-018-0484-8
29. Alvarez-Risco A, Del-Aguila-Arcentales S, Diaz-Risco S (2018) Pharmacovigilance as a tool for sustainable development of healthcare in Peru. PharmacoVigilance Rev 10(2):4–6
30. Alvarez-Risco A, Del-Aguila-Arcentales S, Rosen MA, García-Ibarra V, Maycotte-Felkel S, Martínez-Toro GM (2021) Expectations and interests of university students in covid-19 times about sustainable development goals: evidence from colombia, ecuador, mexico, and peru. Sustainability (Switzerland) 13(6), Article 3306. https://doi.org/10.3390/su13063306
31. Alvarez-Risco A, Del-Aguila-Arcentales S, Stevenson JG (2015) Pharmacists and mass communication for implementing pharmaceutical care. Am J Pharm Benefits 7(3):e125–e126
32. Alvarez-Risco A, Del-Aguila-Arcentales S, Villalobos-Alvarez D, Diaz-Risco S (2022) Leadership for sustainability in crisis time. In: Alvarez-Risco A, Muthu SS, Del-Aguila-Arcentales S (eds) Circular economy: impact on carbon and water footprint. Springer Singapore, pp 41–64. https://doi.org/10.1007/978-981-19-0549-0_3
33. Alvarez-Risco A, Del-Aguila-Arcentales S, Yáñez JA, Alvarez-Risco A (2021) Telemedicine in Peru as a Result of the COVID-19 pandemic: perspective from a country with limited internet access. Am J Trop Med Hyg 105(1):6–11. https://doi.org/10.4269/ajtmh.21-0255

34. Alvarez-Risco A, Del-Aguila-Arcentales S, Yáñez JA, Rosen MA, Mejia CR (2021) Influence of technostress on academic performance of university medicine students in peru during the covid-19 pandemic. Sustainability (Switzerland) 13(16), Article 8949. https://doi.org/10.3390/su13168949
35. Alvarez-Risco A, Delgado-Zegarra J, Yáñez JA, Diaz-Risco S, Del-Aguila-Arcentales S (2018) Predation risk by gastronomic boom—case Peru. J Landscape Ecol (Czech Republic) 11(1):100–103. https://doi.org/10.2478/jlecol-2018-0003
36. Alvarez-Risco A, Estrada-Merino A, Anderson-Seminario MM, Mlodzianowska S, García-Ibarra V, Villagomez-Buele C, Carvache-Franco M (2021) Multitasking behavior in online classrooms and academic performance: case of university students in Ecuador during COVID-19 outbreak. Interact Technol Smart Educ 18(3):422–434. https://doi.org/10.1108/ITSE-08-2020-0160
37. Alvarez-Risco A, Estrada-Merino A, Rosen MA, Vargas-Herrera A, Del-Aguila-Arcentales S (2021) Factors for implementation of circular economy in firms in COVID-19 pandemic times: the case of Peru. Environments 8(9):95. https://doi.org/10.3390/environments8090095
38. Alvarez-Risco A, Mejia CR, Delgado-Zegarra J, Del-Aguila-Arcentales S, Arce-Esquivel AA, Valladares-Garrido MJ, … Yáñez JA (2020) The Peru approach against the COVID-19 infodemic: Insights and strategies. Am J Trop Med Hyg 103(2):583–586. https://doi.org/10.4269/ajtmh.20-0536
39. Alvarez-Risco A, Mlodzianowska S, García-Ibarra V, Rosen MA, Del-Aguila-Arcentales S (2021) Factors affecting green entrepreneurship intentions in business university students in covid-19 pandemic times: case of ecuador. Sustainability (Switzerland) 13(11), Article 6447. https://doi.org/10.3390/su13116447
40. Alvarez-Risco A, Mlodzianowska S, Zamora-Ramos U, Del-Aguila-Arcentales S (2021) Green entrepreneurship intention in university students: the case of Peru. Entrepreneurial Bus Econ Rev 9(4):85–100. https://doi.org/10.15678/EBER.2021.090406
41. Alvarez-Risco A, Quipuzco-Chicata L, Escudero-Cipriani C (2021) Determinantes de la intención de recompra en línea en tiempos de COVID-19: evidencia de una economía emergente. Lecturas de Economia (96):101–143. https://doi.org/10.17533/udea.le.n96a342638
42. Alvarez-Risco A, Quiroz-Delgado D, Del-Aguila-Arcentales S (2016) Pharmaceutical care in hypertension patients in a peruvian hospital [Article]. Indian J Public Health Res Develop 7(3):183–188. https://doi.org/10.5958/0976-5506.2016.00153.4
43. Alvarez-Risco A, Turpo-Cama A, Ortiz-Palomino L, Gongora-Amaut N, Del-Aguila-Arcentales S (2016) Barriers to the implementation of pharmaceutical care in pharmacies in Cusco, Peru. Pharm Care Espana 18(5):194–205
44. Alvarez-Risco A, Van Mil JWF (2007) Pharmaceutical care in community pharmacies: practice and research in Peru. Ann Pharmacother 41(12):2032–2037. https://doi.org/10.1345/aph.1K117
45. Amberg N, Fogarassy C (2019) Green consumer behavior in the cosmetics market. Resources 8(3). https://doi.org/10.3390/resources8030137
46. Amorós JE, Felzensztein C, Etchebarne M (2012) Emprendimiento internacional en Latinoamérica: desafíos para el desarrollo. Esic Market Econ Bus J 43(3):513–529
47. Ananda ARW, Astuty P, Nugroho YC (2018) Role of green supply chain management in embolden competitiveness and performance: Evidence from Indonesian organizations. Int J Supply Chain Manage 7(5):437–442
48. Ang KL, Saw ET, He W, Dong X, Ramakrishna S (2021) Sustainability framework for pharmaceutical manufacturing (PM): a review of research landscape and implementation barriers for circular economy transition. J Clean Prod 280:124264. https://doi.org/10.1016/j.jclepro.2020.124264
49. Apcho-Ccencho LV, Cuya-Velásquez BB, Rodríguez DA, Anderson-Seminario MLM, Alvarez-Risco A, Estrada-Merino A, Mlodzianowska S (2021) The impact of international price on the technological industry in the united states and china during times of crisis: commercial war and covid-19. In: Advances in business and management forecasting, vol 14, pp 149–160

50. Aragon-Correa JA, Marcus AA, Rivera JE, Kenworthy AL (2017) Sustainability management teaching resources and the challenge of balancing planet, people, and profits. Acad Manage Learn Educ 16(3):469–483. https://doi.org/10.5465/amle.2017.0180
51. Arias-Meza M, Alvarez-Risco A, Cuya-Velásquez BB, de las Mercedes Anderson-Seminario M, Del-Aguila-Arcentales S (2022) Fashion and textile circularity and waste footprint. In: Alvarez-Risco A, Muthu SS, Del-Aguila-Arcentales S (eds) Circular economy: impact on carbon and water footprint. Springer Singapore, pp 181–204. https://doi.org/10.1007/978-981-19-0549-0_9
52. Aschemann-Witzel J, Giménez A, Ares G (2018) Convenience or price orientation? Consumer characteristics influencing food waste behaviour in the context of an emerging country and the impact on future sustainability of the global food sector. Glob Environ Chang 49:85–94. https://doi.org/10.1016/j.gloenvcha.2018.02.002
53. Ataman K, Mayowa J-O, Senkan E, Olusola AM (2018) Green entrepreneurship: an opportunity for entrepreneurial development in Nigeria. Covenant J Entrepreneurship (Special Edition), 1(1)
54. Bag S, Gupta S, Telukdarie A (2018) Exploring the relationship between unethical practices, buyer–supplier relationships and green design for sustainability. Int J Sustain Eng 11(2):97–109. https://doi.org/10.1080/19397038.2017.1376723
55. Bamgbade JA, Kamaruddeen AM, Nawi MNM, Adeleke AQ, Salimon MG, Ajibike WA (2019) Analysis of some factors driving ecological sustainability in construction firms. J Clean Prod 208:1537–1545. https://doi.org/10.1016/j.jclepro.2018.10.229
56. Banerjee SB (2002) Organisational strategies for sustainable development: developing a research agenda for the new millennium. Aust J Manage 27(1_suppl):105–117. https://doi.org/10.1177/031289620202701S11
57. Barr S, Gilg A, Shaw G (2011) 'Helping people make better choices': exploring the behaviour change agenda for environmental sustainability. Appl Geogr 31(2):712–720. https://doi.org/10.1016/j.apgeog.2010.12.003
58. Barragán-Ocaña A, Silva-Borjas P, Olmos-Peña S, Polanco-Olguín M (2020) Biotechnology and bioprocesses: their contribution to sustainability. Processes 8(4). https://doi.org/10.3390/pr8040436
59. Barros MV, Salvador R, do Prado GF, de Francisco AC, Piekarski CM (2021) Circular economy as a driver to sustainable businesses. Cleaner Environ Syst 2:100006. https://doi.org/10.1016/j.cesys.2020.100006
60. Batista AA, Francisco AC (2018) Organizational sustainability practices: a study of the firms listed by the corporate sustainability index. Sustainability 10(1). https://doi.org/10.3390/su10010226
61. Baumann H, Boons F, Bragd A (2002) Mapping the green product development field: engineering, policy and business perspectives. J Clean Prod 10(5):409–425. https://doi.org/10.1016/S0959-6526(02)00015-X
62. Beltramello A, Haie-Fayle L, Pilat D (2013) Why new business models matter for green growth. OECD Green Growth Papers, No. 2013/1. Retrieved 01/01/2022 from https://www.oecd-ilibrary.org/environment/why-new-business-models-matter-for-green-growth_5k97gk40v3ln-en
63. Biswas A, Roy M (2016) A study of consumers' willingness to pay for green products. J Adv Manage Sci 4(3). https://doi.org/10.12720/joams.4.3.211-215
64. Bocken NMP, Short SW, Rana P, Evans S (2014) A literature and practice review to develop sustainable business model archetypes. J Clean Prod 65:42–56. https://doi.org/10.1016/j.jclepro.2013.11.039
65. Bokpin GA (2017) Foreign direct investment and environmental sustainability in Africa: the role of institutions and governance. Res Int Bus Financ 39:239–247. https://doi.org/10.1016/j.ribaf.2016.07.038
66. Bombiak E (2019) Green human resource management–the latest trend or strategic necessity? Entrepreneurship Sustain Issues 6(4):1647–1662. https://doi.org/10.9770/jesi.2019.6.4(7)

67. Boons F, Lüdeke-Freund F (2013) Business models for sustainable innovation: state-of-the-art and steps towards a research agenda. J Clean Prod 45:9–19. https://doi.org/10.1016/j.jclepro.2012.07.007
68. Boyko CT, Cooper R, Davey CL, Wootton AB (2006) Addressing sustainability early in the urban design process. Manage Environ Qual Int J 17(6):689–706. https://doi.org/10.1108/14777830610702520
69. Brito RM, Rodríguez C, Aparicio JL (2018) Sustainability in teaching: an evaluation of university teachers and students. Sustainability 10(2):439. https://doi.org/10.3390/su10020439
70. Brown HS, de Jong M, Levy DL (2009) Building institutions based on information disclosure: lessons from GRI's sustainability reporting. J Clean Prod 17(6):571–580. https://doi.org/10.1016/j.jclepro.2008.12.009
71. Brown KA, Harris F, Potter C, Knai C (2020) The future of environmental sustainability labelling on food products. Lancet Planet Health 4(4):e137–e138. https://doi.org/10.1016/S2542-5196(20)30074-7
72. Buffa F, Franch M, Rizio D (2018) Environmental management practices for sustainable business models in small and medium sized hotel enterprises. J Clean Prod 194:656–664. https://doi.org/10.1016/j.jclepro.2018.05.143
73. Bürer MJ, Capezzali M, Lapparent Md, Pallotta V, Carpita M (2019) Blockchain in Industry: review of key use cases and lessons learned. In: 2019 IEEE international conference on engineering, technology and innovation (ICE/ITMC), pp 1–7. https://doi.org/10.1109/ICE.2019.8792674
74. Burzyńska D, Jabłońska M, Dziuba R (2018) Opportunities and conditions for the development of green entrepreneurship in the polish textile sector. Fibres Text Eastern Europe 26(2):13–19. https://doi.org/10.5604/01.3001.0011.5733
75. Carter CR, Rogers DS (2008) A framework of sustainable supply chain management: moving toward new theory. Int J Phys Distrib Logist Manage 38(5):360–387. https://doi.org/10.1108/09600030810882816
76. Carvache-Franco M, Alvarez-Risco A, Carvache-Franco O, Carvache-Franco W, Estrada-Merino A, Villalobos-Alvarez D (2021) Perceived value and its influence on satisfaction and loyalty in a coastal city: a study from Lima, Peru. J Policy Res Tourism, Leisure and Events. https://doi.org/10.1080/19407963.2021.1883634
77. Carvache-Franco M, Alvarez-Risco A, Carvache-Franco W, Carvache-Franco O, Estrada-Merino A, Rosen MA (2021) Coastal cities seen from loyalty and their tourist motivations: a study in Lima, Peru. Sustainability (Switzerland) 13(21), Article 11575. https://doi.org/10.3390/su132111575
78. Carvache-Franco M, Carvache-Franco O, Carvache-Franco W, Alvarez-Risco A, Estrada-Merino A (2021) Motivations and segmentation of the demand for coastal cities: a study in Lima, Peru. Int J Tourism Res 23(4):517–531. https://doi.org/10.1002/jtr.2423
79. Castillo-Benancio S, Alvarez-Risco A, Esquerre-Botton S, Leclercq-Machado L, Calle-Nole M, Morales-Ríos F, … Del-Aguila-Arcentales S (2022) Circular economy for packaging and carbon footprint. In: Alvarez-Risco A, Muthu SS, Del-Aguila-Arcentales S (eds) Circular economy: impact on carbon and water footprint. Springer Singapore, pp 115–138. https://doi.org/10.1007/978-981-19-0549-0_6
80. Chafloque-Cespedes R, Alvarez-Risco A, Robayo-Acuña PV, Gamarra-Chavez CA, Martinez-Toro GM, Vicente-Ramos W (2021) Effect of sociodemographic factors in entrepreneurial orientation and entrepreneurial intention in university students of latin american business schools. In: Contemporary issues in entrepreneurship research, vol 11, pp 151–165. https://doi.org/10.1108/S2040-724620210000011010
81. Chafloque-Céspedes R, Vara-Horna A, Asencios-Gonzales Z, López-Odar D, Alvarez-Risco A, Quipuzco-Chicata L, … Sánchez-Villagomez M (2020) Academic presenteeism and violence against women in schools of business and engineering in Peruvian universities. Lecturas de Economia 93:127–153. https://doi.org/10.17533/udea.le.n93a340726

82. Chan HK (2011) Green process and product design in practice. Procedia Soc Behav Sci 25:398–402. https://doi.org/10.1016/j.sbspro.2012.02.050
83. Chen C (2001) Design for the environment: a quality-based model for green product development. Manage Sci 47(2):250–263. https://doi.org/10.1287/mnsc.47.2.250.9841
84. Chen X, Zhang SX, Jahanshahi AA, Alvarez-Risco A, Dai H, Li J, Ibarra VG (2020) Belief in a COVID-19 conspiracy theory as a predictor of mental health and well-being of health care workers in Ecuador: Cross-sectional survey study. JMIR Public Health Surveill 6(3), Article e20737. https://doi.org/10.2196/20737
85. Chen Y-S, Chang T-W, Lin C-Y, Lai P-Y, Wang K-H (2016) The influence of proactive green innovation and reactive green innovation on green product development performance: the mediation role of green creativity. Sustainability 8(10):966. https://doi.org/10.3390/su8100966
86. Chiarini A, Baccarani C, Mascherpa V (2018) Lean production, Toyota Production System and Kaizen philosophy. TQM Journal 30(4):425–438. https://doi.org/10.1108/TQM-12-2017-0178
87. Chung SA, Olivera S, Rojas Román B, Alanoca E, Moscoso S, Limpias Terceros B, … Yáñez JA (2021) Temáticas de la producción científica de la Revista Cubana de Farmacia indizada en Scopus (1967–2020). Revista Cubana de Farmacia 54(1) (2021): (Enero-Marzo). http://www.revfarmacia.sld.cu/index.php/far/article/view/511
88. Chung SA, Olivera S, Román BR, Alanoca E, Moscoso S, Terceros BL, … Yáñez JA (2021) Themes of scientific production of the cuban journal of pharmacy indexed in scopus (1967–2020). Revista Cubana de Farmacia 54(1), Article e511
89. Chygryn O (2017) Green entrepreneurship: EU experience and Ukraine perspectives. Centre Stud Eur Integr Working Pap Ser (6):6–13
90. ClimateLaunchpad. (n.d.). Ecosaf. Retrieved 01/01/2022 from https://climatelaunchpad.org/finalists/ecosaf
91. Contreras-Taica A, Alvarez-Risco A, Arias-Meza M, Campos-Dávalos N, Calle-Nole M, Almanza-Cruz C, … Del-Aguila-Arcentales S (2022) Virtual education: carbon footprint and circularity. In: Alvarez-Risco A, Muthu SS, Del-Aguila-Arcentales S (eds) Circular economy: impact on carbon and water footprint. Springer Singapore, pp 265–285. https://doi.org/10.1007/978-981-19-0549-0_13
92. Cruz-Torres W, Alvarez-Risco A, Del-Aguila-Arcentales S (2021) Impact of enterprise resource planning (ERP) implementation on performance of an education enterprise: a structural equation modeling (SEM). Stud Bus Econ 16(2):37–52. https://doi.org/10.2478/sbe-2021-0023
93. Cruz JP, Alshammari F, Felicilda-Reynaldo RFD (2018) Predictors of Saudi nursing students' attitudes towards environment and sustainability in health care. Int Nurs Rev 65(3):408–416. https://doi.org/10.1111/inr.12432
94. Cuya-Velásquez BB, Alvarez-Risco A, Gomez-Prado R, Juarez-Rojas L, Contreras-Taica A, Ortiz-Guerra A, … Del-Aguila-Arcentales S (2022) Circular economy for food loss reduction and water footprint. In: Alvarez-Risco A, Muthu SS, Del-Aguila-Arcentales S (eds) Circular economy: impact on carbon and water footprint. Springer Singapore, pp 65–91. https://doi.org/10.1007/978-981-19-0549-0_4
95. D'Amato D, Veijonaho S, Toppinen A (2020) Towards sustainability? Forest-based circular bioeconomy business models in Finnish SMEs. Forest Policy Econ 110:101848. https://doi.org/10.1016/j.forpol.2018.12.004
96. D'Souza C, Taghian M, Sullivan-Mort G, Gilmore A (2015) An evaluation of the role of green marketing and a firm's internal practices for environmental sustainability. J Strateg Mark 23(7):600–615. https://doi.org/10.1080/0965254X.2014.1001866
97. De-la-Cruz-Diaz M, Alvarez-Risco A, Jaramillo-Arévalo M, de las Mercedes Anderson-Seminario M, Del-Aguila-Arcentales S (2022) 3D print, circularity, and footprints. In: Alvarez-Risco A, Muthu SS, Del-Aguila-Arcentales S (eds) Circular economy: impact on carbon and water footprint. Springer Singapore, pp 93–112. https://doi.org/10.1007/978-981-19-0549-0_5

98. de Bucourt M, Busse R, Güttler F, Wintzer C, Collettini F, Kloeters C, … Teichgräber UK (2011) Lean manufacturing and Toyota production system terminology applied to the procurement of vascular stents in interventional radiology. Insights Imaging 2(4):415–423. https://doi.org/10.1007/s13244-011-0097-0
99. de las Mercedes Anderson-Seminario M, Alvarez-Risco A (2022) Better students, better companies, better life: circular learning. In: Alvarez-Risco A, Muthu SS, Del-Aguila-Arcentales S (eds) Circular economy: impact on carbon and water footprint. Springer Singapore, pp 19–40. https://doi.org/10.1007/978-981-19-0549-0_2
100. de Medeiros JF, Ribeiro JLD (2017) Environmentally sustainable innovation: expected attributes in the purchase of green products. J Clean Prod 142:240–248. https://doi.org/10.1016/j.jclepro.2016.07.191
101. Del-Aguila-Arcentales S, Alvarez-Risco A (2013) Human error or burnout as explanation for mistakes in pharmaceutical laboratories. Accred Qual Assur 18(5):447–448. https://doi.org/10.1007/s00769-013-1000-0
102. Del-Aguila-Arcentales S, Alvarez-Risco A, Muthu SS (2022) Measuring circular economy. In: Alvarez-Risco A, Muthu SS, Del-Aguila-Arcentales S (eds) Circular economy: impact on carbon and water footprint. Springer Singapore, pp 3–17. https://doi.org/10.1007/978-981-19-0549-0_1
103. Del Giudice M, Chierici R, Mazzucchelli A, Fiano F (2021) Supply chain management in the era of circular economy: the moderating effect of big data. Int J Logistics Manage 32(2):337–356. https://doi.org/10.1108/IJLM-03-2020-0119
104. Delgado-Zegarra J, Alvarez-Risco A, Yáñez JA (2018) Indiscriminate use of pesticides and lack of sanitary control in the domestic market in Peru. Revista Panamericana de Salud Publica/Pan American Journal of Public Health, 42, Article e3. https://doi.org/10.26633/RPSP.2018.3
105. Desideri U, Proietti S, Sdringola P, Taticchi P, Carbone P, Tonelli F (2010) Integrated approach to a multifunctional complex. Manage Environ Qual Int J 21(5):659–679. https://doi.org/10.1108/14777831011067944
106. Donner M, Gohier R, de Vries H (2020) A new circular business model typology for creating value from agro-waste. Sci Total Environ 716:137065. https://doi.org/10.1016/j.scitotenv.2020.137065
107. dos Santos RA, Mexas MP, Meirino MJ, Sampaio MC, Costa HG (2020) Criteria for assessing a sustainable hotel business. J Clean Prod 262:121347. https://doi.org/10.1016/j.jclepro.2020.121347
108. Du B, Liu Q, Li G (2017) Coordinating leader-follower supply chain with sustainable green technology innovation on their fairness concerns. Int J Environ Res Public Health 14(11). https://doi.org/10.3390/ijerph14111357
109. Duarte Cueva F (2007) Entrepreneurship, enterprise and business growth [Emprendimiento, empresa y crecimiento empresarial]. Contabilidad y Negocios 2(3):46–55
110. Enciso-Zarate A, Guzmán-Oviedo J, Sánchez-Cardona F, Martínez-Rohenes D, Rodríguez-Palomino JC, Alvarez-Risco A, … Diaz-Risco S (2016) Evaluation of contamination by cytotoxic agents in colombian hospitals. Pharm Care Espana 18(6):241–250
111. Eppinger E, Jain A, Vimalnath P, Gurtoo A, Tietze F, Hernandez Chea R (2021) Sustainability transitions in manufacturing: the role of intellectual property. Curr Opin Environ Sustain 49:118–126. https://doi.org/10.1016/j.cosust.2021.03.018
112. Ertuna B, Karatas-Ozkan M, Yamak S (2019) Diffusion of sustainability and CSR discourse in hospitality industry. Int J Contemp Hosp Manage 31(6):2564–2581. https://doi.org/10.1108/IJCHM-06-2018-0464
113. Esmaeilpour M, Bahmiary E (2017) Investigating the impact of environmental attitude on the decision to purchase a green product with the mediating role of environmental concern and care for green products. Manage Mark 12(2):297
114. Esquerre-Botton S, Alvarez-Risco A, Leclercq-Machado L, de las Mercedes Anderson-Seminario M, Del-Aguila-Arcentales S (2022) Food loss reduction and carbon footprint practices worldwide: a benchmarking approach of circular economy. In: Alvarez-Risco A,

Muthu SS, Del-Aguila-Arcentales S (eds) Circular economy: impact on carbon and water footprint. Springer Singapore, pp 161–179. https://doi.org/10.1007/978-981-19-0549-0_8
115. Foo P-Y, Lee V-H, Tan GW-H, Ooi K-B (2018) A gateway to realising sustainability performance via green supply chain management practices: A PLS–ANN approach. Expert Syst Appl 107:1–14. https://doi.org/10.1016/j.eswa.2018.04.013
116. Fuergy. (n.d.). Fuergy. Retrieved 01/01/2022 from https://www.fuergy.com
117. Galleli B, Teles NEB, Santos JARd, Freitas-Martins MS, Hourneaux Junior F (2021) Sustainability university rankings: a comparative analysis of UI green metric and the times higher education world university rankings. Int J Sustain High Educ, ahead-of-print(ahead-of-print). https://doi.org/10.1108/IJSHE-12-2020-0475
118. Geissdoerfer M, Vladimirova D, Evans S (2018) Sustainable business model innovation: a review. J Clean Prod 198:401–416. https://doi.org/10.1016/j.jclepro.2018.06.240
119. Gerdt S-O, Wagner E, Schewe G (2019) The relationship between sustainability and customer satisfaction in hospitality: an explorative investigation using eWOM as a data source. Tour Manage 74:155–172. https://doi.org/10.1016/j.tourman.2019.02.010
120. Gómez-Prado R, Alvarez-Risco A, Sánchez-Palomino J, de las Mercedes Anderson-Seminario M, Del-Aguila-Arcentales S (2022) Circular economy for waste reduction and carbon footprint. In: Alvarez-Risco A, Muthu SS, Del-Aguila-Arcentales S (eds) Circular economy: impact on carbon and water footprint. Springer Singapore, pp 139–159. https://doi.org/10.1007/978-981-19-0549-0_7
121. Grandia J, Voncken D (2019) Sustainable public procurement: the impact of ability, motivation, and opportunity on the implementation of different types of sustainable public procurement. Sustainability 11(19):5215. https://doi.org/10.3390/su11195215
122. Grewal J, Hauptmann C, Serafeim G (2021) Material sustainability information and stock price informativeness. J Bus Ethics 171(3):513–544. https://doi.org/10.1007/s10551-020-04451-2
123. Guo M, Nowakowska-Grunt J, Gorbanyov V, Egorova M (2020) Green technology and sustainable development: assessment and green growth frameworks. Sustainability 12(16):6571. https://doi.org/10.3390/su12166571
124. Guo Y-F (2017) Green innovation design of products under the perspective of sustainable development. IOP Conf Ser Earth Environ Sci 51:012011. https://doi.org/10.1088/1742-6596/51/1/012011
125. Habib MA, Bao Y, Ilmudeen A (2020) The impact of green entrepreneurial orientation, market orientation and green supply chain management practices on sustainable firm performance. Cogent Bus Manage 7(1):1743616. https://doi.org/10.1080/23311975.2020.1743616
126. Hall MR (2019) The sustainability price: expanding environmental life cycle costing to include the costs of poverty and climate change. Int J Life Cycle Assess 24(2):223–236. https://doi.org/10.1007/s11367-018-1520-2
127. Halme M, Anttonen M, Kuisma M, Kontoniemi N, Heino E (2007) Business models for material efficiency services: conceptualization and application. Ecol Econ 63(1):126–137. https://doi.org/10.1016/j.ecolecon.2006.10.003
128. Handfield RB, Walton SV, Seegers LK, Melnyk SA (1997) 'Green' value chain practices in the furniture industry. J Oper Manag 15(4):293–315. https://doi.org/10.1016/S0272-6963(97)00004-1
129. Harmon J, Fairfield KD, Wirtenberg J (2010) Missing an opportunity: HR leadership and sustainability. People Strategy 33(1):16
130. Hoffman AJ, Sandelands LE (2005) Getting right with nature: anthropocentrism, ecocentrism, and theocentrism. Organ Environ 18(2):141–162. https://doi.org/10.1177/1086026605276197
131. Huang X-X, Hu Z-P, Liu C-S, Yu D-J, Yu L-F (2016) The relationships between regulatory and customer pressure, green organizational responses, and green innovation performance. J Clean Prod 112:3423–3433. https://doi.org/10.1016/j.jclepro.2015.10.106
132. Jones P, Comfort D (2020) The COVID-19 crisis and sustainability in the hospitality industry. Int J Contemp Hosp Manage 32(10):3037–3050. https://doi.org/10.1108/IJCHM-04-2020-0357

133. Juarez-Rojas L, Alvarez-Risco A, Campos-Dávalos N, de las Mercedes Anderson-Seminario M, Del-Aguila-Arcentales S (2022) Water footprint in the textile and food supply chain management: trends to become circular and sustainable. In: Alvarez-Risco A, Muthu SS, Del-Aguila-Arcentales S (eds) Circular economy: impact on carbon and water footprint. Springer Singapore, pp 225–243. https://doi.org/10.1007/978-981-19-0549-0_11
134. Junior SSB, da Silva D, Gabriel M, de Oliveira Braga WR (2018) The influence of environmental concern and purchase intent in buying green products. Asian J Behav Stud 3(12):183
135. Kallio L, Heiskanen E, Apajalahti E-L, Matschoss K (2020) Farm power: How a new business model impacts the energy transition in Finland. Energy Res Soc Sci 65:101484. https://doi.org/10.1016/j.erss.2020.101484
136. Kaur J, Sidhu R, Awasthi A, Chauhan S, Goyal S (2018) A DEMATEL based approach for investigating barriers in green supply chain management in Canadian manufacturing firms. Int J Prod Res 56(1–2):312–332. https://doi.org/10.1080/00207543.2017.1395522
137. Khalfallah M, Lakhal L (2021) The impact of lean manufacturing practices on operational and financial performance: the mediating role of agile manufacturing. Int J Qual Reliab Manage 38(1):147–168. https://doi.org/10.1108/IJQRM-07-2019-0244
138. Khan S, Henderson C (2020) How Western Michigan University is approaching its commitment to sustainability through sustainability-focused courses. J Clean Prod 253:119741. https://doi.org/10.1016/j.jclepro.2019.119741
139. Khan SAR, Razzaq A, Yu Z, Miller S (2021) Industry 4.0 and circular economy practices: a new era business strategies for environmental sustainability. Bus Strategy Environ 30(8):4001–4014. https://doi.org/10.1002/bse.2853
140. Kong L, Liu Z, Wu J (2020) A systematic review of big data-based urban sustainability research: state-of-the-science and future directions. J Clean Prod 273:123142. https://doi.org/10.1016/j.jclepro.2020.123142
141. Kozik N (2020) Sustainable packaging as a tool for global sustainable development. SHS Web of Conferences
142. Krishnan V, Ulrich KT (2001) Product development decisions: a review of the literature. Manage Sci 47(1):1–21
143. Kumar M (2008) The role of product design in value creation, transmission and interpretation: Implications for consumer preference. ProQuest
144. Kumar P (2015) Green marketing innovations in small Indian firms. World J Entrepreneurship, Manage Sustain Develop 11(3):176–190. https://doi.org/10.1108/WJEMSD-01-2015-0003
145. Larrán Jorge M, Andrades Peña FJ, Herrera Madueño J (2019) An analysis of university sustainability reports from the GRI database: an examination of influential variables. J Environ Plann Manage 62(6):1019–1044. https://doi.org/10.1080/09640568.2018.1457952
146. Leiva-Martinez MA, Anderson-Seminario MLM, Alvarez-Risco A, Estrada-Merino A, Mlodzianowska S (2021) Price variation in lower goods as of previous economic crisis and the contrast of the current price situation in the context of covid-19 in peru. In: Advances in business and management forecasting, vol 14, pp 161–166
147. Liebowitz J (2010) The role of HR in achieving a sustainability culture. J Sustain Develop 3(4):50–57
148. Lopez-Odar D, Alvarez-Risco A, Vara-Horna A, Chafloque-Cespedes R, Sekar MC (2020) Validity and reliability of the questionnaire that evaluates factors associated with perceived environmental behavior and perceived ecological purchasing behavior in Peruvian consumers. Soc Responsib J 16(3):403–417. https://doi.org/10.1108/SRJ-08-2018-0201
149. Lotfi M, Yousefi A, Jafari S (2018) The effect of emerging green market on green entrepreneurship and sustainable development in knowledge-based companies. Sustainability 10(7):2308. https://doi.org/10.3390/su10072308
150. Lozano R, Barreiro-Gen M, Lozano FJ, Sammalisto K (2019) Teaching sustainability in European Higher Education Institutions: assessing the connections between competences and pedagogical approaches. Sustainability 11(6):1602. https://doi.org/10.3390/su11061602

151. Mampra M (2013) Green HRM: Does it help to build a competitive service sector? A study. In: Proceedings of tenth AIMS international conference on management, vol 3, pp 1273–1281
152. Manavalan E, Jayakrishna K (2019) An analysis on sustainable supply chain for circular economy. Procedia Manuf 33:477–484. https://doi.org/10.1016/j.promfg.2019.04.059
153. Mani V, Gunasekaran A, Delgado C (2018) Supply chain social sustainability: standard adoption practices in Portuguese manufacturing firms. Int J Prod Econ 198:149–164. https://doi.org/10.1016/j.ijpe.2018.01.032
154. Manzoor SR, Ho JSY, Al Mahmud A (2021) Revisiting the 'university image model' for higher education institutions' sustainability. J Mark High Educ 31(2):220–239. https://doi.org/10.1080/08841241.2020.1781736
155. Martinez-Martinez A, Cegarra-Navarro J-G, Garcia-Perez A, Wensley A (2019) Knowledge agents as drivers of environmental sustainability and business performance in the hospitality sector. Tour Manage 70:381–389. https://doi.org/10.1016/j.tourman.2018.08.030
156. Martínez AN, Porcelli AM (2018) Study on the circular economy as a sustainable alternative to the decline of the traditional economy (first part) [Estudio sobre la economía circular como una alternativa sustentable frente al ocaso de la economía tradicional (primera parte)]. Lex: Revista de la Facultad de Derecho y Ciencia Política de la Universidad Alas Peruanas, 16(22):301–334
157. Masudin I, Wastono T, Zulfikarijah F (2018) The effect of managerial intention and initiative on green supply chain management adoption in Indonesian manufacturing performance. Cogent Bus Manage 5(1):1485212. https://doi.org/10.1080/23311975.2018.1485212
158. Maware C, Okwu MO, Adetunji O (2021) A systematic literature review of lean manufacturing implementation in manufacturing-based sectors of the developing and developed countries. Int JLean Six Sigma, ahead-of-print(ahead-of-print). https://doi.org/10.1108/IJLSS-12-2020-0223
159. Mejía-Acosta N, Alvarez-Risco A, Solís-Tarazona Z, Matos-Valerio E, Zegarra-Arellano E, Del-Aguila-Arcentales S (2016) Adverse drug reactions reported as a result of the implementation of pharmaceutical care in the Institutional Pharmacy DIGEMID—Ministry of Health. Pharma Care Espana 18(2):67–74
160. Melissen F, van Ginneken R, Wood RC (2016) Sustainability challenges and opportunities arising from the owner-operator split in hotels. Int J Hosp Manage 54:35–42. https://doi.org/10.1016/j.ijhm.2016.01.005
161. Melton T (2005) The benefits of lean manufacturing: What lean thinking has to offer the process industries. Chem Eng Res Des 83(6):662–673. https://doi.org/10.1205/cherd.04351
162. Mengelkamp E, Notheisen B, Beer C, Dauer D, Weinhardt C (2018) A blockchain-based smart grid: towards sustainable local energy markets. Comput Sci Res Dev 33(1):207–214. https://doi.org/10.1007/s00450-017-0360-9
163. Minniti M (2012) El emprendimiento y el crecimiento económico de las naciones. Economía industrial 383:23–30
164. Mital A, Desai A, Subramanian A, Mital A (2014) Product development: a structured approach to consumer product development, design, and manufacture. Elsevier
165. Mohamued EA, Ahmed M, Pypłacz P, Liczmańska-Kopcewicz K, Khan MA (2021) Global oil price and innovation for sustainability: the impact of R&D spending, oil price and oil price volatility on GHG emissions. Energies 14(6). https://doi.org/10.3390/en14061757
166. Morales-Ríos F, Alvarez-Risco A, Castillo-Benancio S, de las Mercedes Anderson-Seminario M, Del-Aguila-Arcentales S (2022) Material selection for circularity and footprints. In: Alvarez-Risco A, Muthu SS, Del-Aguila-Arcentales S (eds) Circular economy: impact on carbon and water footprint. Springer Singapore, pp 205–221. https://doi.org/10.1007/978-981-19-0549-0_10
167. Moser AK (2016) Consumers' purchasing decisions regarding environmentally friendly products: an empirical analysis of German consumers. J Retail Consum Serv 31:389–397. https://doi.org/10.1016/j.jretconser.2016.05.006
168. Mzembe AN, Lindgreen A, Idemudia U, Melissen F (2020) A club perspective of sustainability certification schemes in the tourism and hospitality industry. J Sustain Tour 28(9):1332–1350. https://doi.org/10.1080/09669582.2020.1737092

169. Nair S, Paulose H (2014) Emergence of green business models: the case of algae biofuel for aviation. Energy Policy 65:175–184. https://doi.org/10.1016/j.enpol.2013.10.034
170. Ndubisi NO, Nair SR (2009) Green entrepreneurship (GE) and green value added (GVA): a conceptual framework. Int J Entrepreneurship 13:21
171. Nikolaou IE, Tasopoulou K, Tsagarakis K (2018) A typology of green entrepreneurs based on institutional and resource-based views. J Entrepreneurship 27(1):111–132. https://doi.org/10.1177/0971355717738601
172. Nordin R, Hassan RA (2019) The role of opportunities for green entrepreneurship towards investigating the practice of green entrepreneurship among SMEs in Malaysia. Rev Integr Bus Econ Res 8:99–116
173. Norström AV, Cvitanovic C, Löf MF, West S, Wyborn C, Balvanera P, … Österblom H (2020) Principles for knowledge co-production in sustainability research. Nature Sustain 3(3):182–190. https://doi.org/10.1038/s41893-019-0448-2
174. Nueno P (2009) Entrepreneurship towards 2020. A renewed global perspective on the art of entrepreneurship and its artists [Emprendiendo hacia el 2020. Una renovada perspectiva global del arte de crear empresas y sus artistas] https://assets-libr.cantook.net/assets/publications/11729/medias/excerpt.pdf
175. Otache I, Mahmood R (2015) Corporate entrepreneurship and business performance: The role of external environment and organizational culture: A proposed framework, vol 6. https://www.mcser.org/journal/index.php/mjss/article/view/7318
176. Ozturk M (2019) Two samples of enviromental and green procurement at Turkcell. Procedia Comput Sci 158:921–928. https://doi.org/10.1016/j.procs.2019.09.132
177. Ozturkoglu Y, Sari FO, Saygili E (2021) A new holistic conceptual framework for sustainability oriented hospitality innovation with triple bottom line perspective. J Hosp Tour Technol 12(1):39–57. https://doi.org/10.1108/JHTT-02-2019-0022
178. Peeters P (2018) Why space tourism will not be part of sustainable tourism. Tour Recreat Res 43(4):540–543. https://doi.org/10.1080/02508281.2018.1511942
179. Peng H, Li B, Zhou C, Sadowski BM (2021) How does the appeal of environmental values influence sustainable entrepreneurial intention? Int J Environ Res Public Health 18(3). https://doi.org/10.3390/ijerph18031070
180. Peña Miguel N, Corral Lage J, Mata Galindez A (2020) Assessment of the development of professional skills in university students: Sustainability and serious games. Sustainability 12(3). https://doi.org/10.3390/su12031014
181. Pham H, Kim S-Y (2019) The effects of sustainable practices and managers' leadership competences on sustainability performance of construction firms. Sustain Prod Consumption 20:1–14. https://doi.org/10.1016/j.spc.2019.05.003
182. Plewnia F, Guenther E (2018) Mapping the sharing economy for sustainability research. Manage Decis 56(3):570–583. https://doi.org/10.1108/MD-11-2016-0766
183. Przychodzen J, Przychodzen W (2015) Relationships between eco-innovation and financial performance—evidence from publicly traded companies in Poland and Hungary. J Clean Prod 90:253–263. https://doi.org/10.1016/j.jclepro.2014.11.034
184. Quispe-Cañari JF, Fidel-Rosales E, Manrique D, Mascaró-Zan J, Huamán-Castillón KM, Chamorro–Espinoza SE, … Mejia CR (2021) Self-medication practices during the COVID-19 pandemic among the adult population in Peru: a cross-sectional survey. Saudi Pharm J 29(1):1–11. https://doi.org/10.1016/j.jsps.2020.12.001
185. Rajesh R (2020) Exploring the sustainability performances of firms using environmental, social, and governance scores. J Clean Prod 247:119600. https://doi.org/10.1016/j.jclepro.2019.119600
186. Ramísio PJ, Pinto LMC, Gouveia N, Costa H, Arezes D (2019) Sustainability strategy in higher education institutions: lessons learned from a nine-year case study. J Clean Prod 222:300–309. https://doi.org/10.1016/j.jclepro.2019.02.257
187. Ransfield AK, Reichenberger I (2021) Māori Indigenous values and tourism business sustainability. AlterNative: Int J Indigenous Peoples 17(1):49–60. https://doi.org/10.1177/1177180121994680

188. Rau H, Goggins G, Fahy F (2018) From invisibility to impact: Recognising the scientific and societal relevance of interdisciplinary sustainability research. Res Policy 47(1):266–276. https://doi.org/10.1016/j.respol.2017.11.005
189. Reinhardt R, Christodoulou I, García BA, Gassó-Domingo S (2020) Sustainable business model archetypes for the electric vehicle battery second use industry: towards a conceptual framework. J Clean Prod 254:119994. https://doi.org/10.1016/j.jclepro.2020.119994
190. Ritala P, Huotari P, Bocken N, Albareda L, Puumalainen K (2018) Sustainable business model adoption among S&P 500 firms: a longitudinal content analysis study. J Clean Prod 170:216–226. https://doi.org/10.1016/j.jclepro.2017.09.159
191. Rojas-Osorio M, Alvarez-Risco A (2019) Intention to use smartphones among Peruvian university students. Int J Interact Mobile Technol 13(3):40–52. https://doi.org/10.3991/ijim.v13i03.9356
192. Román BR, Moscoso S, Chung SA, Terceros BL, Álvarez-Risco A, Yáñez JA (2020) Treatment of COVID-19 in peru and bolivia, and self-medication risks. Revista Cubana de Farmacia 53(2):1–20, Article e435
193. Rosenbaum E (2017) Green growth—magic bullet or damp squib? Sustainability 9(7). https://doi.org/10.3390/su9071092
194. Rossignoli F, Lionzo A (2018) Network impact on business models for sustainability: case study in the energy sector. J Clean Prod 182:694–704. https://doi.org/10.1016/j.jclepro.2018.02.015
195. Rouhollahi Z, Ebrahimi-Nik M, Ebrahimi SH, Abbaspour-Fard MH, Zeynali R, Bayati MR (2020) Farm biogas plants, a sustainable waste to energy and bio-fertilizer opportunity for Iran. J Clean Prod 253:119876. https://doi.org/10.1016/j.jclepro.2019.119876
196. Saleem F, Adeel A, Ali R, Hyder S (2018) Intentions to adopt ecopreneurship: moderating role of collectivism and altruism. Entrepreneurship Sustain Issues 6(2):517–537. https://doi.org/10.9770/jesi.2018.6.2(4)
197. Salmerón-Manzano E, Manzano-Agugliaro F (2018) The higher education sustainability through virtual laboratories: the spanish university as case of study. Sustainability 10(11). https://doi.org/10.3390/su10114040
198. Schaltegger S, Lüdeke-Freund F, Hansen EG (2012) Business cases for sustainability: the role of business model innovation for corporate sustainability. Int J Innov Sustain Develop 6(2):95–119
199. Schaper M, Volery T, Weber P, Lewis K (2010) Entrepreneurship and small business. Wiley
200. Sdrolia E, Zarotiadis G (2019) A comprehensive review for green product term: from definition to evaluation. J Econ Surv 33(1):150–178. https://doi.org/10.1111/joes.12268
201. Seay JR (2015) Education for sustainability: developing a taxonomy of the key principles for sustainable process and product design. Comput Chem Eng 81:147–152. https://doi.org/10.1016/j.compchemeng.2015.03.010
202. Serafeim G (2020) Public sentiment and the price of corporate sustainability. Financ Anal J 76(2):26–46. https://doi.org/10.1080/0015198X.2020.1723390
203. Seyfang G, Gilbert-Squires A (2019) Move your money? Sustainability transitions in regimes and practices in the UK retail banking sector. Ecol Econ 156:224–235. https://doi.org/10.1016/j.ecolecon.2018.09.014
204. Sharma N, Kushwaha G (2015) Emerging green market as an opportunity for green entrepreneurs and sustainable development in India. J Entrepreneurship Organ Manage 4(2):2–7
205. Shuqin C, Minyan L, Hongwei T, Xiaoyu L, Jian G (2019) Assessing sustainability on Chinese university campuses: development of a campus sustainability evaluation system and its application with a case study. J Build Eng 24:100747. https://doi.org/10.1016/j.jobe.2019.100747
206. Silajdžić I, Kurtagić SM, Vučijak B (2015) Green entrepreneurship in transition economies: a case study of Bosnia and Herzegovina. J Clean Prod 88:376–384. https://doi.org/10.1016/j.jclepro.2014.07.004

207. Soenarto S, Rahmawati R, Suprapti A, Handayani R, Sudira P (2018) Green entrepreneurship development strategy based on local characteristic to support power eco-tourism continuous at Lombok. J Tourism Hospitality 7(6):2–15. https://doi.org/10.4172/2167-0269.1000394
208. Sroufe R, Gopalakrishna-Remani V (2018) Management, social sustainability, reputation, and financial performance relationships: an empirical examination of U.S. firms. Organ Environ 32(3):331–362. https://doi.org/10.1177/1086026618756611
209. Stahel WR (2016) The circular economy. Nature 531(7595):435–438. https://doi.org/10.1038/531435a
210. Statista (2021) Global startups—statistics & facts. Retrieved 01/01/2022 from https://www.statista.com/topics/4733/startups-worldwide/#dossierKeyfigures
211. Susanty A, Sari DP, Rinawati DI, Setiawan L (2019) The role of internal and external drivers for successful implementation of GSCM practices. J Manuf Technol Manage 30(2):391–420. https://doi.org/10.1108/JMTM-07-2018-0217
212. Teufel B, Sentic A, Barmet M (2019) Blockchain energy: blockchain in future energy systems. J Electron Sci Technol 17(4):100011. https://doi.org/10.1016/j.jnlest.2020.100011
213. Tippayawong KY, Niyomyat N, Sopadang A, Ramingwong S (2016) Factors affecting green supply chain operational performance of the Thai auto parts industry. Sustainability 8(11):1161. https://doi.org/10.3390/su8111161
214. Todeschini BV, Cortimiglia MN, Callegaro-de-Menezes D, Ghezzi A (2017) Innovative and sustainable business models in the fashion industry: Entrepreneurial drivers, opportunities, and challenges. Bus Horiz 60(6):759–770. https://doi.org/10.1016/j.bushor.2017.07.003
215. Topleva SA, Prokopov TV (2020) Integrated business model for sustainability of small and medium-sized enterprises in the food industry. Brit Food J 122(5):1463–1483. https://doi.org/10.1108/BFJ-03-2019-0208
216. Trapp CTC, Kanbach DK (2021) Green entrepreneurship and business models: deriving green technology business model archetypes. J Clean Prod 297:126694. https://doi.org/10.1016/j.jclepro.2021.126694
217. Trosper RL (2002) Northwest coast indigenous institutions that supported resilience and sustainability. Ecol Econ 41(2):329–344. https://doi.org/10.1016/S0921-8009(02)00041-1
218. Tseng M-L, Islam MS, Karia N, Fauzi FA, Afrin S (2019) A literature review on green supply chain management: trends and future challenges. Resour Conserv Recycl 141:145–162. https://doi.org/10.1016/j.resconrec.2018.10.009
219. Tunn V, Dekoninck E (2016) How does sustainability help or hinder innovation? In: Setchi R, Howlett RJ, Liu Y, Theobald P (eds) Sustainable design and manufacturing 2016. Springer International Publishing, pp 73–83. https://doi.org/10.1007/978-3-319-32098-4_7
220. Upadhyay P, Kumar A (2020) A house of sustainability-based approach for green product design. Manage Environ Qual Int J 31(4):819–846. https://doi.org/10.1108/MEQ-03-2019-0057
221. Vanalle RM, Ganga GMD, Godinho Filho M, Lucato WC (2017) Green supply chain management: an investigation of pressures, practices, and performance within the Brazilian automotive supply chain. J Clean Prod 151:250–259. https://doi.org/10.1016/j.jclepro.2017.03.066
222. Vijayvargy L, Thakkar J, Agarwal G (2017) Green supply chain management practices and performance. J Manuf Technol Manage 28(3):299–323. https://doi.org/10.1108/JMTM-09-2016-0123
223. Vizcardo D, Salvador LF, Nole-Vara A, Dávila KP, Alvarez-Risco A, Yáñez JA, Mejia CR (2022) Sociodemographic predictors associated with the willingness to get vaccinated against COVID-19 in Peru: a cross-sectional survey. Vaccines 10(1), Article 48. https://doi.org/10.3390/vaccines10010048
224. Weber T, Brain D, Mitchell D, Xu S, Espley J, Halekas J, … Jakosky B (2019) The influence of solar wind pressure on martian crustal magnetic field topology. Geophys Res Lett 46(5):2347–2354. https://doi.org/10.1029/2019GL081913
225. Xihua X (1999) Ecological design: The inevitable chioce of the development of the industrial design. Journal of Zhejiang University, 4. https://en.cnki.com.cn/Article_en/CJFDTotal-ZJDX199904022.htm

226. Yahaya AS, Javaid N, Alzahrani FA, Rehman A, Ullah I, Shahid A, Shafiq M (2020) Blockchain based sustainable local energy trading considering home energy management and demurrage mechanism. Sustainability 12(8):3385. https://doi.org/10.3390/su12083385
227. Yan J, Kim S, Zhang SX, Foo MD, Alvarez-Risco A, Del-Aguila-Arcentales S, Yáñez JA (2021) Hospitality workers' COVID-19 risk perception and depression: a contingent model based on transactional theory of stress model. Int J Hospitality Manage 95, Article 102935. https://doi.org/10.1016/j.ijhm.2021.102935
228. Yáñez JA, Alvarez-Risco A, Delgado-Zegarra J (2020) Covid-19 in Peru: from supervised walks for children to the first case of Kawasaki-like syndrome. The BMJ, 369, Article m2418. https://doi.org/10.1136/bmj.m2418
229. Yáñez JA, Chung SA, Román BR, Hernández-Yépez PJ, Garcia-Solorzano FO, Del-Aguila-Arcentales S, … Alvarez-Risco A (2021) Prescription, over-the-counter (OTC), herbal, and other treatments and preventive uses for COVID-19. In: Hadi Dehghani M, Karri RR, Roy S (eds) Environmental and health management of novel coronavirus disease (COVID-19), chap 14. Academic Press, pp 379–416. https://doi.org/10.1016/B978-0-323-85780-2.00001-9
230. Yáñez JA, Jahanshahi AA, Alvarez-Risco A, Li J, Zhang SX (2020) Anxiety, distress, and turnover intention of healthcare workers in Peru by their distance to the epicenter during the COVID-19 crisis. Am J Trop Med Hyg 103(4):1614–1620. https://doi.org/10.4269/ajtmh.20-0800
231. Yáñez S, Uruburu Á, Moreno A, Lumbreras J (2019) The sustainability report as an essential tool for the holistic and strategic vision of higher education institutions. J Clean Prod 207:57–66. https://doi.org/10.1016/j.jclepro.2018.09.171
232. Yarime M, Tanaka Y (2012) The issues and methodologies in sustainability assessment tools for higher education institutions: a review of recent trends and future challenges. J Educ Sustain Dev 6(1):63–77. https://doi.org/10.1177/097340821100600113
233. Yasmeen H, Wang Y, Zameer H, Ismail H (2020) Modeling the role of government, firm, and civil society for environmental sustainability. In: Developing eco-cities through policy, planning, and innovation: can it really work?. IGI Global, pp 62–83. https://doi.org/10.4018/978-1-7998-0441-3.ch003
234. Yip AWH, Bocken NMP (2018) Sustainable business model archetypes for the banking industry. J Clean Prod 174:150–169. https://doi.org/10.1016/j.jclepro.2017.10.190
235. Yung WKC, Chan HK, Choi ACK, Yue TM, Mazhar MI (2008) An environmental assessment framework with respect to the requirements of energy-using products directive. Proc Inst Mech Eng, Part B: J Eng Manuf 222(5):643–651. https://doi.org/10.1243/09544054JEM977
236. Zameer H, Wang Y, Yasmeen H (2020) Reinforcing green competitive advantage through green production, creativity and green brand image: implications for cleaner production in China. J Clean Prod 247:119119. https://doi.org/10.1016/j.jclepro.2019.119119
237. Zamora-Polo F, Sánchez-Martín J (2019) Teaching for a better world. sustainability and sustainable development goals in the construction of a Change-Maker University. Sustainability 11(15). https://doi.org/10.3390/su11154224
238. Zhang SX, Chen J, Afshar Jahanshahi A, Alvarez-Risco A, Dai H, Li J, Patty-Tito RM (2021) Succumbing to the COVID-19 pandemic—healthcare workers not satisfied and intend to leave their jobs. Int J Ment Heal Addict. https://doi.org/10.1007/s11469-020-00418-6
239. Zhang SX, Chen J, Jahanshahi AA, Alvarez-Risco A, Dai H, Li J, Patty-Tito RM (2021) Correction to: succumbing to the COVID-19 pandemic—healthcare workers not satisfied and intend to leave their jobs (international journal of mental health and addiction. Int J Ment Heal Addict. https://doi.org/10.1007/s11469-021-00502-5
240. Zhang SX, Sun S, Afshar Jahanshahi A, Alvarez-Risco A, Ibarra VG, Li J, Patty-Tito RM (2020) Developing and testing a measure of COVID-19 organizational support of healthcare workers—results from Peru, Ecuador, and Bolivia. Psychiatry Res 291. https://doi.org/10.1016/j.psychres.2020.113174
241. Zhou W, Su D, Yang J, Tao D, Sohn D (2021) When do strategic orientations matter to innovation performance of green-tech ventures? The moderating effects of network positions. J Clean Prod 279:123743. https://doi.org/10.1016/j.jclepro.2020.123743

Effectiveness of Renewable Energy Policies in Promoting Green Entrepreneurship: A Global Benchmark Comparison

Luis Juarez-Rojas, Aldo Alvarez-Risco, Nilda Campos-Dávalos, María de las Mercedes Anderson-Seminario, and Shyla Del-Aguila-Arcentales

Abstract Renewable energy policies in some countries are essential for creating a sustainable energy system and reducing carbon emissions. The countries apply different policies to encourage the use of renewable energy whose primary purpose is the promotion, transition, and use of renewable energy. These policies provide incentives that allow companies to benefit, as long as they choose to use renewable sources in their economic activities. This chapter has evaluated renewable energy policies in 20 countries (Liechtenstein, Norway, Sweden, Iceland, Uruguay, Denmark, Niue, Tajikistan, Portugal, Costa Rica, United States, China, Japan, Germany, United Kingdom, India, France, Italy, Canada, and Russia). The analysis focused on the influence of these policies on creating green entrepreneurship. In general, it has been determined that, in many cases, renewable energy policies contribute to creating green entrepreneurship.

Keywords Renewable energy · Renewable energy policies · Green entrepreneurship · Government policies · Sustainability · Sustainable development goals · Environment · Carbon emissions · Renewable sources · Entrepreneurship · Developed countries · Circularity · Entrepreneurs

1 Introduction

With the emergence of globalization, the world has become increasingly industrialized, leading to a deterioration of the environment. Consequently, with the arrival of the twenty-first century, awareness was generated by the population. This problem generated new opportunities for entrepreneurs as new markets were created

L. Juarez-Rojas · A. Alvarez-Risco · N. Campos-Dávalos ·
M. de las Mercedes Anderson-Seminario
Universidad de Lima, Lima, Peru

S. Del-Aguila-Arcentales (✉)
Escuela de Posgrado, Universidad San Ignacio de Loyola, Lima, Peru
e-mail: sdelaguila@usil.edu.pe

A. Alvarez-Risco et al. (eds.), *Footprint and Entrepreneurship*, Environmental Footprints and Eco-design of Products and Processes,
https://doi.org/10.1007/978-981-19-8895-0_3

with consumers oriented to sustainability, motivating the creation of ideas, innovations, and projects related to the care of the environment, also known as green entrepreneurship.

The management of the energy is part of sustainability which is important for people [11, 13, 15, 21, 28, 33, 42–44, 53, 58, 74, 104, 127, 144, 188, 200, 207, 217, 222], institutions [5, 22, 26, 35, 62–64, 134, 141, 170, 191, 194, 202, 219, 220] and firms [6, 14, 18, 32, 57, 61, 85, 90, 143, 173, 176, 198] and activities as tourism [9, 12, 19, 67–69, 158, 169, 178], education [4, 15, 22, 31, 35, 38, 39, 41, 73, 136, 177], circular economy approach [2, 10, 24, 40, 60, 92, 128, 140, 142], prices [51, 52, 55, 56, 112, 116, 136, 154, 192], hospitality [3, 20, 99, 106, 124, 145, 167, 216], intellectual property [7, 25, 30, 50, 59, 98, 148], health [1, 8, 16, 17, 23, 27–29, 34, 36, 45–47, 65, 76, 86, 92, 95, 147, 175, 189, 218, 225–227], and research [79, 130, 161, 163, 164, 174, 179, 190].

The current chapter analyzes how renewable energy policies contribute to green entrepreneurship in 20 selected countries. The first group corresponds to the top 10 countries with the most efficient renewable energy policies, which were extracted from the research by [105]. The second group comprises the 10 countries with the highest economic development based on their GDP. The result of the research focuses on the effect of renewable energy policies in creating green entrepreneurship. However, there are still some factors, mainly social and economic, that hinder the task.

2 Green Entrepreneurship

First of all, the definition of green entrepreneurship includes concepts related to innovation and development of products that, apart from catering to consumers' needs, also help in the process of Sustainable Development but in the long term [115]. Now, there is a growing awareness about the problem of climate change, which can be seen through consumers' behavior, which means new opportunities to do business for entrepreneurs and for companies to bring innovation into their products or development for new ones based on the green market scenario.

Haldar [115] also mentioned that green entrepreneurship is relatively new because it got the attention in the 1990s, which means the concept is far less researched than other similar concepts such as entrepreneurship. The studies are based on empirical research on actual markets, however, there has been a recent shift, and more studies help set a base for developing businesses. Considering that businesses now have a more exigent environment than a few years ago, these studies want to set a basis on what the exigencies may be for future ones.

2.1 Factors of Green Entrepreneurship

The world urgently addresses a climate crisis through economic context and strong political leadership. The planet must consider creating new opportunities to transition to a future with a cleaner economy as a whole and a product life cycle (production, introduction, growth, maturity, and decline) that is entirely environmentally friendly. Environmental changes, driven primarily by human demand for resources, rising living standards, and world population growth, are accelerating and pose new challenges. Environmental trends bring threats and vulnerabilities and provide new sustainable business opportunities for the green economy. For this reason, green entrepreneurship has been increasing worldwide as awareness of caring for the planet has grown. The existence of a strengthened ecosystem is a factor that affects the competitiveness of a country; in this sense, the generation of an entrepreneurial culture requires the presence of a public policy on entrepreneurship that constitutes the basis for the formation of an entrepreneurial ecosystem [110].

The motivation of globalization is seen in global events; a consumer has become more and more advertised and aware of their responsibilities with the globalization of the environment. The motivation in global events has forced the production industry to have a more strategic time. Time to plan its processes and positions. Their goods and services. We saw significant growth in green companies respecting the environment, especially in a country rich in biodiversity. Green businesses related to economic activities are based on the provision of services and products to achieve positive environmental impacts and, at the same time, allow good development of environmental practices in organizations.

The value of business is at risk from developing global environmental conditions. High risks, but at the same time there are interesting new business opportunities for companies. Assessing the significance of environmental trends throughout the product life cycle, begins with the training of raw materials for recycling or entertainment [204] and combine these trends in activities and fabrics. The figure plans, improves the competitiveness of companies, reducing the competitiveness of environmentally degrading companies and improving the happy people [205]. Companies, cities, and countries are more aware that when they subject themselves to unsustainable economic sanctions, they can also gain economic benefits if they seriously embrace sustainability [117]. There is evidence from international business experience in the green economy, people with leading characteristics are companies that are environment and sustainable development concerns in management practice, mission, and surname strategy. These are cases of companies with concerns that do not come from a legal attitude or a legal attitude of environmental benefits, which are a strategic or creative position, related to the lips, the school is as an essential element of the strategy and development of vision, commitment, and the environment. Leadership and environment consistent with its business value [137].

2.2 *Factors to the Development of Renewable Energy*

It is essential to mention the factors that help the growth of this type of energy to understand how effective energy policies create green entrepreneurship based on renewable energies. The main factors to consider are economic, technological, social, and political. As [215] explain, the most important factors among these four would be the economic and policy ones, Through the purpose of this paper, we delve into them.

Economic Factors
Renewable energy and its development directly correlate with the development of a country [215]. The economic situation of a country has a relation to its GDP because the economic income of its population determines the consumption of renewable energy. Additionally, the promotion of economic development is the basis for developing clean energy [129], which requires the corresponding financial support, which requires different strategies, so that said investment is sustainable and that it can be sustained due to the savings generated for the eco-efficiency achieved [146]. As a result, we can analyze the impact of economic growth on the growth of renewable energy. The economic situation includes various factors, however, we use GDP as a representative indicator of this factor. As Fig. 1 describes the United States and China lead in GDP, which would mean they have more financial ground to invest in renewable energy and promote green entrepreneurship; however, this is not necessarily the case.

Policy Factors
Political factors promote legislation incorporating legislation that points toward renewable energy in a country [215]. The ecological regulation that may be generated directly influences business activities [114]. For the regulation obtained to have

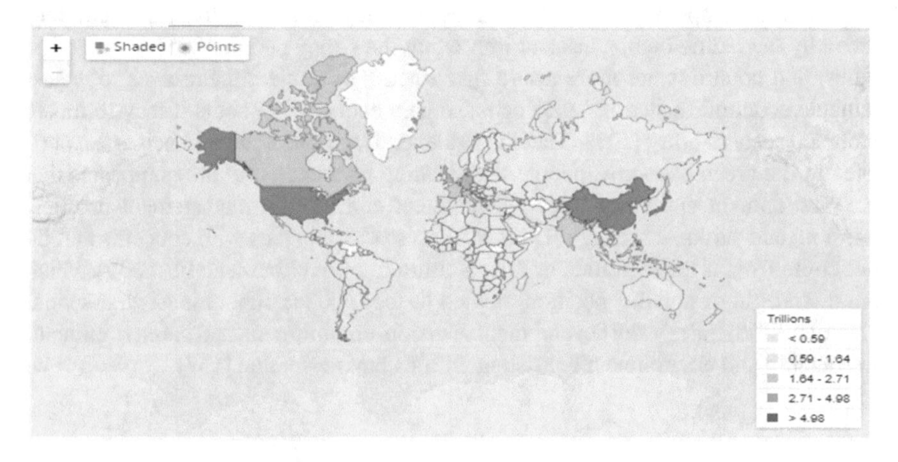

Fig. 1 World GDP (Current US$). *Source* World Bank [212]

results, the government's active role is needed to generate tax facilities for companies and generate programs that help efficiently transmit information to consumers [223]. Globally, different strategies are established to encourage more institutions to generate ecological impacts [215].

Renewable energy policy performance, which relates to its effectiveness and efficiency, depends on several essential elements. For example, policies with a sound design have no outstanding performance of renewable energy development if other critical factors are not considered, such as a sustainable growing cost recuperation mechanism (purchased through sustainable subsidy sources or a surcharge on the tariffs of the consumer). Additionally, transmission infrastructure with renewable energy integration and clear connection and transmission access [212]. Furthermore, developing countries tried various instruments to develop renewable energy, which has shifted over the years. Additionally, most of them are currently using both quantity and price positioning instruments to support different renewable energy market segments. However, policy shifts should be managed through systems that permit stakeholders to control the risks to keep regulatory reliability [212].

Other Factors

On the one hand, *technical factors* include renewable energy technologies, which are crucial to ensure a competitive and inexpensive energy supply [183]. Encouraging the development of these renewable energy technologies reduces maintenance costs and improves the efficiency needed to promote power generation [93]. On the other hand, *social factors* refer to the population's awareness of sustainable energy. Above all, these are also related to the sustainability indicator system developed in Agenda 21. [215] concluded that most countries are aware of the importance of caring for the environment, so this factor rarely is a problem or boundary for developing renewable energy worldwide.

Table 1 shows a brief overview of the importance and meaning of every factor; later in this investigation, we see these factors involved in the policies made by each country.

2.3 Green Entrepreneurship Cases Linked to Renewable Energy

There are several terms to identify the concept of "green entrepreneurship". The most common terms are eco entrepreneurship, environmental entrepreneurship, sustainable entrepreneurship, enviro-preneurship, ecological entrepreneurship, or sustainopreneurship [162]. Although some of these concepts seem the same, there are specific differences in objectives, ideals, or structure. In short, green entrepreneurship is an economic activity that aims to reduce the adverse effects of products, services, or production methods on the environment [107]. Alternatively, entrepreneurship provides renewable energy services.

Table 1 Factors influencing the development of renewable energy policies

Factors	Framework
Economic factor	With economic growth, it is feasible to finance investment, resist the financial risks brought about by the development and utilization of renewable energy
Political factor	Governments should also improve welfare through renewable energy incentives, including subsidies or tariffs
Social factor	Most of the countries are aware of the importance of caring for the environment, so this factor is not a boundary for the development of renewable energy
Technical factor	Encouraging renewable energy technologies developed to reduce its maintenance costs and improve reliability, applicability, and energy conversion efficiency

Source [215]

The examples shown below are entrepreneurship that provides innovations through renewable energies. It is essential to mention that these are cases of entrepreneurship that are not necessarily related to the countries that be analyzed later.

1. **Rathi Solar (India)**: A entrepreneurship that emerged in 2016 to supply solar installations that fit the optimal space required by customers. It is a company that has 2–10 employees, including independent contractors, energy professionals, among others. It is a green enterprise because it contributes to renewable sources, mainly solar energy. Its impact as a company has been significant because it has been able to bring energy to places where its government has not reached. At the environmental level, it has managed to stop customers from using carbon-based energy sources [193].
2. **Bonergie (Senegal)**: It is a company that offers solar energy-based products capable of providing electricity, water, and light. Currently, this company is increasing its market share in Senegal. A market that had not been attractive to some companies. The company's main objective is to offer solutions to social problems, thus contributing directly to improving environmental conditions [208].
3. **Eavor (Canada)**: It is a closed-loop geothermal energy company. Its goal is to create sustainable energy systems that preserve the environment. The company was founded in 2017 and, to date, has 11–50 employees. It is an excellent example of promoting renewable energies to generate economic benefits [138].

3 Renewable Energy Policies

First, renewable energy can be defined as a relevant energy supply complement able to protect the ecological territory and balance supply and optimize structures of

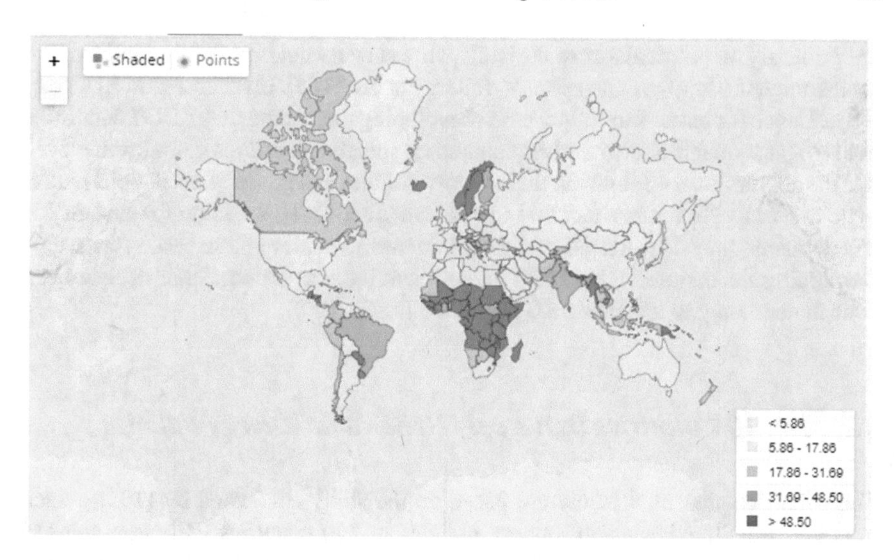

Fig. 2 Renewable energy consumption (% of total final energy consumption). *Source* World Bank [213]

this energy. The International Renewable Energy Agency explained that this type of energy must be almost 60% of the global energy supply by 2050. Promoting the utilization and growth of renewable energy has become a usual discussion and needs the effort of every country globally [215]. Figure 2 shows an overview of renewable energy consumption in the world.

It is essential to acknowledge that implementing renewable energy policies does not only mean setting boundaries or obligations. These policies make high expectations in entrepreneurs and are supposed to be met once in the field [78]. They are made to promote green entrepreneurship, create opportunities and incentives, and make this business easier and accessible.

3.1 Renewable Energy Policies' Contribution to the Accomplishment of SDGs

The United Nations' Sustainable Development Goal (SDG) 7 aims to reach energy security, primarily affordable and clean energy parameters, by 2030. The objective of the SDG 7 is to "endure access to affordable, reliable, sustainable, and modern energy for all" [203]. One of its targets is focused on renewable energy, with the reach to "substantially escalate the share of renewable energy in the global energy mix" [105].

The importance of the implementation of policies relies heavily on some countries for the achievement of the SDG 7, for example, energy problems in Africa are primarily on political terms; basically, almost every country in this continent has issues about the supply, expansion, and energy prices [165]. In other words, SDGs are the guidance for nations to follow when developing their strategies. SDG7 details the goals for sustainable energy and explains some specific objectives globally, however, it is the country's choice how far their energy policies and government actions participate in them. While it is a fact that SDGs lead the global discussion for sustainable development, they do not necessarily explain how to achieve these goals concretely. Combining the nations' efforts and giving them tools to support their decision can contribute mainly to achieving SDG7's 2030 targets [66].

3.2 Top 10 Countries that Apply Renewable Energy Policies

The countries presented below are based on the study conducted by [105], which evaluates the effectiveness of energy policies in 230 countries. It is important to highlight the term "effectiveness", which is not directly related to renewable energy production. If we were to compare energy production in Norway and Liechtenstein, Norway would have a significant advantage over Liechtenstein. The efficiency of energy policies has been evaluated using the single composite interval-based indicator. The ranking of the countries with the most efficient energy policies is presented below:

Liechtenstein
There is no scientific research analyzing Liechtenstein's energy policies, only descriptive reports, and Internet information from the companies in charge of energy supply in Liechtenstein. Liechtenstein's relations with Switzerland and Austria are crucial [131]. These two countries–especially Switzerland, with 98% of electricity imports–are predominant energy suppliers in the country of reference [96]. Despite this, strategies are being implemented at the government level to generate more independence regarding electricity production. Liechtenstein's energy policies are based on three critical pillars: (1) increased energy efficiency, (2) increasing the use of renewable energy sources, and (3) reducing energy consumption [131]. All this is based on the Law on the Promotion of Energy Efficiency and Renewable Energy. Currently, energy access in society is 100% [80].

To encourage the use of renewable energies, the government provides incentives such as financial subsidies for renewable energy sources and improved insulation in buildings [111]. In addition, it is constantly searching for alternatives to increase the consumption of renewable energy sources. To such an extent that, in mid-2015, it became the country with the most prominent photovoltaic installations in the world [181]. Despite its limited geographical extension, Liechtenstein is above countries such as Germany in terms of PV installations [111]. In general, this country has shown that despite being a small country of other European countries, it is leading one of the most significant efforts to implement sustainable energy policies. Despite this, the government continues to encounter limitations mainly at the geographical level since installing renewable energy sources requires space, which this country does not have [131].

Norway

The Nordic countries pursue aggressive energy and climate policies to reduce carbon emissions by decarbonizing electricity, heat, and buildings [196]. According to the [152], 98% of electricity production comes from renewable energy sources, mainly from hydropower sources from rivers and waterfalls. In ancient times, Norway was supplied with electricity using oil and gas, which focused its energy policies on these resources. However, Norway is developing as a hydropower nation [156]. Electricity consumption per capita has increased over the last few years. Household consumption, for example, uses mainly hydropower, with 95.8% recorded in 2016 [49].

Norway's energy policies have achieved effective results based on the increased participation of renewable sources over fossil fuels. Figure 3 details the share of each renewable source in electricity production.

Hydropower in Norway accounts for approximately 91.82% of the electricity produced in this country. Policies are aimed at increasing the energy production capacity through this source. However, a slight decrease can be observed due to the energy crisis occurring in Norway. Current renewable energy policies in Norway aim at fulfilling four priority areas. Table 2 presents what these main areas are.

The four primary areas of Norway's energy policies are explained in The White Paper Meld. St. 25 (2015–2016) [180]. First, the security of energy supply refers to the ability of the Norwegian electricity system to provide continuous electricity. Both

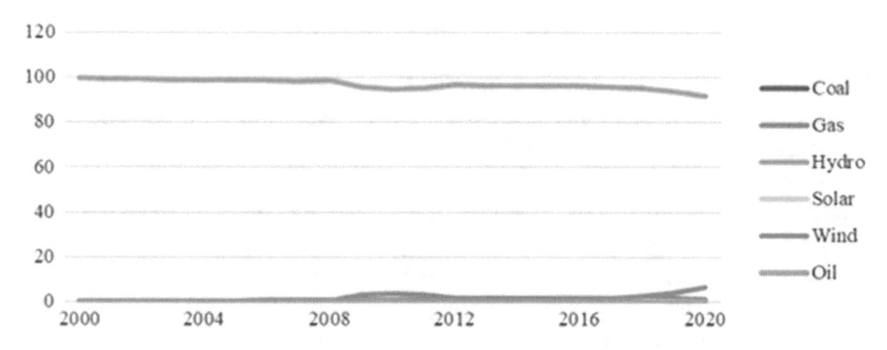

Fig. 3 Share of electricity production by source. *Source* Our World in Data [166]

Table 2 Four main priority areas for the Norwegian energy policy

Scope 1	Scope 2	Scope 3	Scope 4
Improving the security of supply	Profitable development of renewable energy	More efficient and climate-friendly energy use	Value creation based on Norway's renewable energy resources

Source [153]

households and businesses depend on a reliable power supply. Second, Norway seeks to produce energy cost-effectively. In this way, more energy is provided at a lower price. Thirdly, the implementation of energy policies must be aligned with reducing carbon emissions. However, some economic activities such as transportation or oil extraction continue to generate emissions. Finally, it is estimated that the renewable energy industry employs 20,000 people in the country. Therefore, energy policies should evaluate the economic benefits that renewable energy companies can generate and the social welfare in Norway.

Another set of policies related to the use of renewable energy in Norway is linked to technical regulations for building construction. This sector has great potential for carbon emission reduction through increased energy efficiency [49]. Building policies aim to reduce up to 50% of carbon emissions by 2030. Even though technology, emission-free construction equipment is developed and supplied to companies in the industry. Since 2020, it has been mandatory for non-residential buildings more significant than 1000 m^2 to have an energy performance certificate detailing information on the energy needs of the buildings. In addition, it is helpful to generate awareness about energy use in them [153].

Implementing renewable energy policies does not always ensure the country's economic sustainability. The main consequence of energy policies in Norway is the difficulty in raising private capital for renewable energy investments [211]. Another necessary consequence is the clash of interests between maximizing natural resources to generate profitability and reducing climate change. For example, a major political issue in Norway is exploring and extracting oil outside Lofoten (an archipelago in Norway). However, in a country where an emerging discourse on climate change prevails, the interests of oil entrepreneurs are affected [120].

Sweden

Sweden leads countries seeking a transition to a low-carbon society [121]. The primary energy sources in this country come from hydropower (45%) and nuclear (30%) [201]. Currently, many researchers are evaluating the possibility that Sweden's electricity system could be supplied exclusively from renewable energy sources without nuclear sources [126]. However, this possibility continues to be evaluated as replacing nuclear energy with renewables does not generate economic or ecological benefits concerning climate [119]. This issue involves the participation of the political environment, even some political groups agree to stop subsidizing nuclear energy to initiate a transition to renewable sources [121].

Sweden's energy policies continue to decrease the use of nuclear sources toward a transition to renewable sources. Figure 4 details the share of each renewable source in electricity production.

Electricity production from nuclear sources has decreased by 9% between 2019 and 2020, which hints at the effectiveness of finding solutions and management alternatives for the sustainable energy transition. In general, the abundance of energy from hydropower sources ensures a low-cost renewable energy system without using nuclear energy sources [126], which could secure the goal of decarbonizing the electricity system in Sweden. Biomass and wind energy are the main options for

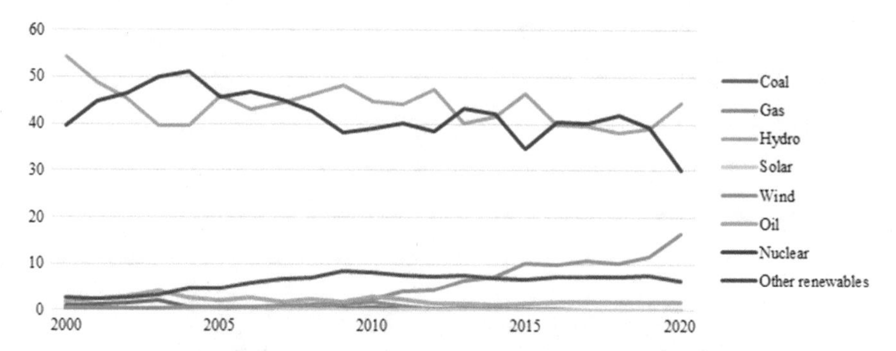

Fig. 4 Share of electricity production by source. *Source* Our World in Data [166]

renewable energy production, mainly, the latter has the highest growth in Sweden [210]. The installation of wind farms has increased considerably in Sweden during the last decade [97]. In this country, the government can designate those spaces with the necessary characteristics for installing wind plants as a "national interest" [135].

Since 2003, Sweden has introduced a green certificate system to secure renewable energy sources [135]. This certificate is a support system that replaces the previous public subsidies and grants systems [201]. Sweden's policies are not only focused on providing benefits for the promotion of renewable energy use, but also, on a structural level, they have managed to improve their electricity systems making them more effective. They have achieved high electrification in industry and buildings, increased use of biomass in industries, and limited the use of oil to activities in the transport sector [150].

Iceland

Currently, about 85% of primary energy [149] and about 98% of electricity [197] in Iceland comes from renewable energy sources, mainly hydro and geothermal. The government of this country has been developing policies involving the development and transition of the country to renewable energy throughout the entire energy supply [149]. For many years, the primary energy source in Iceland was supplied by coal and oil [139]. Today, however, Iceland has integrated a sustainable energy system that is essential for moving away from fossil fuels and reducing greenhouse gases generated by transportation [197].

Iceland's policies effectively decrease dependence on nonrenewable sources such as fossil fuels for electricity production. Figure 5 details the share of each renewable source in electricity production.

Energy dependence on nonrenewable energy sources seems to be decreasing. Most of the electricity comes from renewable sources. The "Other renewables" category includes geothermal energy, whose share in 2019 was 30.79%. This trend appears to continue to increase. In Iceland, 100% of the population has access to energy. Electricity and heating prices are low. However, prices are not equal [197]. About 80% of the electricity in this country supplies export-oriented heavy industry [113]. Figure 6 shows the amount of energy consumed by the sector in 2012.

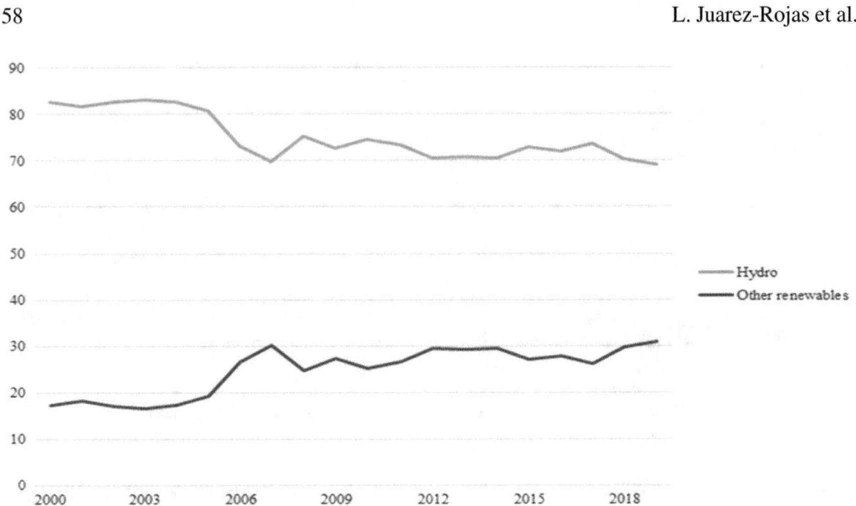

Fig. 5 Share of electricity production by source. *Source* Our World in Data [166]

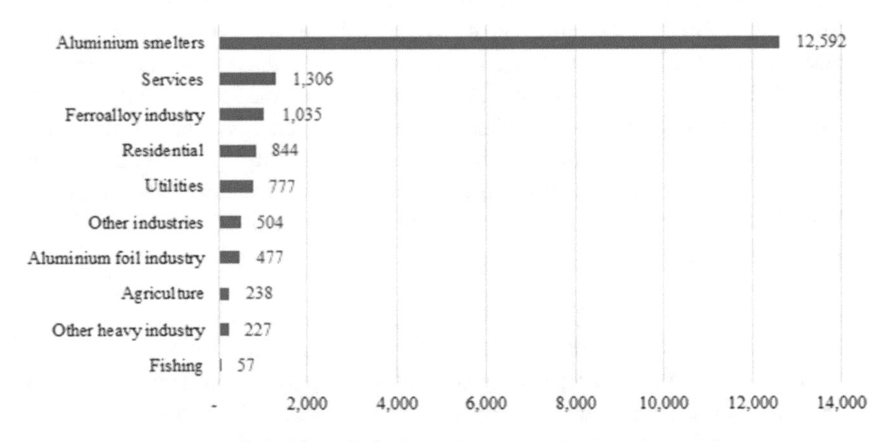

Fig. 6 Energy consumed by activity, industry, and economic sector. *Source* [113]

Figure 6 shows the energy needed for different activities, sectors, and industries. Indeed, the effectiveness of renewable energy use in Iceland is significant. However, questions about the social and economic purposes for which this energy serves to remain questionable [113]. The search for new renewable energy sources may be critical to achieving a sustainable electricity system. The role of the government to achieve the above focuses on ongoing studies to determine the possibility of promoting the production of renewable energy without affecting the environment [159].

Uruguay

Uruguay has become a world leader in the use of renewable energies [103]. The promotion of renewable energies in this country started in 2002 with Law 17.567 on the Promotion of Renewable Energies [221]. At the business level, the Government of Uruguay provides Energy Efficiency Certificates through the Ministry of Industry, Energy and Mining, which encourage the participation of households, industries, public and private organizations, also, it promotes the implementation of renewable energies in different economic activities [151]. In Uruguay, renewable energies represented 98% of the energy matrix in 2019 [206].

Uruguay's policies have proven to be one of the most effective in South America. The trend of renewable energy as a source of electricity continues to be the primary energy supply for this country. Figure 7 details the share of each renewable source in electricity production.

According to Fig. 7, a considerable decrease of approximately 40% of the energy produced by hydroelectric sources is evident. However, the growing trend of wind energy, with a share of 40.31% in 2020, demonstrates the excellent availability of other energy sources. These data are results of the effectiveness of policies in this country. The inclusion of renewable energy policies in Uruguay responds mainly to three factors shown in Table 3.

Traditionally, Uruguay's total energy consumption has relied heavily on fossil fuel imports [100]. Therefore, implementing renewable energy policies allowed maintaining some energy sovereignty, decreasing the risks of commercial vulnerability

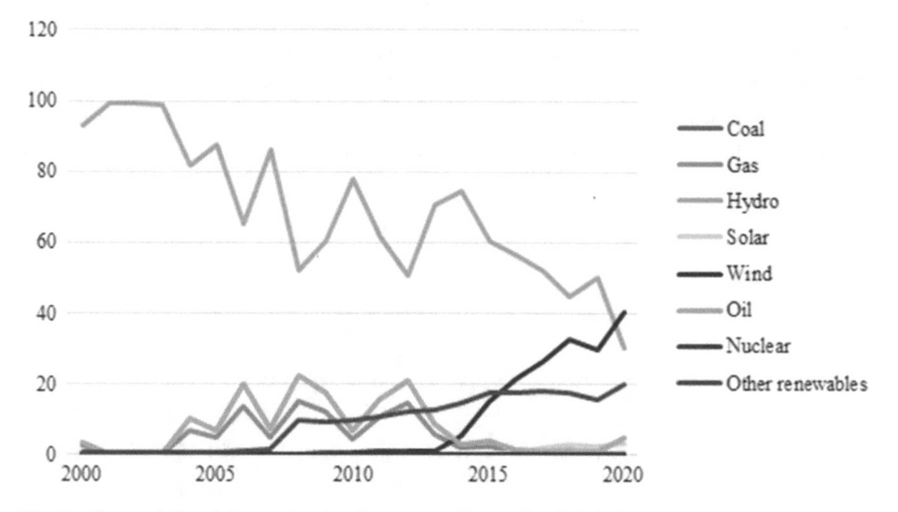

Fig. 7 Share of electricity production by source. *Source* Our World in Data [166]

Table 3 Factors influencing the promotion of renewable energy policies

Factor 1	Factor 2	Factor 3
Dependence on fossil fuel imports	Uruguay lacked energy stability due to a long-standing structural deficit	A plan to increase the share of gas in the mix became unfeasible

Source [103]

resulting from oil imports [103]. The second factor refers to structural failures in the energy supply system in Uruguay during 1990 and 2000. Hydroelectric energy was already being generated during those years, with shallow production peaks in certain seasons. As a result, oil imports were again used [103]. Therefore, it was essential to implement policies that would increase production capacity through other types of renewable energies. The third factor responds to problems in the supply of gas from Argentina. The latter ceased to be a reliable supplier for Uruguay [103].

Denmark

Nowadays, Denmark has focused on developing renewable energy policies that have successfully overcome its high dependence on fossil fuels [209]. This country is searching for solutions to reduce carbon emissions, thus generating economic development, energy security, and environmental preservation. It is focused on developing new technologies to meet the objectives set out in its Energy Strategy 2050 [157]. The success of energy policies in Denmark is based on its focus on decentralized strategies [209]. The policies are required at the local level, i.e., at the community level, to achieve the objectives set in this country regarding the use of renewable energies [172].

Renewable energy sources are increasingly increasing, leaving aside those nonrenewable sources that limit the possibility of achieving decarbonization in Denmark. Figure 8 details the share of each renewable source in electricity production.

The significant reduction of coal as a source of energy production is evident. Between 2019 and 2020, there was a decrease of approximately 7%. In turn, the use of wind energy as the primary source of supply shows the effectiveness of the policies implemented by this country. As part of its strategic direction toward

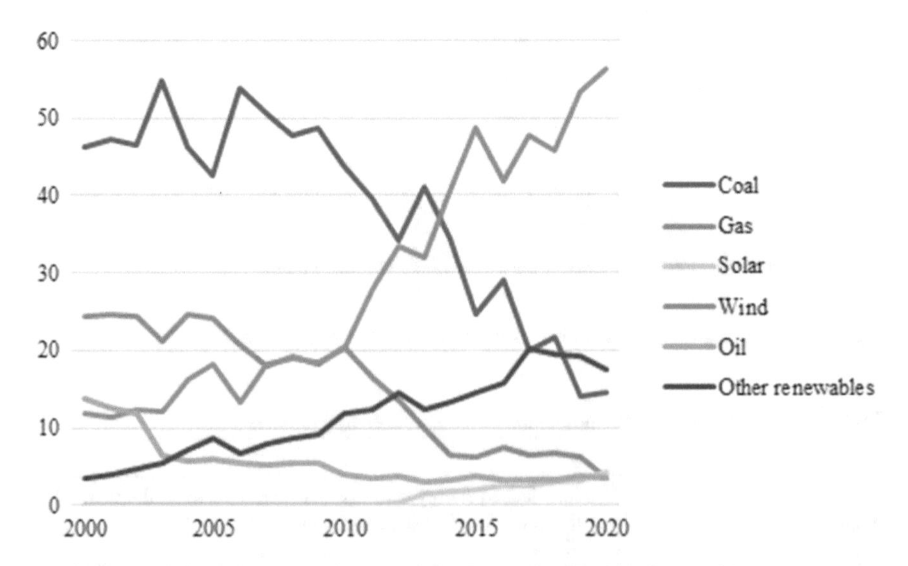

Fig. 8 Share of electricity production by source. *Source* Our World in Data [166]

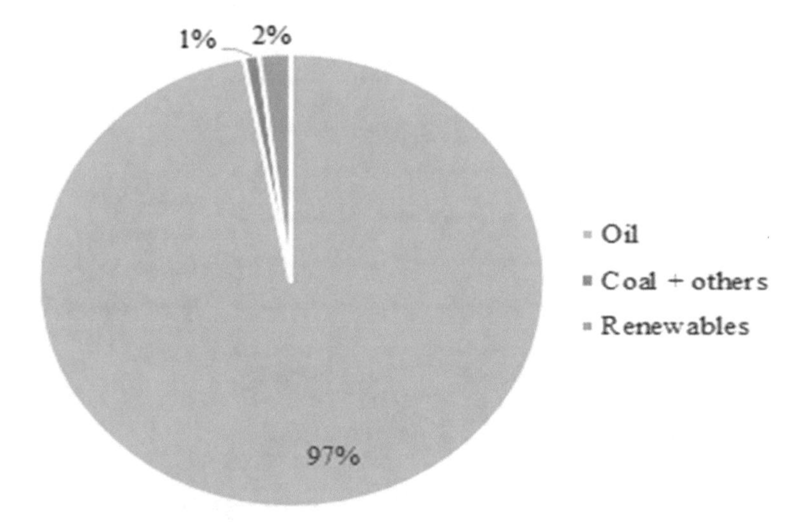

Fig. 9 Main energy sources in Niue. *Source* [122]

2050, Denmark's initial policies focus on achieving more than half of its electricity consumption from wind energy [121]. It is essential to achieve energy efficiency at the business level as it has direct benefits by achieving greater competitiveness and productivity [157]. Finally, like other countries analyzed, the government provides economic subsidies to promote renewable energies in different productive activities [121].

Niue

The case of Niue seems to be an exception to the relationship presented in the reference research. Most of Niue's electricity comes from nonrenewable sources, especially fossil fuels, and gases. Only 2% of its primary energy comes from solar energy [122]. The present research does not address in depth the analysis of this country. Figure 9 shows its main energy sources.

Indeed, unlike other countries, this small territory still retains its strong dependence on energy from nonrenewable sources. Despite this, Niue has integrated a series of policies to reverse this situation.

Costa Rica

Costa Rica has integrated policies to achieve decarbonization like many countries under analysis. Currently, almost 100% of energy is supplied by renewable sources [72]. However, achieving carbon emission reductions still encounters limitations, mainly due to transportation [108]. In 2017, renewable energy sources such as hydroelectric, geothermal, wind, solar, and biomass represented 99.7% of the national electricity mix [75]. By 2019, this figure rose to 99.89%. Costa Rica has one of the best indicators of energy use from renewable sources in its national energy mix at the level of Central American countries. Figure 10 details the share of each renewable source in electricity production.

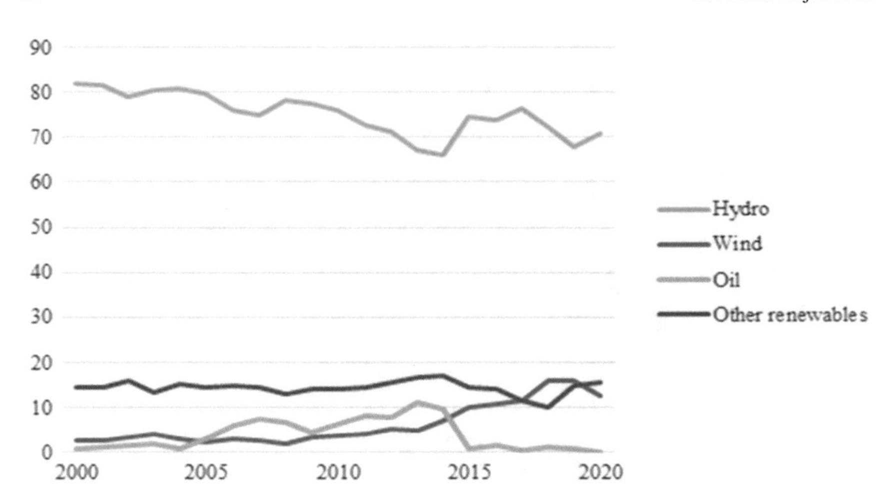

Fig. 10 Share of electricity production by source. *Source* Our World in Data [166]

Energy dependence on nonrenewable energy sources has decreased considerably over the last five years. Meanwhile, the participation of hydropower appears to be important in the country's energy system. By 2020, this source represents 0.16% of total electricity production. The central policies regarding renewable energies are based on Costa Rica's Energy Expansion Plan 2016–2035. This plan is renewed every two years and sets the strategic basis for promoting a sustainable energy chain [121].

Tajikistan

Tajikistan is one of the countries with the best water availability indicators globally. It even holds second place in terms of water resources per capita, mainly due to the favorable geographical location that allows it to have the right conditions (land and climate) for hydropower development and production. Although Tajikistan has significant hydropower development compared to other Central Asian countries, it has many barriers to achieving Sustainable Development (SD). The strengthening and promotion for the development of green entrepreneurship are limited by issues such as Lack of legal policies to encourage the use of renewable energy, low tariffs for electricity, and lack of knowledge regarding the use of other renewable energy sources [94].

Currently, the promotion of policies to encourage entrepreneurship continues to develop favorably. The State Program for Supporting Entrepreneurship for 2012–2020 is an ideal option for creating entrepreneurship. This program includes economic incentives such as subsidies or loans to developing business models. In addition, reductions in tax and customs clearance fees are set mainly in construction and goods processing businesses [195]. However, as expressed above, this country has certain limitations that disfavor investment attraction to develop green entrepreneurship. Government regulations generate increases in costs and this, in turn, decreases the attention of entrepreneurs. The situation of foreign-owned Sangtuda-1 and

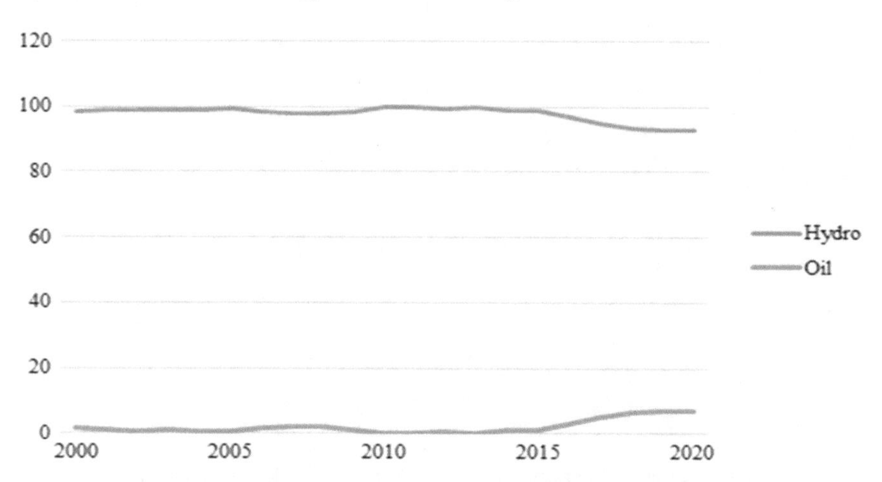

Fig. 11 Share of electricity production by source. *Source* Our World in Data [166]

Sangtuda-2 hydropower plant companies with non-payments and lack of private investments in Rogun hydropower plant are examples of disadvantageous business conditions [133].

According to Fig. 11, hydropower in Tajikistan accounts for approximately 92.88% of the electricity produced in this country. This country focuses its efforts on constructing hydroelectric power plants to take advantage of the potential of the water resource. The Rogun Hydropower Plant is the primary example of development constructions that boost the use of renewable energy [133].

Portugal

Portugal's policies focus on Feed-in tariffs to promote renewable energy sources. However, although hydroelectric power plants have played an important role in Portugal since the 1950s, they operate mainly outside feed-in tariff schemes. Among the most widely used renewable energy sources in this country, wind power stands out. At the end of 2012, a production of 4194 MW was reached, and, according to the Portuguese Renewable Action Plan, this number would increase to 5300 MW in 2020 [184] (Fig. 12). Renewable energies are expected to become the most representative source of energy production by 2040. In this way, they create a sustainable energy system and reduce greenhouse gas emissions from the electricity sector [84].

In general terms, Portugal shows a well-shared consumption of energy, led by gas electricity. As seen above, even though hydroelectric power is an essential source of energy in the country and wind energy, they are still not above gas electricity, which would mean there are still changes that need to be made on its policies for them to be more effective.

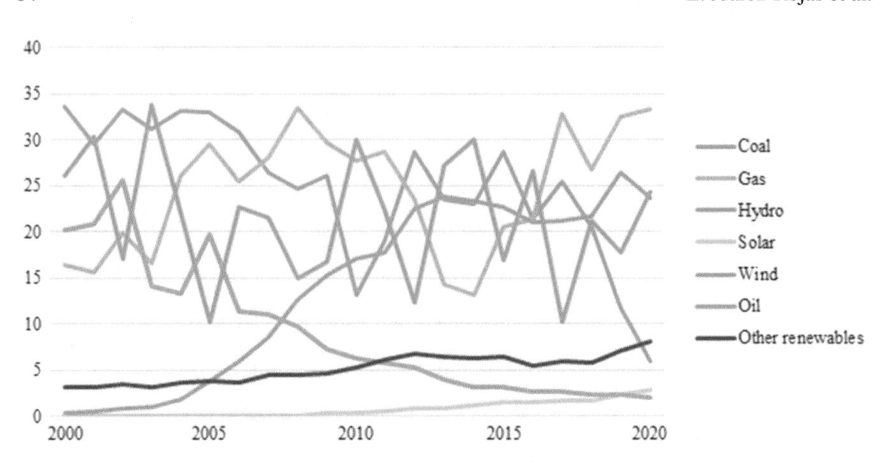

Fig. 12 Share of electricity production by source. *Source* Our World in Data [166]

3.3 Top 10 Developed Countries Based on Their GDP

Next are the 10 countries with the highest gross domestic product, and we described energy management efforts.

United States

The United States is one of the world's largest energy consumers and one of the largest producers. But although in recent years, it has not been reflected in its renewable energy policies, they are not used in the best way, which is because according to [123], the United States is a federation consisting of fifty states. Energy policy is conducted at the state and federal levels, assuming that state power is dominant. However, it should be remembered that the United States is a very economically free country, so despite regulations, the main limiting factor is the competitive market. In addition, investments in the energy infrastructure of all types require a long development period, so it is an area of great inertia where changes do not occur immediately. In the early twentieth century, state and federal regulations did not exist. Energy companies or utilities (which include much electricity) appear distributed as small local companies at the city level.

But despite having this type of process for its realization, it does have an Office of Energy Efficiency and Renewable Energy, which is responsible for seeking that the whole country comes to be supplied by renewable or clean energy. It is for this reason, considering the opinion of Jiménez and [123], it could be said that the policy of improving these capacities differs from the traditional policy of most European countries, since in the United States, there is no guaranteed price for all renewable generators. However, development targets have been set for these technologies. In 37 of the 50 states, a minimum target for renewable electricity consumption has been established following the Renewable Portfolio Standard (RPS). RPSs range from 15% by 2021 in states such as Michigan to 100% in California or Hawaii by 2045.

China

China is the world's largest energy consumer and the country's geopolitical significance is not without stress. According to Xinhuanet [214], the energy problem cannot be separated from enabling conversion issues. China's growing economy generates a significant impact on the environment. With the emergence of the COVID-19 pandemic, China has grown in using renewable energy sources to reduce carbon emissions and promote green development. The installed capacity of renewable energy production nationwide is approximately 930 million kilowatts by the end of 2020, accounting for 42.4% of the country. The installation capacity of hydropower capacity has been created for the first time worldwide for 16 consecutive years, while the main installation capacity of wind and photovoltaic installation in eleven and six consecutive years.

In 2020, the use of renewable energy in China was 680 million tons of coal equivalent standard, creating similar energy for nearly one billion tons of coal, reducing emissions of dioxide. Carbon, sulfur dioxide, and nitrous oxide are approximately 1.79 billion tons. Equivalent to 864,000 tons and 798,000 tons. Power generated by renewable energy sources in China reached 2.2 billion kilowatt-hours, for 29.5% of the country's total electricity consumption, 9.5% compared to 2012. In addition, the government is taking many steps to promote the development of high-quality regenerative energy. Development and positive market orientation build an energy system representing new energy pillars. New efforts are being implemented to accelerate wind power and solar power, improve the creation of hydropower in local conditions and promotions, and order the safety security of the security power system. Meanwhile, stricter energy consumption standards are applied to support and promote the use of fossil energy and modify energy use in essential areas and regions, such as industry, construction, and transport.

Japan

The Fukushima nuclear accident marked a turning point in Japanese energy policy. The country must compensate for energy by starting up thermal power plants and importing more oil and liquefied natural gas in the short term. In the medium term, according to [70], it helped reform the electricity market through further liberalization and the introduction of grid-linked tariffs in 2012 to make renewable energy investments more affordable. Japan now aims to increase its use of renewable energy and reduce fossil fuel consumption over the next decade. However, there are still obstacles for renewables, such as instability of supply, technological challenges for storage, and high costs, which play a vital role in the medium to long-term future. According to [185], there is no obligation on the part of the Japanese government. Based on the Fifth Strategic Energy Plan, Japan proposes that renewable energy should account for between 22 and 24% of Japanese production, renewable energy accounts for 19% of the country's total electricity production. Previously, around 8% came from wind and solar power.

From a policy point of view, the Japanese authorities are promoting its development by addressing institutional and regulatory barriers and improving the competitiveness of the energy market. From an investment point of view, many local and international investors are interested in the energy sector, especially renewables. On the other hand, Japan's energy diversification strategy seeks to find a balance between different energy sources to compensate for energy shortages quickly in case of supply disruptions. It, therefore, supports the decarbonization of other sources, such as nuclear power, thermal power with CCS technology, and new clean energy sources, such as hydrogen used in traffic.

Germany

According to [132], energy transformation means transforming an energy economy based on oil, coal, gas, and nuclear power to renewable energy. Germany's goal is to be essentially climate neutral by 2045. By this year, at least 80% of Germany's electricity supply and 60% of Germany's energy supply is expected to come from renewables. Until 2022, all nuclear power plants are shut down in a row. As of 2019, there are only six nuclear power plants in operation. As a result, the German federal government continued the sustainable restructuring of the energy system, starting in 2000 with the first decision to abandon atomic energy and law on renewable energies. Support for renewable energies in 2000, this was supplemented by the Renewable Energy.

Sources Act. Reducing dependence on imports: at the beginning of the decade, the federal government agreed with the energy companies to forgo nuclear power until 2022. The German government's 2011 energy transition decisions were part of a tradition of transitioning energy supply from sustainable sources. The German energy system began a rapid transformation in 2011, along the lines of the Fukushima nuclear catastrophe, Japan, based on a decision of the German Senate and with the explicit broad support from most of the population. For the German government, it is "a necessary step on the way to an industrial society committed to the idea of sustainability and nature conservation." The energy transition and the German economy benefit the environment and the climate in the future. Specifically, it aims to reduce dependence on oil and natural gas imports. Germany has imported coal, oil, and natural gas worth approximately 45 billion euros annually. The domestic value creation in the renewable energy sector will decrease in the coming years. In addition, these actions open new export opportunities and generate employment. Use energy more efficiently.

Another central task, according to [75], is to promote the "second pillar" of the energy transition: using energy more economically and efficiently. At the end of 2019, the German government approved the Energy Efficiency Strategy 2050. The goal is to reduce primary energy consumption by 30% by 2030 compared to 2008. In cooperation with professional associations and civil society groups, the provinces and representatives of the scientific community carry out the analysis. Measures to halve consumption by 2050 and propose specific action plans. For the periods up to 2030 and 2050. Climate protection–the world's first binding law.

For this reason, the German government's climate protection program, adopted at the end of 2019 and revised in 2021, sets emission ceilings for buildings and other areas. In addition, the Climate Protection Act sets a constant rate for carbon dioxide emissions from vehicles and buildings, which has already been applied, in the context of emissions trading in Europe, to the energy and energy-intensive sectors.

The energy transition aims to reduce risks and reduce climate impacts and secure supply. With the dynamic expansion of renewable energies, it is possible to significantly increase the proportion of carbon dioxide-free energy in the energy mix. Green Power had 42.1% in 2019. Depending on weather conditions, photovoltaics and wind power can meet up to 90% of Germany's electricity demand at peak generation. More than 66% of new residential buildings were heated with renewable energy. By the end of 2020, there were 1.7 million PV installations, with a nominal capacity of about 49.5 gigawatts. With this installed capacity, Germany ranks third after China and Japan. Renewable energy law as an international role model.

The Renewable Energy Law is considered a model in many countries. In 2014, it was revised to ensure that energy remains affordable and secure. The "EEG contribution", which allocates a higher proportion of green energy expansion costs to consumers, has increased significantly since 2009 due to the expansion of solar installations and a new calculation method, which sparked a heated debate in the community about the cost of the green sector and the energy transition. As of 2014, EEG's contribution has remained relatively constant: with slight fluctuations between six and seven cents per kilowatt-hour.

United Kingdom

The United Kingdom, according to Comunicar [82], has become the first G7 country to reach a historic agreement to support the transition of the oil and gas industry to clean and green energy, compared to the work of 40,000 years ago. On March 31 2021, the UK government stopped supporting the offshore fossil fuel energy industry. The industrial agreement between the UK government and the oil and gas industry supports workers, businesses, and the supply chain during this transition, harnessing existing industrial skills, infrastructure, and potential to exploit new and emerging technologies, such as cooking and storage hydrogen production. This agreement helps businesses currently prepare for a 2050 grid and create a complete business environment to attract new UK-based industries, it develops new opportunities to export UK businesses and secure long-term jobs. Coal, as well as naval wind energy, decommissioned. Oil extraction for the UK continental shelf is directly responsible for approximately 3.5% of greenhouse gas emissions. Thanks to the packaging measures, the agreements should reduce pollution by 60 million tons by 2030, 15 million tons of oil and gas production on the UK continental shelf, like the annual emissions of 90% of UK households, and support 40,000 jobs along the supply chain. According to [186], the 40 GW Energy target can help cover some 20 billion books of private investment by 2030, creating 60,000 new jobs and economic growth. In areas like the Northeast of England, Yorkshire and the Humber are East Anglia and Scotland. It is estimated that 60% of their wind energy expenditure will be depreciated by 2030. The UK sets carbon use, and storage has recovered in the two industry clusters between 2020, reducing UCI 1 billion floor funding. Reaching £1 billion provide the industry with the one needed to implement this technology following the scale and winds. The government's target is for the UK to develop 5 GW of hydropower production capacity by 2030, which can benefit the UK with around 8000 jobs at industrial sites, to be supported by many different measures, including a net fund of £240 million.

In addition, BP has developed a plan for the highest hydropower production facility in England at Teesside, which can produce 1 GW HYDROPOTER (20% hydropower lens) by 2030 and capture and send the maximum capacity of two million tons of carbon dioxide each year. The economic measures are consistent with the recent budget in which the Prime Minister funded the oil and gas industry and green energy development companies. Ensure that the British economy is always competitive and resilient to net zero, that companies in all areas must be willing to face their risks and take advantage of the family name of the queen gas opportunities. For this reason, the British government also announced the current proposals to establish that the UK's largest utility and professional company, which accounts for 4 billion in revenues, must be ready to change the economy.

Companies are encouraged to reduce emissions on request. Companies that reduce their emissions do not pay taxes. On the other hand, it can be noted that according to [102], there are other policies, such as the climate change tax introduced in 2001 on energy use in industry. The tax was adopted based on Lord Marshall's October 1998 report "Economic Instruments for Business and Energy Use". The UK Parliament has just passed the "Green Deal" for housing, stating that 43% of inefficient housing emissions are wholly replaced. Green deals allow consumers to make their homes energy efficient at no extra cost. Loans are granted and repaid through energy savings. Since 2015, the Green Investment Bank has financed loans for low-carbon investments focusing on wind energy and offshore waste. And finally, energy market reform: introducing a subsidized renewable energy pricing scheme, including long-term contracts, ensuring stability and predictability, and encouraging investment in low-carbon energy production. The reform includes carbon dioxide emissions from the most polluting power plants. The subsidy is valid until around 2025, when the market is sufficient for renewables to compete.

India

In recent years, according to [118], India has been widely recognized for its commitment to renewable energy. Since signing the Paris Agreement in 2015, the world's fifth-largest economy has increased its solar capacity almost ninefold and has committed to reducing emissions drastically. At the same time, however, the country still relies on highly polluting fossil fuels. But this does not mean that there is a policy in India to eliminate polluting waste energy. Improving people's quality of life, cleaning the air and water, and, of course, combating climate change.

Everything changed in 2015 with the Paris climate agreement. Since the agreement, India has been one of the countries that have made the most progress in complying with the Paris Agreement. As an Asian country, the second-most populous country in the world, and the third-largest producer of greenhouse gases after

China and the United States, this is the reason why the Asian country is in such a hurry to switch to green energy. It is worth noting that the country has a population of 1.3 billion, 300 million of whom do not have access to electricity, and the average annual income is just over 1400 euros.

According to [48], they aim to decouple economic growth from environmental influences and leave a better world. Today, coal accounts for 75% of India's total energy. However, the Asian country shows signs of new plans to replace this polluting energy production. According to National [160] in Clean Energy XXI, renewable energy in India has doubled its energy production (mainly solar) since 2017. Thus, the country's target is to reach the installed capacity of renewable energy.

Currently, the country has an installed renewable energy capacity of 82.6 GW, which is 23% of the total energy matrix. By 2022, it is expected to reach 175 GW, and by 2030, 450 GW. The country has around 1.339 billion people, so the amount of electricity required is enormous. As a result, the Indian government has begun to invest in extensive renewable energy facilities, including some of the world's most famous photovoltaic parks.

France

Continuous increase in technological advances in terms of reliability and predictability of production, coupled with staggering cost reductions, has allowed French public authorities to look toward renewable energy. It is seen as a sustainable solution and plays a legitimate role in the national energy mix. The French government has begun to realize the enormous potential of renewable energies; to catch up with other European countries and achieve the objectives of the European Union, France is committed to the comprehensive development of the means of generating electricity from renewable energy sources renewable resources; onshore and offshore solar and wind energy.

Based on information acquired from [81], the implementation of its Energy Conversion Act in August 2015, just four months from the Paris Climate Summit, France showed that the international community changed its current energy model for a more sustainable company and continued economic development. The French government's adoption in October 2016 of Plurian Energy Programming 2016–2023 allows this national road to make this conversion. It is characterized by its central ambition but also by its uniqueness. Ambition because it wants beyond the commitments to be acquired by the country at the international level and initial, because it understands that the conversion must overcome a set of economic areas and horizontal energy origin, face each of them, all potentials can create a new model of growth. The law's implementation is the outcome of collaborative work between government, business, and French civil society. Its route was announced in late October 2016 and analyzed the third part of this work, the Ordinance, in the period 2016–2023, the main actions to achieve the objectives set by the law.

The President of France, Emmanuel Macron, presented, according to [187], the Energy and Climate Strategy 2030. You can see things like the coal shutdown in 2022, where the last plants finally stopped operating after 4 years, and this is happening. No new power plants using fossil fuels are allowed to be planned. Double the installed capacity of renewable energy by 2028. The target is to reach 45% renewable energy by 2035. The French government has also set a 50% share of nuclear power by 2035. In addition, France explores various options to ensure the long-term security of supply, especially the possibility of building new nuclear reactors. This study also increases renewable heat production from 40 to 59% and reduces renewable gas share in gas consumption to 10%, assuming a sharp reduction in costs. Thus, it is also intended to significantly improve the energy efficiency of buildings and reduce their emissions. And as a last resort, consider stopping the sale of fossil fuel cars by 2040.

Italy
According to [182], based on the statement of the Minister of Environment, Gian Luca Galletti, the country's energy future is renewable (wind, solar, geothermal), and to motivate them have been targeted investments of 13.5 billion euros. Italy is the second-largest producer of solar photovoltaic energy and the seventh in terms of wind energy, showing a solid increase in the use of renewable energy in recent years to reduce dependence on imports, oil, and natural gas. The wind industry generates 3.9% of the electrical energy needed, and wind farms are mainly established in Sardinia and Sicily and south of the Apennines.

Information gathered from [171] tells us that the Italian government is working these days on a plan to be submitted by April 30 to the European Commission to raise 200 billion euros in the coming months that may come from the European Fund for the Recovery of Epidemic Affected Countries. In the field of environmental transformation, the minister wants Italy to invest in decarbonization technology so that in the future, it does not have to rely on other countries in this strategic area. His goal is to make Italy a "world champion" in environmental transformation "and invest in at least four areas: sustainable agriculture and circular economy, renewable energy, hydrogen, and sustainable mobility; energy efficiency and building restoration; land and water protection.

In addition, it should be noted that according to Parlamento [168], the Commission adopted the Alliance's 2030 Climate Target Plan (COM/2020/562), which includes an emissions reduction target that is updated to 2030 for 55% compared to 1990 levels, up from the current target of 40%.

Canada
Renewable energy policy operations here include quantity-based instruments and price-based operations. Real studies compare both operations on how effective they are in distributing technology, reducing economic costs, and composing a stable investment environment that could attract capital. There is a debate about governments' role in supporting renewable energy technologies and the choice of policy instruments to increase the share of renewables in the electricity supply mix [199].

However, it is not clear whether the Canadian government allocated funds for energy innovation to mechanisms of technological impulse, through the creation of knowledge and research and development (R&D), or to mechanisms of attraction of demand, through market creation through subsidies or guaranteed markets. While there are a variety of motivations for implementing policies to support renewables, little research is done on their effectiveness, and it is precisely through the negotiation process between political actors, including political unions, the state, and regulators, that the ultimate structure of renewable energies. Energetic mechanisms are developed. For example, in Ontario, feed-in tariffs are created to encourage green entrepreneurship and renewable energy [199].

Russia

There is insufficient information on the renewable energy policies in this country; most of this information focuses on using technology to develop electricity from renewable sources. The country's efforts to implement renewable energy policies have not significantly impacted electricity production in the short and long term. Specific schemes have been developed to improve renewable energy policies, such as the Power and Fuel Complex Support Program, the Alternative Energy Program, the Non-Conventional Renewable Energy Application Program, and the Environmental Energy Technology Program [224].

This country faces a series of limitations that impede the use and consumption of renewable energy as the primary source of electricity. For example, there is a lack of technology promotion agencies, incentive programs, financing for companies in the energy sector, and a lack of knowledge about renewable energy sources [224]. Although Russia has the potential and availability of renewable energy sources ideal for developing electricity production, renewable energy production in the country's energy balance is less than 1%. In general, the country's ideals to achieve a transition to renewable energy are affected by the lack of funding from state governments. Despite this, the government achieves a 20% share of renewable energy by 2024. Solar energy, the renewable source, increases its share despite the previously described limitations [77].

The main element determining the difference between the alternative energy indicators is the little technology development in this country. Additionally, the production capacity of the needed materials does not have an increasing development, obtaining a lag. The proposed plan on the capacity of wind and hydroelectric plants in 2017–2019 is not feasible. The most likely scenario is that the share of renewable energies in the energy balance does not exceed 4–6%, which would be 3.5–5 times lower than the state's objectives [77].

Recent Suggested Lectures About Green Approach

- Waste reduction and carbon footprint [109]
- Material selection for circularity and footprints [155]
- 3D Print, circularity, and footprints [88]
- Virtual tourism [88]
- Leadership for sustainability in crisis time [37]
- Virtual education and circularity [83]
- Circular economy for packaging [71]
- Students oriented to circular learning [89]

- Food and circular economy [87]
- Water footprint and food supply chain management [125]
- Waste footprint [54]
- Measuring circular economy [91]
- Carbon footprint [101].

4 Closing Remarks

As part of the review of previous concepts on green entrepreneurship, it is evident that different factors influence the creation of green entrepreneurship. These factors can be political, social-cultural, and economical, resulting from renewable energy as an initial variable. Figure 2.4 represents how the factors above contribute to green entrepreneurship (Fig. 13).

The analysis is limited to the evaluation of political factors. We leave the possibility of evaluating how social-cultural and economic factors, whose initial variable is renewable energy, may or may not contribute to creating green entrepreneurship. Liechtenstein, Norway, Sweden, Iceland, Uruguay, Denmark, Niue, Costa Rica, Tajikistan, and Portugal. First, the countries with the most efficient renewable energy policies were evaluated concerning the baseline research mentioned above. Except for Niue, most countries mentioned applying three types of policies: incentive-based, decarbonization, and linked to the search for a sustainable energy system.

1. **Incentive-based policies**: As part of the process of promotion, development, and use of renewable energies, most countries apply economic incentives through subsidies, tax rate reductions, feed-in tariffs, among others. These types of policies directly affect the creation of green businesses. The possibility of introducing new products and services related to renewable energies by reducing the tax rate is feasible, considering that this reduction can contribute to increasing or receiving higher profits. On the other hand, obtaining subsidies from public agencies promotes the competitiveness of companies and allows them to use this economic incentive as capital to improve production levels.

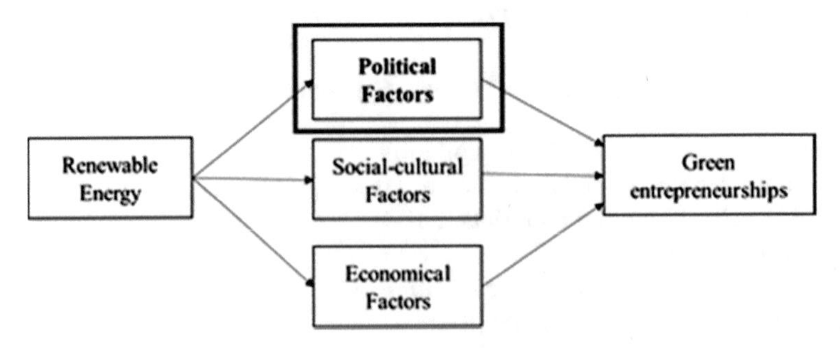

Fig. 13 Model for creating green enterprises based on political, social-cultural, and economic factors

2. **Decarbonization policies**: The implementation of renewable energy policies are based, in most cases, on achieving a decrease in carbon emissions–a product of using nonrenewable sources such as carbon and fossil fuels–over a certain period. Some of the countries evaluated issue Energy Efficiency Certificates that bring a series of benefits that create green businesses. For example, through this type of certificate, companies can save energy, reduce various manufacturing costs, generate economic benefits for the enterprise, and provide a positive business image to customers. The purpose of this certificate–among other reasons–is based on the efficient transition toward the use of renewable energies as primary sources of energy production.
3. **Sustainable energy system**: Failures in energy systems can affect the productive activities of companies. Therefore, ensuring continuity in the constant supply of energy is fundamental. Implementing renewable energy policies in the countries evaluated improves their energy systems, contributing to the better development of green entrepreneurship.

However, some barriers need to be evaluated. Some cases, such as Tajikistan, do not create green entrepreneurship due to the absence of legislative support, lack of funding for new technologies, and lack of awareness and promotion of other renewable energy sources. Therefore, although the policies of some countries seem to encourage business development, some others have not yet succeeded in creating an enabling environment for such entrepreneurship. In the second point, the countries with the highest Gross Domestic Product were analyzed since they are the countries that have the most significant influence in the world. The countries considered are the United States, China, Japan, Germany, United Kingdom, India, France, Italy, Canada, and Russia. Most of the countries described seeking the improvement of the environment seeks the implementation, monitoring, or compliance of the different renewable energy policies. Many of them began their work since the Paris treaty in 2015, and all have both short and long-term visions. It can be observed that they all have a struggle among them to become the country with the minor carbon emissions by 2050.

Despite all the good things, it can be observed that these countries are not among the first list not because they are not applying their policies correctly but because their population is much larger than that of the countries in the first list; therefore, the level of carbon footprint in these countries is much larger. Apart from that, other factors influenced some countries compared to the policies made, for example, Russia, in which even though there was an incentive of renewable energy policies, it did not go as planned because of the technological factor, an element explained before.

References

1. Aaron L (2020) The oft forgotten part of nutrition: lessons from an integrated approach to understand consumer food safety behaviors. Curr Develop Nutrition 4(Supplement_2), 1322–1322. https://doi.org/10.1093/cdn/nzaa059_039

2. Abad-Segura E, Batlles-delaFuente A, González-Zamar M-D, Belmonte-Ureña LJ (2021) Implications for sustainability of the joint application of bioeconomy and circular economy: a worldwide trend study. Sustainability 13(13):7182. https://doi.org/10.3390/su13137182
3. Abad-Segura E, Cortés-García FJ, Belmonte-Ureña LJ (2019) The sustainable approach to corporate social responsibility: a global analysis and future trends. Sustainability 11(19):5382. https://doi.org/10.3390/su11195382
4. Abad-Segura E, González-Zamar MD (2021) Sustainable economic development in higher education institutions: a global analysis within the SDGs framework. J Cleaner Prod 294:126133. https://doi.org/10.1016/j.jclepro.2021.126133
5. Abad-Segura E, González-Zamar M-D, Infante-Moro JC, Ruipérez García G (2020) Sustainable management of digital transformation in higher education: global research trends. Sustainability 12(5):2107. https://doi.org/10.3390/su12052107
6. Abad-Segura E, Morales ME, Cortés-García FJ, Belmonte-Ureña LJ (2020) Industrial processes management for a sustainable society: global research analysis. Processes 8(5). https://doi.org/10.3390/pr8050631
7. Abdaljaleel M, Singer EJ, Yong WH (2019) Sustainability in biobanking. In: Yong WH (ed) Biobanking: methods and protocols. Springer, New York, pp 1–6. https://doi.org/10.1007/978-1-4939-8935-5_1
8. Aberilla JM, Gallego-Schmid A, Stamford L, Azapagic A (2020) Environmental sustainability of cooking fuels in remote communities: Life cycle and local impacts. Sci Total Environ 713:136445. https://doi.org/10.1016/j.scitotenv.2019.136445
9. Acevedo-Duque Á, Vega-Muñoz A, Salazar-Sepúlveda G (2020) Analysis of hospitality, leisure, and tourism studies in Chile. Sustainability 12(18):7238. https://doi.org/10.3390/su12187238
10. Ada E, Sagnak M, Mangla SK, Kazancoglu Y (2021) A circular business cluster model for sustainable operations management. Int J Logistics Res Appl 1–19. https://doi.org/10.1080/13675567.2021.2008335
11. Adam A (2015) Challenges of public finance sustainability in the European Union. Procedia Econ Finance 23:298–302. https://doi.org/10.1016/S2212-5671(15)00507-9
12. Adam I, Agyeiwaah E, Dayour F (2021) Understanding the social identity, motivations, and sustainable behaviour among backpackers: a clustering approach. J Travel Tour Mark 38(2):139–154. https://doi.org/10.1080/10548408.2021.1887053
13. Adam JN, Adams T, Gerber JD, Haller T (2021) Decentralization for increased sustainability in natural resource management? two cautionary cases from Ghana. Sustainability, 13(12). https://doi.org/10.3390/su13126885
14. Adam, M. (2018). The role of human resource management (HRM) for the implementation of sustainable product-service systems (PSS)—an analysis of fashion retailers. Sustainability, 10(7). https://doi.org/10.3390/su10072518
15. Adams R, Martin S, Boom K (2018) University culture and sustainability: designing and implementing an enabling framework. J Clean Prod 171:434–445. https://doi.org/10.1016/j.jclepro.2017.10.032
16. Adesogan AT, Havelaar AH, McKune SL, Eilittä M, Dahl GE (2020) Animal source foods: Sustainability problem or malnutrition and sustainability solution? Perspective matters. Global Food Secur 25:100325. https://doi.org/10.1016/j.gfs.2019.100325
17. Adomah-Afari A, Chandler JA (2018) The role of government and community in the scaling up and sustainability of mutual health organisations: an exploratory study in Ghana. Soc Sci Med 207:25–37. https://doi.org/10.1016/j.socscimed.2018.04.044
18. Adomako S, Amankwah-Amoah J, Danso A, Konadu R, Owusu-Agyei S (2019) Environmental sustainability orientation and performance of family and nonfamily firms. Bus Strateg Environ 28(6):1250–1259. https://doi.org/10.1002/bse.2314
19. Adongo CA, Taale F, Adam I (2018) Tourists' values and empathic attitude toward sustainable development in tourism. Ecol Econ 150:251–263. https://doi.org/10.1016/j.ecolecon.2018.04.013

20. Agyeiwaah E (2019) Exploring the relevance of sustainability to micro tourism and hospitality accommodation enterprises (MTHAEs): evidence from home-stay owners. J Clean Prod 226:159–171. https://doi.org/10.1016/j.jclepro.2019.04.089
21. Al-Naqbi AK, Alshannag Q (2018) The status of education for sustainable development and sustainability knowledge, attitudes, and behaviors of UAE University students. Int J Sustain High Educ 19(3):566–588. https://doi.org/10.1108/IJSHE-06-2017-0091
22. Aleixo AM, Leal S, Azeiteiro UM (2018) Conceptualization of sustainable higher education institutions, roles, barriers, and challenges for sustainability: an exploratory study in Portugal. J Clean Prod 172:1664–1673. https://doi.org/10.1016/j.jclepro.2016.11.010
23. Ali M, de Azevedo ARG, Marvila MT, Khan MI, Memon AM, Masood F, Haq IU et al (2021) The influence of COVID-19-induced daily activities on health parameters—a case study in Malaysia. Sustainability 13(13):7465. https://doi.org/10.3390/su13137465
24. Alkhuzaim L, Zhu Q, Sarkis J (2021) Evaluating emergy analysis at the nexus of circular economy and sustainable supply chain management. Sustainable Production and Consumption 25:413–424. https://doi.org/10.1016/j.spc.2020.11.022
25. Alonso-Fradejas A (2021) The resource property question in climate stewardship and sustainability transitions. Land Use Policy 108:105529. https://doi.org/10.1016/j.landusepol.2021.105529
26. Alshuwaikhat HM, Adenle YA, Saghir B (2016) Sustainability assessment of higher education institutions in Saudi Arabia. Sustainability, 8(8). https://doi.org/10.3390/su8080750
27. Álvarez-Risco A, Arellano EZ, Valerio EM, Acosta NM, Tarazona ZS (2013) Pharmaceutical care campaign as a strategy for implementation of pharmaceutical services: experience Peru. Pharmaceutical Care Espana 15(1):35–37
28. Alvarez-Risco A, Dawson J, Johnson W, Conteh-Barrat M, Aslani P, Del-Aguila-Arcentales S, Diaz-Risco S (2020) Ebola virus disease outbreak: a global approach for health systems. Revista Cubana de Farmacia, 53(4), 1–13, Article e491
29. Alvarez-Risco A, Del-Aguila-Arcentales S (2015) Prescription errors as a barrier to pharmaceutical care in public health facilities: experience Peru. Pharmaceutical Care Espana 17(6):725–731
30. Alvarez-Risco A, Del-Aguila-Arcentales S (2021a) A note on changing regulation in international business: the world intellectual property organization (wipo) and artificial intelligence. In: Progress in international business research, vol 15, pp 363–371. https://doi.org/10.1108/S1745-886220210000015020
31. Alvarez-Risco A, Del-Aguila-Arcentales S (2021b) Public policies and private efforts to increase women entrepreneurship based on STEM background. In: Contributions to management science, pp 75–87. https://doi.org/10.1007/978-3-030-83792-1_5
32. Alvarez-Risco A, Del-Aguila-Arcentales S (2022) Sustainable initiatives in international markets. In: Contributions to management science, pp 181–191. https://doi.org/10.1007/978-3-030-85950-3_10
33. Alvarez-Risco A, Del-Aguila-Arcentales S, Delgado-Zegarra J, Yáñez JA, Diaz-Risco S (2019) Doping in sports: findings of the analytical test and its interpretation by the public. Sport Sciences for Health 15(1):255–257. https://doi.org/10.1007/s11332-018-0484-8
34. Alvarez-Risco A, Del-Aguila-Arcentales S, Diaz-Risco S (2018) Pharmacovigilance as a tool for sustainable development of healthcare in Peru. PharmacoVigilance Review 10(2):4–6
35. Alvarez-Risco A, Del-Aguila-Arcentales S, Rosen MA, García-Ibarra V, Maycotte-Felkel S, Martínez-Toro GM (2021) Expectations and interests of university students in covid-19 times about sustainable development goals: evidence from Colombia, Ecuador, Mexico, and Peru. Sustainability (Switzerland), 13(6), Article 3306. https://doi.org/10.3390/su13063306
36. Alvarez-Risco A, Del-Aguila-Arcentales S, Stevenson JG (2015) Pharmacists and mass communication for implementing pharmaceutical care. Am J Pharm Benefits 7(3):e125–e126
37. Alvarez-Risco A, Del-Aguila-Arcentales S, Villalobos-Alvarez D, Diaz-Risco S (2022) Leadership for sustainability in crisis time. In: Alvarez-Risco A, Muthu SS, Del-Aguila-Arcentales S (eds) Circular economy: impact on carbon and water footprint. Springer, Singapore, pp 41–64. https://doi.org/10.1007/978-981-19-0549-0_3

38. Alvarez-Risco A, Del-Aguila-Arcentales S, Yáñez JA, Alvarez-Risco A (2021) Telemedicine in Peru as a result of the COVID-19 pandemic: perspective from a country with limited internet access. Am J Trop Med Hyg 105(1):6–11. https://doi.org/10.4269/ajtmh.21-0255
39. Alvarez-Risco A, Del-Aguila-Arcentales S, Yáñez JA, Rosen MA, Mejia CR (2021) Influence of technostress on academic performance of university medicine students in Peru during the covid-19 pandemic. Sustainability (Switzerland), 13(16), Article 8949. https://doi.org/10.3390/su13168949
40. Alvarez-Risco A, Delgado-Zegarra J, Yáñez JA, Diaz-Risco S, Del-Aguila-Arcentales S (2018) Predation risk by gastronomic boom—case Peru. J Landscape Ecol (Czech Republic) 11(1):100–103. https://doi.org/10.2478/jlecol-2018-0003
41. Alvarez-Risco A, Estrada-Merino A, Anderson-Seminario MM, Mlodzianowska S, García-Ibarra V, Villagomez-Buele C, Carvache-Franco M (2021) Multitasking behavior in online classrooms and academic performance: case of university students in ecuador during COVID-19 outbreak. Interactive Technol Smart Educ 18(3):422–434. https://doi.org/10.1108/ITSE-08-2020-0160
42. Alvarez-Risco A, Mejia CR, Delgado-Zegarra J, Del-Aguila-Arcentales S, Arce-Esquivel AA, Valladares-Garrido MJ, Yáñez JA et al (2020) The Peru approach against the COVID-19 infodemic: insights and strategies. Am J Trop Med Hyg 103(2):583–586. https://doi.org/10.4269/ajtmh.20-0536
43. Alvarez-Risco A, Mlodzianowska S, García-Ibarra V, Rosen MA, Del-Aguila-Arcentales S (2021) Factors affecting green entrepreneurship intentions in business university students in covid-19 pandemic times: case of ecuador. Sustainability (Switzerland), 13(11), Article 6447. https://doi.org/10.3390/su13116447
44. Alvarez-Risco A, Mlodzianowska S, Zamora-Ramos U, Del-Aguila-Arcentales S (2021) Green entrepreneurship intention in university students: the case of Peru. Entrepreneurial Bus Econ Rev 9(4):85–100. https://doi.org/10.15678/EBER.2021.090406
45. Alvarez-Risco A, Quiroz-Delgado D, Del-Aguila-Arcentales S (2016) Pharmaceutical care in hypertension patients in a peruvian hospital [Article]. Indian J Public Health Res Develop 7(3):183–188. https://doi.org/10.5958/0976-5506.2016.00153.4
46. Alvarez-Risco A, Turpo-Cama A, Ortiz-Palomino L, Gongora-Amaut N, Del-Aguila-Arcentales S (2016) Barriers to the implementation of pharmaceutical care in pharmacies in Cusco Peru. Pharmaceutical Care Espana 18(5):194–205
47. Alvarez-Risco A, Van Mil JWF (2007) Pharmaceutical care in community pharmacies: practice and research in Peru. Ann Pharmacother 41(12):2032–2037. https://doi.org/10.1345/aph.1K117
48. Ambientum (2020) What is the situation of renewables in India? Retrieved 12/09/2021 from https://www.ambientum.com/ambientum/energia/como-es-la-situacion-de-las-renovables-en-india.asp
49. Amoruso G, Donevska N, Skomedal G (2018) German and Norwegian policy approach to residential buildings' energy efficiency—a comparative assessment. Energ Effi 11(6):1375–1395. https://doi.org/10.1007/s12053-018-9637-5
50. Ang KL, Saw ET, He W, Dong X, Ramakrishna S (2021) Sustainability framework for pharmaceutical manufacturing (PM): a review of research landscape and implementation barriers for circular economy transition. J Clean Prod 280:124264. https://doi.org/10.1016/j.jclepro.2020.124264
51. Angulo-Mosquera LS, Alvarado-Alvarado AA, Rivas-Arrieta MJ, Cattaneo CR, Rene ER, García-Depraect O (2021) Production of solid biofuels from organic waste in developing countries: A review from sustainability and economic feasibility perspectives. Sci Total Environ 795:148816. https://doi.org/10.1016/j.scitotenv.2021.148816
52. Apcho-Ccencho LV, Cuya-Velásquez BB, Rodríguez DA, Anderson-Seminario MLM, Alvarez-Risco A, Estrada-Merino A, Mlodzianowska S (2021) The impact of international price on the technological industry in the united states and china during times of crisis: commercial war and covid-19. Adv Business Manage Forecasting, 14:149–160

53. Aragon-Correa JA, Marcus AA, Rivera JE, Kenworthy AL (2017) Sustainability management teaching resources and the challenge of balancing planet, people, and profits. Acad Manage Learn Educ 16(3):469–483. https://doi.org/10.5465/amle.2017.0180
54. Arias-Meza M, Alvarez-Risco A, Cuya-Velásquez BB, de las Mercedes Anderson-Seminario M, Del-Aguila-Arcentales S (2022) Fashion and textile circularity and waste footprint. In: Alvarez-Risco A, Muthu SS, Del-Aguila-Arcentales S (eds) Circular economy: impact on carbon and water footprint. Springer, Singapore, pp 181–204. https://doi.org/10.1007/978-981-19-0549-0_9
55. Aschemann-Witzel J, Giménez A, Ares G (2018) Convenience or price orientation? Consumer characteristics influencing food waste behaviour in the context of an emerging country and the impact on future sustainability of the global food sector. Glob Environ Chang 49:85–94. https://doi.org/10.1016/j.gloenvcha.2018.02.002
56. Aydın B, Alvarez MD (2020) Understanding the tourists' perspective of sustainability in cultural tourist destinations. Sustainability 12(21):8846. https://doi.org/10.3390/su12218846
57. Bamgbade JA, Kamaruddeen AM, Nawi MNM, Adeleke AQ, Salimon MG, Ajibike WA (2019) Analysis of some factors driving ecological sustainability in construction firms. J Clean Prod 208:1537–1545. https://doi.org/10.1016/j.jclepro.2018.10.229
58. Barr S, Gilg A, Shaw G (2011) 'Helping People Make Better Choices': Exploring the behaviour change agenda for environmental sustainability. Appl Geogr 31(2):712–720. https://doi.org/10.1016/j.apgeog.2010.12.003
59. Barragán-Ocaña A, Silva-Borjas P, Olmos-Peña S, Polanco-Olguín M (2020) Biotechnology and bioprocesses: their contribution to sustainability. Processes 8(4). https://doi.org/10.3390/pr8040436
60. Barros MV, Salvador R, do Prado GF, de Francisco AC, Piekarski CM (2021) Circular economy as a driver to sustainable businesses. Cleaner Environ Syst 2:100006. https://doi.org/10.1016/j.cesys.2020.100006
61. Batista AA, Francisco AC (2018) Organizational sustainability practices: a study of the firms listed by the corporate sustainability index. Sustainability 10(1). https://doi.org/10.3390/su10010226
62. Bokpin GA (2017) Foreign direct investment and environmental sustainability in Africa: the role of institutions and governance. Res Int Bus Finance 39:239–247. https://doi.org/10.1016/j.ribaf.2016.07.038
63. Brito RM, Rodríguez C, Aparicio JL (2018) Sustainability in teaching: an evaluation of university teachers and students. Sustainability 10(2):439. https://doi.org/10.3390/su10020439
64. Brown HS, de Jong M, Levy DL (2009) Building institutions based on information disclosure: lessons from GRI's sustainability reporting. J Clean Prod 17(6):571–580. https://doi.org/10.1016/j.jclepro.2008.12.009
65. Brown KA, Harris F, Potter C, Knai C (2020) The future of environmental sustainability labelling on food products. Lancet Planetary Health 4(4):e137–e138. https://doi.org/10.1016/S2542-5196(20)30074-7
66. Büyüközkan Feyzioğlu G, Karabulut Y, Mukul E (2018) A novel renewable energy selection model for United Nations' sustainable development goals
67. Carvache-Franco M, Alvarez-Risco A, Carvache-Franco O, Carvache-Franco W, Estrada-Merino A, Villalobos-Alvarez D (2021) Perceived value and its influence on satisfaction and loyalty in a coastal city: a study from Lima, Peru. J Policy Res Tourism Leisure and Events. https://doi.org/10.1080/19407963.2021.1883634
68. Carvache-Franco M, Alvarez-Risco A, Carvache-Franco W, Carvache-Franco O, Estrada-Merino A, Rosen MA (2021) Coastal cities seen from loyalty and their tourist motivations: a study in Lima, Peru. Sustainability (Switzerland) 13(21), Article 11575. https://doi.org/10.3390/su132111575
69. Carvache-Franco M, Carvache-Franco O, Carvache-Franco W, Alvarez-Risco A, Estrada-Merino A (2021) Motivations and segmentation of the demand for coastal cities: a study in Lima Peru. Int J Tourism Res 23(4):517–531. https://doi.org/10.1002/jtr.2423

70. Casado M (2021) Japan's energy transition a decade after Fukushima. Retrieved 12 Jul 2021 from https://www.esglobal.org/la-transicion-energetica-de-japon-una-decada-despues-de-fukushima
71. Castillo-Benancio S, Alvarez-Risco A, Esquerre-Botton S, Leclercq-Machado L, Calle-Nole M, Morales-Ríos F, Del-Aguila-Arcentales S et al (2022) Circular economy for packaging and carbon footprint. In: Alvarez-Risco A, Muthu SS, Del-Aguila-Arcentales S (eds) Circular economy: impact on carbon and water footprint. Springer, Singapore, pp 115–138. https://doi.org/10.1007/978-981-19-0549-0_6
72. Centro Nacional de Control de Energía (2020) Generación y demanda. Informe anual. Retrieved 12Jul 2021 from https://apps.grupoice.com/CenceWeb/CenceDescargaArchivos.jsf?init=true&categoria=3&codigoTipoArchivo=3008
73. Chafloque-Cespedes R, Alvarez-Risco A, Robayo-Acuña PV, Gamarra-Chavez CA, Martinez-Toro GM, Vicente-Ramos W (2021) Effect of sociodemographic factors in entrepreneurial orientation and entrepreneurial intention in university students of Latin American business schools. Contemp Issues Entrepreneurship Res 11:151–165. https://doi.org/10.1108/S2040-724620210000011010
74. Chafloque-Céspedes R, Vara-Horna A, Asencios-Gonzales Z, López-Odar D, Alvarez-Risco A, Quipuzco-Chicata L, Sánchez-Villagomez M et al (2020) Academic presenteeism and violence against women in schools of business and engineering in Peruvian universities. Lecturas de Economia 93:127–153. https://doi.org/10.17533/udea.le.n93a340726
75. Chavez J (2021) Renewables in Germany to outperform fossil fuels in 2020. Retrieved 12 Aug 2021 from https://energiahoy.com/2021/01/05/renovables-en-alemania-superaron-al-combustible-fosil-en-2020
76. Chen X, Zhang SX, Jahanshahi AA, Alvarez-Risco A, Dai H, Li J, Ibarra VG (2020) Belief in a COVID-19 conspiracy theory as a predictor of mental health and well-being of health care workers in Ecuador: Cross-sectional survey study. JMIR Public Health Surveillance 6(3), Article e20737. https://doi.org/10.2196/20737
77. Cherepovitsyn A, Tcvetkov P (2017) Overview of the prospects for developing a renewable energy in Russia. In: 2017 international conference on green energy and applications (ICGEA)
78. Chu F, Zhang W, Jiang Y (2021) How does policy perception affect green entrepreneurship behavior? An empirical analysis from China. Discret Dyn Nat Soc 2021:7973046. https://doi.org/10.1155/2021/7973046
79. Chung SA, Olivera S, Román BR, Alanoca E, Moscoso S, Terceros BL, Yáñez JA et al (2021) Themes of scientific production of the Cuban journal of pharmacy indexed in scopus (1967–2020). Revista Cubana de Farmacia 54(1), Article e511.
80. CIA (2021) Liechtenstein. Retrieved 12 Aug 2021 from https://www.cia.gov/the-world-factbook/countries/liechtenstein
81. Collin J (2017) The French energy transition law for green growth and the multiannual energy programming 2016–2023. Retrieved 12 Sept 2021 from http://www.realinstitutoelcano.org/wps/portal/rielcano_es/contenido?WCM_GLOBAL_CONTEXT=/elcano/elcano_es/zonas_es/ari18-2017-collin-ley-transicion-energetica-francia-crecimiento-verde
82. Comunicar Se (2021) Landmark UK agreement to boost clean energy transition. Retrieved 12 Aug 2021 from https://www.comunicarseweb.com/noticia/acuerdo-historico-del-reino-unido-para-impulsar-la-transicion-hacia-energias-limpias
83. Contreras-Taica A, Alvarez-Risco A, Arias-Meza M, Campos-Dávalos N, Calle-Nole M, Almanza-Cruz C, Del-Aguila-Arcentales S et al (2022) Virtual education: carbon footprint and circularity. In: Alvarez-Risco A, Muthu SS, Del-Aguila-Arcentales S (eds) Circular economy: impact on carbon and water footprint. Springer, Singapore, pp 265–285. https://doi.org/10.1007/978-981-19-0549-0_13
84. Coren MJ (2018) Portugal generated enough renewable energy to power the whole country in March. Retrieved 12 Aug 2021 from https://qz.com/1245048/portugal-generated-enough renewable-energy-to-power-the-whole-country-in-march

85. Cruz-Torres W, Alvarez-Risco A, Del-Aguila-Arcentales S (2021) Impact of enterprise resource planning (ERP) implementation on performance of an education enterprise: a structural equation modeling (SEM) [Article]. Stud Bus Econ 16(2):37–52. https://doi.org/10.2478/sbe-2021-0023
86. Cruz JP, Alshammari F, Felicilda-Reynaldo RFD (2018) Predictors of Saudi nursing students' attitudes towards environment and sustainability in health care. Int Nurs Rev 65(3):408–416. https://doi.org/10.1111/inr.12432
87. Cuya-Velásquez BB, Alvarez-Risco A, Gomez-Prado R, Juarez-Rojas L, Contreras-Taica A, Ortiz-Guerra A, Del-Aguila-Arcentales S et al (2022) Circular economy for food loss reduction and water footprint. In: Alvarez-Risco A, Muthu SS, Del-Aguila-Arcentales S (eds) Circular economy: impact on carbon and water footprint. Springer, Singapore, pp 65–91. https://doi.org/10.1007/978-981-19-0549-0_4
88. De-la-Cruz-Diaz M, Alvarez-Risco A, Jaramillo-Arévalo M, de las Mercedes Anderson-Seminario M, Del-Aguila-Arcentales S (2022) 3D Print, circularity, and footprints. In: Alvarez-Risco A, Muthu SS, Del-Aguila-Arcentales S (eds) Circular economy: impact on carbon and water footprint. Springer, Singapore, pp 93–112. https://doi.org/10.1007/978-981-19-0549-0_5
89. de las Mercedes Anderson-Seminario M, Alvarez-Risco A (2022) Better students, better companies, better life: circular learning. In: Alvarez-Risco A, Muthu SS, Del-Aguila-Arcentales S (eds) Circular economy: impact on carbon and water footprint. Springer, Singapore, pp 19–40. https://doi.org/10.1007/978-981-19-0549-0_2
90. Del-Aguila-Arcentales S, Alvarez-Risco A (2013) Human error or burnout as explanation for mistakes in pharmaceutical laboratories. Accred Qual Assur 18(5):447–448. https://doi.org/10.1007/s00769-013-1000-0
91. Del-Aguila-Arcentales S, Alvarez-Risco A, Muthu SS (2022) Measuring circular economy. In: Alvarez-Risco A, Muthu SS, Del-Aguila-Arcentales S (eds) Circular economy: impact on carbon and water footprint. Springer, Singapore, pp 3–17. https://doi.org/10.1007/978-981-19-0549-0_1
92. Delgado-Zegarra J, Alvarez-Risco A, Yáñez JA (2018) Indiscriminate use of pesticides and lack of sanitary control in the domestic market in Peru. Revista Panamericana de Salud Publica/Pan Am J Public Health 42, Article e3. https://doi.org/10.26633/RPSP.2018.3
93. Dincer I (2001) Environmental issues: II-potential solutions. Energy Sources 23(1):83–92. https://doi.org/10.1080/00908310151092218
94. Doukas H, Marinakis V, Karakosta C, Psarras J (2012) Promoting renewables in the energy sector of Tajikistan. Renew Energy 39(1):411–418. https://doi.org/10.1016/j.renene.2011.09.007
95. Enciso-Zarate A, Guzmán-Oviedo J, Sánchez-Cardona F, Martínez-Rohenes D, Rodríguez-Palomino JC, Alvarez-Risco A, Diaz-Risco S et al (2016) Evaluation of contamination by cytotoxic agents in colombian hospitals. Pharm Care Espana 18(6):241–250
96. Energiestrategie (2020) Energy practice. Retrieved 12 Aug 2021 from https://www.energiebuendel.li/EnergiePraxis.aspx
97. Enevoldsen P, Permien F-H (2018) Mapping the wind energy potential of Sweden: a sociotechnical Wind Atlas. J Renew Energy 2018:1650794. https://doi.org/10.1155/2018/1650794
98. Eppinger E, Jain A, Vimalnath P, Gurtoo A, Tietze F, Hernandez Chea R (2021) Sustainability transitions in manufacturing: the role of intellectual property. Curr Opin Environ Sustain 49:118–126. https://doi.org/10.1016/j.cosust.2021.03.018
99. Ertuna B, Karatas-Ozkan M, Yamak S (2019) Diffusion of sustainability and CSR discourse in hospitality industry. Int J Contemp Hosp Manag 31(6):2564–2581. https://doi.org/10.1108/IJCHM-06-2018-0464
100. Espinasa R, Bonzi A, Anaya F (2017) Energy dossier: Uruguay. Retrieved 12 Aug 2021 from https://publications.iadb.org/es/publications/spanish/document/Dossier-energ%C3%A9tico-Uruguay.pdf

101. Esquerre-Botton S, Alvarez-Risco A, Leclercq-Machado L, de las Mercedes Anderson-Seminario M, Del-Aguila-Arcentales S (2022) Food loss reduction and carbon footprint practices worldwide: a benchmarking approach of circular economy. In: Alvarez-Risco A, Muthu SS, Del-Aguila-Arcentales S (eds) Circular economy: impact on carbon and water footprint. Springer, Singapore, pp 161–179. https://doi.org/10.1007/978-981-19-0549-0_8
102. FCDO (2012) UK national and international energy and climate change policies to 2050. Retrieved 12Jul 2021 from https://blogs.fcdo.gov.uk/es/ukinperu/2012/07/05/politicas-nacionales-e-internacionales-del-reino-unido-respecto-de-energia-y-cambio-climatico-al-2050
103. Fornillo B (2021) Transición energética en Uruguay:¿ dominio del mercado o potencia público-social? Ambiente and Sociedade 24
104. Galleli B, Teles NEB, Santos JARD, Freitas-Martins MS, Hourneaux Junior F (2021) Sustainability university rankings: a comparative analysis of UI green metric and the times higher education world university rankings. Int J Sustain Higher Educ Ahead-of-print(ahead-of-print). https://doi.org/10.1108/IJSHE-12-2020-0475
105. Gatto A, Drago C (2021) When renewable energy, empowerment, and entrepreneurship connect: measuring energy policy effectiveness in 230 countries. Energy Res Soc Sci 78:101977. https://doi.org/10.1016/j.erss.2021.101977
106. Gerdt S-O, Wagner E, Schewe G (2019) The relationship between sustainability and customer satisfaction in hospitality: an explorative investigation using eWOM as a data source. Tour Manage 74:155–172. https://doi.org/10.1016/j.tourman.2019.02.010
107. Gevrenova T (2015) Nature and characteristics of green entrepreneurship. Trakia J Sci 13(Suppl 2):321–323
108. Godínez-Zamora G, Victor-Gallardo L, Angulo-Paniagua J, Ramos E, Howells M, Usher W, Quirós-Tortós J et al (2020) Decarbonising the transport and energy sectors: technical feasibility and socioeconomic impacts in Costa Rica. Energ Strat Rev 32:100573. https://doi.org/10.1016/j.esr.2020.100573
109. Gómez-Prado R, Alvarez-Risco A, Sánchez-Palomino J, de las Mercedes Anderson-Seminario M, Del-Aguila-Arcentales S (2022) Circular economy for waste reduction and carbon footprint. In: Alvarez-Risco A, Muthu SS, Del-Aguila-Arcentales S (eds) Circular economy: impact on carbon and water footprint. Springer, Singapore, pp 139–159. https://doi.org/10.1007/978-981-19-0549-0_7
110. Gonzaga Añazco SJ, Alaña Castillo TP (2017) Competitividad y emprendimiento: herramientas de crecimiento económico de un país. INNOVA Res J 2(8.1):322–328
111. Government of the Principality of Liechtenstein (2019) Sustainability in Liechtenstein. Retrieved 12 Aug 2021 from https://sustainabledevelopment.un.org/content/documents/23369Full_VNR_Liechtenstein_June_2019.pdf
112. Grewal J, Hauptmann C, Serafeim G (2021) Material sustainability information and stock price informativeness. J Bus Ethics 171(3):513–544. https://doi.org/10.1007/s10551-020-04451-2
113. Guðmundsdóttir H, Carton W, Busch H, Ramasar V (2018) Modernist dreams and green sagas: the neoliberal politics of Iceland's renewable energy economy. Environ Plan E: Nat Space 1(4):579–601. https://doi.org/10.1177/2514848618796829
114. Haar N, Theyel G (2006) U.S. electric utilities and renewable energy: drivers for adoption. Int J Green Energy 3(3):271–281. https://doi.org/10.1080/01971520600704043
115. Haldar S (2019) Green entrepreneurship in theory and practice: insights from India. Int J Green Econ 13(2):99–119
116. Hall MR (2019) The sustainability price: expanding environmental life cycle costing to include the costs of poverty and climate change. Int J Life Cycle Assess 24(2):223–236. https://doi.org/10.1007/s11367-018-1520-2
117. Hargroves K, Smith M (2005) The natural advantage of nations: business opportunities, innovation and governance in the 21st Century. Routledge
118. Hemalatha K (2021) India: between solar and coal. India leads the world's largest clean energy project. Retrieved 12 Oct 2021 from https://www.dw.com/es/india-entre-la-energ%C3%ADa-solar-y-el-carbón/a-54688951

119. Hong S, Qvist S, Brook BW (2018) Economic and environmental costs of replacing nuclear fission with solar and wind energy in Sweden. Energy Policy 112:56–66. https://doi.org/10.1016/j.enpol.2017.10.013
120. Houeland C, Jordhus-Lier DC, Angell FH (2021) Solidarity tested: the case of the Norwegian confederation of trade unions (LO-Norway) and its contradictory climate change policies. Area 53(3):413–421. https://doi.org/10.1111/area.12608
121. International Energy Agency (2019) Sweden 2019. Retrieved 12 Aug 20212 from https://iea.blob.core.windows.net/assets/abf9ceee-2f8f-46a0-8e3b-78fb93f602b0/Energy_Policies_of_IEA_Countries_Sweden_2019_Review.pdf
122. International Renewable Energy Agency (2018) Energy profile—Niue. Retrieved 12 Aug 2021 from https://www.irena.org/IRENADocuments/Statistical_Profiles/Oceania/Niue_Oceania_RE_SP.pdf
123. Jiménez DG, Oliva JS (2019) Energy policy in the United States today [La política energética en Estados Unidos en la actualidad]. Boletín Económico de ICE(3110)
124. Jones P, Comfort D (2020) The COVID-19 crisis and sustainability in the hospitality industry. Int J Contemp Hosp Manag 32(10):3037–3050. https://doi.org/10.1108/IJCHM-04-2020-0357
125. Juarez-Rojas L, Alvarez-Risco A, Campos-Dávalos N, de las Mercedes Anderson-Seminario M, Del-Aguila-Arcentales S (2022) Water footprint in the textile and food supply chain management: trends to become circular and sustainable. In: Alvarez-Risco A, Muthu SS, Del-Aguila-Arcentales S (eds) Circular economy: impact on carbon and water footprint. Springer, Singapore, pp 225–243. https://doi.org/10.1007/978-981-19-0549-0_11
126. Kan X, Hedenus F, Reichenberg L (2020) The cost of a future low-carbon electricity system without nuclear power—the case of Sweden. Energy 195:117015. https://doi.org/10.1016/j.energy.2020.117015
127. Khan S, Henderson C (2020) How Western Michigan University is approaching its commitment to sustainability through sustainability-focused courses. J Clean Prod 253:119741. https://doi.org/10.1016/j.jclepro.2019.119741
128. Khan SAR, Razzaq A, Yu Z, Miller S (2021) Industry 4.0 and circular economy practices: a new era business strategies for environmental sustainability. Bus Strategy Environ 30(8):4001–4014. https://doi.org/10.1002/bse.2853
129. Khoshnevis Yazdi S, Shakouri B (2017) Renewable energy, nonrenewable energy consumption, and economic growth. Energy Sources Part B 12(12):1038–1045. https://doi.org/10.1080/15567249.2017.1316795
130. Kong L, Liu Z, Wu J (2020) A systematic review of big data-based urban sustainability research: state-of-the-science and future directions. J Clean Prod 273:123142. https://doi.org/10.1016/j.jclepro.2020.123142
131. Kucharska A (2019) Energy policy of Liechtenstein. Energy Policy Stud. http://yadda.icm.edu.pl/yadda/element/bwmeta1.element.baztech-8cf3a2cb-5ede-4add-bd5a-90d5b7c0c6e0/c/Kucharska_A_Energy_EPS_No.2_4__2019.pdf
132. La actualidad energética (2021) The energy transition. Retrieved 12 Aug 2021 from https://www.tatsachen-ueber-deutschland.de/es/clima-y-energia/la-transicion-energetica
133. Laldjebaev M, Morreale SJ, Sovacool BK, Kassam K-AS (2018) Rethinking energy security and services in practice: National vulnerability and three energy pathways in Tajikistan. Energy Policy 114:39–50. https://doi.org/10.1016/j.enpol.2017.11.058
134. Larrán Jorge M, Andrades Peña FJ, Herrera Madueño J (2019) An analysis of university sustainability reports from the GRI database: an examination of influential variables. J Environ Planning Manage 62(6):1019–1044. https://doi.org/10.1080/09640568.2018.1457952
135. Lauf T, Ek K, Gawel E, Lehmann P, Söderholm P (2020) The regional heterogeneity of wind power deployment: an empirical investigation of land-use policies in Germany and Sweden. J Environ Planning Manage 63(4):751–778. https://doi.org/10.1080/09640568.2019.1613221
136. Leiva-Martinez MA, Anderson-Seminario MLM, Alvarez-Risco A, Estrada-Merino A, Mlodzianowska S (2021) Price variation in lower goods as of previous economic crisis and the contrast of the current price situation in the context of covid-19 in peru. In: Advances in business and management forecasting, vol 14, pp 161–166

137. Lizarralde E, Ferro E, Guasch Murillo D, Berbegal Mirabent J, Villalta Boix M, Abbad-Jaime de Aragón González F, Gainzarain Sastre M et al (2013) Sectors of the new 20+ 20 economy [Sectores de la nueva economía 20+20. Economía de la accesibilidad]. http://repositori.uic.es/bitstream/handle/20.500.12328/2957/Berbegal%20Mirabent%2C%20Jasmina%20%5Bet%20al.%5D_Economia%20Accesibilidad_2013.pdf?sequence=1&isAllowed=y
138. Llorens D, Jha C, Salmon M, De Paula J, Del Castillo J (2021) Clean, scalable, reliable geothermal power. Retrieved 12 Aug 2021 from https://aim2flourish.com/innovations/clean-scalable-reliable-geothermal-power-1
139. Logadóttir H (n.d.) Iceland's sustainable energy story: a model for the world? Retrieved 12 Aug 2021 from https://www.un.org/en/chronicle/article/icelands-sustainable-energy-story-model-world
140. Lopez-Odar D, Alvarez-Risco A, Vara-Horna A, Chafloque-Cespedes R, Sekar MC (2020) Validity and reliability of the questionnaire that evaluates factors associated with perceived environmental behavior and perceived ecological purchasing behavior in Peruvian consumers [Article]. Soc Respons J 16(3):403–417. https://doi.org/10.1108/SRJ-08-2018-0201
141. Lozano R, Barreiro-Gen M, Lozano FJ, Sammalisto K (2019) Teaching sustainability in European higher education institutions: assessing the connections between competences and pedagogical approaches. Sustainability 11(6):1602. https://doi.org/10.3390/su11061602
142. Manavalan E, Jayakrishna K (2019) An analysis on sustainable supply chain for circular economy. Procedia Manuf 33:477–484. https://doi.org/10.1016/j.promfg.2019.04.059
143. Mani V, Gunasekaran A, Delgado C (2018) Supply chain social sustainability: standard adoption practices in Portuguese manufacturing firms. Int J Prod Econ 198:149–164. https://doi.org/10.1016/j.ijpe.2018.01.032
144. Manzoor SR, Ho JSY, Al Mahmud A (2021) Revisiting the 'university image model' for higher education institutions' sustainability. J Mark High Educ 31(2):220–239. https://doi.org/10.1080/08841241.2020.1781736
145. Martinez-Martinez A, Cegarra-Navarro J-G, Garcia-Perez A, Wensley A (2019) Knowledge agents as drivers of environmental sustainability and business performance in the hospitality sector. Tour Manage 70:381–389. https://doi.org/10.1016/j.tourman.2018.08.030
146. Masini A, Menichetti E (2012) The impact of behavioural factors in the renewable energy investment decision making process: conceptual framework and empirical findings. Energy Policy 40:28–38. https://doi.org/10.1016/j.enpol.2010.06.062
147. Mejía-Acosta N, Alvarez-Risco A, Solís-Tarazona Z, Matos-Valerio E, Zegarra-Arellano E, Del-Aguila-Arcentales S (2016) Adverse drug reactions reported as a result of the implementation of pharmaceutical care in the institutional pharmacy DIGEMID—ministry of health. Pharm Care Espana 18(2):67–74
148. Michelino F, Cammarano A, Celone A, Caputo M (2019) The linkage between sustainability and innovation performance in IT hardware sector. Sustainability 11(16):4275. https://doi.org/10.3390/su11164275
149. Mikhaylov A (2020) Geothermal energy development in Iceland. Int J Energy Econ Policy 10(4):31
150. Millot A, Krook-Riekkola A, Maïzi N (2020) Guiding the future energy transition to net-zero emissions: lessons from exploring the differences between France and Sweden. Energy Policy 139:111358. https://doi.org/10.1016/j.enpol.2020.111358
151. Ministerio de Industria E y. M (2021) Renewable energy sources. Retrieved 12 Aug 2021 from http://www.eficienciaenergetica.gub.uy/-que-es-la-energia-?p_p_id=101&p_p_lifecycle=0&p_p_state=maximized&_101_struts_action=%2Fasset_publisher%2Fview_content&_101_assetEntryId=63094&_101_type=content&_101_urlTitle=fuentes-de-energia-renovables&inheritRedirect=true
152. Ministry of Petroleum and Energy (2016) Renewable energy production in Norway. Retrieved 12 Aug 2021 from https://www.regjeringen.no/en/topics/energy/renewable-energy/renewable-energy-production-in-norway/id2343462
153. Ministry of Petroleum and Energy (2021) The aim of Norwegian energy policy is to provide a suitable framework for maintaining an efficient, climate-friendly and reliable energy supply

system. Retrieved 12 Aug 2021 from https://energifaktanorge.no/en/om-energisektoren/verdt-a-vite-om-norsk-energipolitikk
154. Mohamued EA, Ahmed M, Pypłacz P, Liczmańska-Kopcewicz K, Khan MA (2021) Global oil price and innovation for sustainability: the impact of r&D spending, oil price and oil price volatility on GHG emissions. Energies, 14(6). https://doi.org/10.3390/en14061757
155. Morales-Ríos F, Alvarez-Risco A, Castillo-Benancio S, de las Mercedes Anderson-Seminario M, Del-Aguila-Arcentales S (2022) Material selection for circularity and footprints. In: Alvarez-Risco A, Muthu SS, Del-Aguila-Arcentales S (eds) Circular economy: impact on carbon and water footprint. Springer, Singapore, pp 205–221. https://doi.org/10.1007/978-981-19-0549-0_10
156. Movik S, Allouche J (2020) States of power: energy imaginaries and transnational assemblages in Norway, Nepal and Tanzania. Energy Res Soc Sci 67:101548. https://doi.org/10.1016/j.erss.2020.101548
157. Murad MW, Alam MM, Noman AHM, Ozturk I (2019) Dynamics of technological innovation, energy consumption, energy price and economic growth in Denmark. Environ Prog Sustainable Energy 38(1):22–29. https://doi.org/10.1002/ep.12905
158. Mzembe AN, Lindgreen A, Idemudia U, Melissen F (2020) A club perspective of sustainability certification schemes in the tourism and hospitality industry. J Sustain Tour 28(9):1332–1350. https://doi.org/10.1080/09669582.2020.1737092
159. National Energy Authority (n.d.) Master plan for hydro and geothermal energy resources in Iceland. Retrieved 12 Aug 2021 from https://nea.is/geothermal/master-plan/nr/103
160. National Geographic (2021) India leads the world's largest clean energy project. Retrieved 12 Aug 2021 from https://www.nationalgeographic.es/medio-ambiente/2017/05/india-lidera-el-mayor-proyecto-de-energia-limpia-del-mundo
161. Norström AV, Cvitanovic C, Löf MF, West S, Wyborn C, Balvanera P, Österblom H et al (2020) Principles for knowledge co-production in sustainability research. Nat Sustain 3(3):182–190. https://doi.org/10.1038/s41893-019-0448-2
162. OECD (2011) Working party on SMEs and entrepreneurship (WPSMEE)—Green entrepreneurship, eco-innovation and SMEs. Retrieved 12 Aug 2021 from https://one.oecd.org/document/CFE/SME(2011)9/FINAL/en/pdf
163. Olawumi TO, Chan DWM (2018) A scientometric review of global research on sustainability and sustainable development. J Cleaner Prod 183:231–250. https://doi.org/10.1016/j.jclepro.2018.02.162
164. Omoloso O, Mortimer K, Wise WR, Jraisat L (2021) Sustainability research in the leather industry: a critical review of progress and opportunities for future research. J Cleaner Prod 285:125441. https://doi.org/10.1016/j.jclepro.2020.125441
165. Opoku EEO, Kufuor NK, Manu SA (2021) Gender, electricity access, renewable energy consumption and energy efficiency. Technol Forecast Soc Chang 173:121121. https://doi.org/10.1016/j.techfore.2021.121121
166. Our World in Data (2020) Research and data to make progress against the world's largest problems. Retrieved 12 Aug 2021 from https://ourworldindata.org
167. Ozturkoglu Y, Sari FO, Saygili E (2021) A new holistic conceptual framework for sustainability oriented hospitality innovation with triple bottom line perspective. J Hosp Tour Technol 12(1):39–57. https://doi.org/10.1108/JHTT-02-2019-0022
168. Parlamento Europeo (2021) Energy policy: general principles. Retrieved 12 Jul 2021 from https://www.europarl.europa.eu/factsheets/es/sheet/68/la-politica-energetica-principios-generales
169. Peeters P (2018) Why space tourism will not be part of sustainable tourism. Tour Recreat Res 43(4):540–543. https://doi.org/10.1080/02508281.2018.1511942
170. Peña Miguel N, Corral Lage J, Mata Galindez A (2020) Assessment of the development of professional skills in university students: sustainability and serious games. Sustainability, 12(3). https://doi.org/10.3390/su12031014
171. Periodico de la Energía (2021) Italy wants to 'copy' Spain's renewables auction model. Retrieved 12 Aug 2021 from https://elperiodicodelaenergia.com/italia-quiere-copiar-el-modelo-de-subasta-de-renovables-de-espana

172. Petersen J-P (2018) The application of municipal renewable energy policies at community level in Denmark: a taxonomy of implementation challenges. Sustain Cities Soc 38:205–218. https://doi.org/10.1016/j.scs.2017.12.029
173. Pham H, Kim S-Y (2019) The effects of sustainable practices and managers' leadership competences on sustainability performance of construction firms. Sustain Prod Consum 20:1–14. https://doi.org/10.1016/j.spc.2019.05.003
174. Plewnia F, Guenther E (2018) Mapping the sharing economy for sustainability research. Manag Decis 56(3):570–583. https://doi.org/10.1108/MD-11-2016-0766
175. Quispe-Cañari JF, Fidel-Rosales E, Manrique D, Mascaró-Zan J, Huamán-Castillón KM, Chamorro-Espinoza SE, Mejia CR et al (2021) Self-medication practices during the COVID-19 pandemic among the adult population in Peru: a cross-sectional survey. Saudi Pharm J 29(1):1–11. https://doi.org/10.1016/j.jsps.2020.12.001
176. Rajesh R (2020) Exploring the sustainability performances of firms using environmental, social, and governance scores. J Clean Prod 247:119600. https://doi.org/10.1016/j.jclepro.2019.119600
177. Ramísio PJ, Pinto LMC, Gouveia N, Costa H, Arezes D (2019) Sustainability strategy in higher education institutions: lessons learned from a nine-year case study. J Clean Prod 222:300–309. https://doi.org/10.1016/j.jclepro.2019.02.257
178. Ransfield AK, Reichenberger I (2021) Māori Indigenous values and tourism business sustainability. AlterNative: Int J Indigenous Peoples 17(1):49–60. https://doi.org/10.1177/1177180121994680
179. Rau H, Goggins G, Fahy F (2018) From invisibility to impact: Recognising the scientific and societal relevance of interdisciplinary sustainability research. Res Policy 47(1):266–276. https://doi.org/10.1016/j.respol.2017.11.005
180. Regjeringen (2016) Meld. St. 25 (2015–2016). Retrieved 12 Aug 2021 from https://www.regjeringen.no/no/dokumenter/meld.-st.-25-20152016/id2482952
181. Report RGS (2016) Global status report. Retrieved 12 Aug 2021 from https://www.ren21.net/wp-content/uploads/2019/05/REN21_GSR2016_FullReport_en_11.pdf
182. Reve (2014) Renewable energies in Italy: second in photovoltaic solar energy and seventh in wind energy. Retrieved 12 Feb 2021 from https://www.evwind.com/2014/05/24/italia-en-los-primeros-lugares-de-produccion-de-energias-renovables
183. Rexhäuser S, Löschel A (2015) Invention in energy technologies: Comparing energy efficiency and renewable energy inventions at the firm level. Energy Policy 83:206–217. https://doi.org/10.1016/j.enpol.2015.02.003
184. Ribeiro F, Ferreira P, Araújo M, Braga AC (2014) Public opinion on renewable energy technologies in Portugal. Energy 69:39–50. https://doi.org/10.1016/j.energy.2013.10.074
185. Roca J (2020) Japan to invest $100 billion in solar and wind energy to reach 27% renewables target by 2030. Retrieved 12 Aug 2021 from https://elperiodicodelaenergia.com/japon-invertira-100-000-millones-de-dolares-en-energia-solar-y-eolica-para-alcanzar-un-objetivo-de-renovables-del-27-en-2030
186. Roca J (2021) Renewable energy dominated UK power generation in 2020. Retrieved 12 Jul 2021 from https://elperiodicodelaenergia.com/las-energias-renovables-dominaron-la-generacion-de-energia-del-reino-unido-en-2020
187. Roca R (2018) France's energy strategy for 2020 focuses on nuclear plus renewables: here are the 10 key points. Retrieved 12 Jul 2021 from https://elperiodicodelaenergia.com/la-estrategia-energetica-de-francia-para-2030-apuesta-por-el-binomio-nuclearrenovables-estos-son-sus-10-puntos-clave
188. Rojas-Osorio M, Alvarez-Risco A (2019) Intention to use smartphones among Peruvian university students. Int J Interactive Mobile Technol 13(3):40–52. https://doi.org/10.3991/ijim.v13i03.9356
189. Román BR, Moscoso S, Chung SA, Terceros BL, Álvarez-Risco A, Yáñez JA (2020) Treatment of COVID-19 in peru and bolivia, and self-medication risks. Revista Cubana de Farmacia, 53(2):1–20, Article e435

190. Sakao T, Brambila-Macias, S. A. (2018). Do we share an understanding of transdisciplinarity in environmental sustainability research? *Journal of Cleaner Production, 170*, 1399–1403. https://doi.org/10.1016/j.jclepro.2017.09.226
191. Salmerón-Manzano E, Manzano-Agugliaro F (2018) The higher education sustainability through virtual laboratories: the Spanish university as case of study. Sustainability 10(11). https://doi.org/10.3390/su10114040
192. Serafeim G (2020) Public sentiment and the price of corporate sustainability. Financ Anal J 76(2):26–46. https://doi.org/10.1080/0015198X.2020.1723390
193. Shantanu P, Khoa TT (2020) Insight of entrepreneurship in Indian context. J Glob Bus Adv 13(3):321–335
194. Shuqin C, Minyan L, Hongwei T, Xiaoyu L, Jian G (2019) Assessing sustainability on Chinese university campuses: development of a campus sustainability evaluation system and its application with a case study. J Build Eng 24:100747. https://doi.org/10.1016/j.jobe.2019.100747
195. Skakova D, Livny E (2020) Tajikistan diagnostic. European Bank of reconstruction and development. Retrieved 12 Sept 2021 from https://www.ebrd.com/documents/strategy-and-policy-coordination/tajikistan-diagnostic.pdf
196. Sovacool BK, Noel L, Kester J, Zarazua de Rubens G (2018) Reviewing Nordic transport challenges and climate policy priorities: expert perceptions of decarbonisation in Denmark, Finland, Iceland, Norway, Sweden. Energy 165:532–542. https://doi.org/10.1016/j.energy.2018.09.110
197. Spittler N, Davidsdottir B, Shafiei E, Leaver J, Asgeirsson EI, Stefansson H (2020) The role of geothermal resources in sustainable power system planning in Iceland. Renewable Energy 153:1081–1090. https://doi.org/10.1016/j.renene.2020.02.046
198. Sroufe R, Gopalakrishna-Remani V (2018) Management, social sustainability, reputation, and financial performance relationships: an empirical examination of U.S. firms. Organ Environ 32(3):331–362. https://doi.org/10.1177/1086026618756611
199. Stokes LC (2013) The benefits and challenges of using feed-in tariff policies to encourage renewable energy. Energy Policy 56:490–500
200. Sung E, Kim H, Lee D (2018) Why do people consume and provide sharing economy accommodation?—a sustainability perspective. Sustainability 10(6):2072. https://doi.org/10.3390/su10062072
201. Swedish Energy Agency (2021) The electricity certificate system. Retrieved 12 Aug 2021 from http://www.energimyndigheten.se/en/sustainability/the-electricity-certificate-system
202. Trosper RL (2002) Northwest coast indigenous institutions that supported resilience and sustainability. Ecol Econ 41(2):329–344. https://doi.org/10.1016/S0921-8009(02)00041-1
203. United Nations (2021) Sustainable development goal 7. Retrieved 12 Aug 2021 from https://www.un.org/sustainabledevelopment/energy
204. United Nations Environment Programme (2012) GEO-5 global environmental outlook. Environment for the future we want [GEO-5 perspectivas del medio ambiente mundial. Medio ambiente para el futuro que queremos]. UNEP. Retrieved 12 Aug 2021 from https://wedocs.unep.org/rest/bitstreams/12914/retrieve
205. United Nations Environment Programme (2013) GEO-5 for business. Impacts of a changing environment on the corporate sector. Nairobi: UNEP. Retrieved 12 Aug 2021 from https://www.unep.org/resources/report/geo-5-business-impacts-changing-environment-corporate-sector
206. Uruguay XXI (2020) Investment opportunities. Renewable energies. Retrieved 12 Aug 2021 from https://www.uruguayxxi.gub.uy/uploads/informacion/29a569e6547662d108368a8ec0e57eae2dd1bbb1.pdf
207. Vizcardo D, Salvador LF, Nole-Vara A, Dávila KP, Alvarez-Risco A, Yáñez JA, Mejia CR (2022) Sociodemographic predictors associated with the willingness to get vaccinated against COVID-19 in Peru: a cross-sectional survey. Vaccines, 10(1), Article 48. https://doi.org/10.3390/vaccines10010048

208. Wagmann A, Widsen E, Barner R, Westling V (2020) Solar energy for the African future. Retrieved 12 Aug 2021 from https://aim2flourish.com/innovations/solar-energy-for-the-african-future-1
209. Wang WH, Moreno-Casas V, Huerta de Soto J (2021) A free-market environmentalist transition toward renewable energy: the cases of Germany, Denmark, and the United Kingdom. Energies, 14(15). https://doi.org/10.3390/en14154659
210. Wang Y (2006) Renewable electricity in Sweden: an analysis of policy and regulations. Energy Policy 34(10):1209–1220. https://doi.org/10.1016/j.enpol.2004.10.018
211. White W, Lunnan A, Nybakk E, Kulisic B (2013) The role of governments in renewable energy: the importance of policy consistency. Biomass Bioenerg 57:97–105. https://doi.org/10.1016/j.biombioe.2012.12.035
212. World Bank (2010) Design and performance of policy instruments to promote the development of renewable energy: emerging experience in selected developing countries. Final report, Energy Anchor Unit, sustainable energy department, World Bank, Washington, DC. Retrieved 12 Sept 2021 from https://openknowledge.worldbank.org/handle/10986/9379
213. World Bank (2018) Renewable energy consumption (% of total final energy consumption). Retrieved 12 Aug 2021 from https://data.worldbank.org/indicator/EG.FEC.RNEW.ZS?view=map
214. Xinhuanet Español (2021) China intensifies renewable energy development and utilization. http://spanish.xinhuanet.com/2021-03/30/c_139847197.htm
215. Xu X, Wei Z, Ji Q, Wang C, Gao G (2019) Global renewable energy development: influencing factors, trend predictions and countermeasures. Resour Policy 63:101470. https://doi.org/10.1016/j.resourpol.2019.101470
216. Yan J, Kim S, Zhang SX, Foo MD, Alvarez-Risco A, Del-Aguila-Arcentales S, Yáñez JA (2021) Hospitality workers' COVID-19 risk perception and depression: a contingent model based on transactional theory of stress model. Int J Hospital Manage 95, Article 102935. https://doi.org/10.1016/j.ijhm.2021.102935
217. Yáñez JA, Alvarez-Risco A, Delgado-Zegarra J (2020) Covid-19 in Peru: from supervised walks for children to the first case of Kawasaki-like syndrome. BMJ 369, Article m2418. https://doi.org/10.1136/bmj.m2418
218. Yáñez JA, Jahanshahi AA, Alvarez-Risco A, Li J, Zhang SX (2020) Anxiety, distress, and turnover intention of healthcare workers in Peru by their distance to the epicenter during the COVID-19 crisis. Am J Trop Med Hyg 103(4):1614–1620. https://doi.org/10.4269/ajtmh.20-0800
219. Yáñez S, Uruburu Á, Moreno A, Lumbreras J (2019) The sustainability report as an essential tool for the holistic and strategic vision of higher education institutions. J Clean Prod 207:57–66. https://doi.org/10.1016/j.jclepro.2018.09.171
220. Yarime M, Tanaka Y (2012) The issues and methodologies in sustainability assessment tools for higher education institutions: a review of recent trends and future challenges. J Educ Sustain Dev 6(1):63–77. https://doi.org/10.1177/097340821100600113
221. Zabaloy MF, Guzowski C (2018) Energy transition policy from fossil fuels to renewable energies: the case of Argentina, Brazil and Uruguay in the period 1970–2016 [La política de transición energética de combustibles fósiles a energías renovables: el caso de argentina, Brasil y Uruguay en el periodo 1970–2016]. Economía Coyuntural 3(3):02–34
222. Zamora-Polo F, Sánchez-Martín J (2019) Teaching for a better World. Sustainability and sustainable development goals in the construction of a change-maker university. Sustainability, 11(15). https://doi.org/10.3390/su11154224
223. Zeng M, Li C, Zhou L (2013) Progress and prospective on the police system of renewable energy in China. Renew Sustain Energy Rev 20:36–44. https://doi.org/10.1016/j.rser.2012.11.048
224. Zhang D, Zhang X, He J, Chai Q (2011) Offshore wind energy development in China: current status and future perspective. Renew Sustain Energy Rev 15(9):4673–4684. https://doi.org/10.1016/j.rser.2011.07.084

225. Zhang SX, Chen J, Afshar Jahanshahi A, Alvarez-Risco A, Dai H, Li J, Patty-Tito RM (2021) Succumbing to the COVID-19 pandemic—healthcare workers not satisfied and intend to leave their jobs. Int J Mental Health Addict. https://doi.org/10.1007/s11469-020-00418-6
226. Zhang SX, Chen J, Jahanshahi AA, Alvarez-Risco A, Dai H, Li J, Patty-Tito RM (2021) Correction to: succumbing to the COVID-19 Pandemic—healthcare workers not satisfied and intend to leave their jobs (International Journal of Mental Health and Addiction, 2021). Int J Mental Health Addict. https://doi.org/10.1007/s11469-021-00502-5
227. Zhang SX, Sun S, Jahanshahi AA, Alvarez-Risco A, Ibarra VG, Li J, Patty-Tito RM (2020) Developing and testing a measure of COVID-19 organizational support of healthcare workers—results from Peru, Ecuador, and Bolivia. Psychiatry Res 291. https://doi.org/10.1016/j.psychres.2020.113174

Theory of Sustainable Paths for Entrepreneurship Associated with Fashion and Practical Examples

Marián Arias-Meza, Aldo Alvarez-Risco, Berdy Briggitte Cuya-Velásquez, Romina Gómez-Prado, María de las Mercedes Anderson-Seminario, and Shyla Del-Aguila-Arcentales

Abstract The fashion industry is one sector that generates the most pollution globally due to the overproduction generated by the high demand for products. From its production to the end of its life cycle, it is negatively impacted by chemicals, water pollution, and solid waste; often, workers are exploited, and the likelihood that their health is affected in the long term. New entrepreneurs have emerged seeking to reduce the environmental impact, provide decent work, and offer sustainable alternatives for fashionable market niches, which is the reason why this paper analyzes the theory of sustainable paths for entrepreneurship associated with fashion, and in addition to that, brings real cases that aim to reduce the impact of the fashion industry by providing sustainable fashion products.

Keywords Entrepreneurship · Sustainability · Sustainable fashion · Theory · Effect · Circularity · Circular · Entrepreneur · Sustainable development goals

1 Introduction

The global textile production industry costs $ 3 trillion, which is 2% of the entire world's Gross Domestic Product (GDP). In this sector, several 33.0 million people are hired, which was reflected in the increase of 19.7 million in the last two decades [243]. According to Chen and Burns [98], the textile sector consumes more than 30 million tons per year worldwide, causing harmful effects on the environment and society derived from the supply chain [227], which causes concern about the small use of production and consumption systems that are sustainable [1].

Before delving into the subject of sustainability, it is essential to understand well the meaning of sustainable development, which is determined as a process that is

M. Arias-Meza · A. Alvarez-Risco · B. B. Cuya-Velásquez · R. Gómez-Prado ·
M. de las Mercedes Anderson-Seminario
Universidad de Lima, Lima, Peru

S. Del-Aguila-Arcentales (✉)
Escuela de Posgrado, Universidad San Ignacio de Loyola, Lima, Peru
e-mail: sdelaguila@usil.edu.pe

A. Alvarez-Risco et al. (eds.), *Footprint and Entrepreneurship*, Environmental Footprints and Eco-design of Products and Processes,
https://doi.org/10.1007/978-981-19-8895-0_4

growing and at the same time satisfies the current needs of the consumer without risking the possibility that future generations may satisfy your own needs in the Brundtland report [157]. Therefore, taking care of resources and keeping the environment healthy and clean is vital to ensure the long life of individuals. In this context, sustainability can be applied in the textile sector with new biological systems that allow companies to be competitive and original while maintaining a good position in the market. All dimensions must be considered for this: economic, environmental, and social [97, 144, 233].

If we talk about the textile sector, a fundamental concept in international business is likely related, such as fashion. The drawbacks of sustainability are of particular importance in this industry due to their effect on the environmental, social, and economic aspects [77]. Fashion has been linked to sustainability through ethical, green, ecological, and sustainable fashion [185]. Sustainable fashion was born in the 60s for the first time when consumers noticed a profound effect produced by the production of clothes on the environment, for which a change towards the use of sustainable practices was demanded [155]. Anti-fur campaigns characterized ecological fashion during the 1980s and 1990s, coupled with ethical fashion. The latter groups together fair working conditions, respectful and appropriate use of organic inputs, certifications, and traceability [145, 151]. It is necessary to consider the damage generated in the environment through emissions related to the garment factory and the product distribution process, which is even more relevant due to the great demand in the market. In this way, it is shown that the fashion industry has faced difficult situations and, at the same time, opportunities in terms of reducing environmental consequences throughout the world [192]. Likewise, fashion produces a negative social effects such as the use of labor, the excessive use of resources, and the creation of waste in large quantities [130]. In the economic aspect, investment in machinery and more factories are promoted, which consequently causes environmental pollution.

Consequently, a "megatrend" is born connected to sustainability within the fashion industry that shows signs of gaining more and more strength and importance in the consumer's mind. Similarly, as was cited by Shen [227] and de Brito et al. [111], stated that fashion customers develop a greater social awareness and care for the environment and as a result of this, it affects their purchase decision. For this reason, an increasing number of companies notice the value and opportunity that is generated in trade and integrate greener processes into their supply chain, they understand that, if they use more natural resources, the emission of CO_2 is reduced; also include new marketing strategies related to this trend.

Additionally, there are past studies that reveal the interest and search of fashion customers for sustainable fashion products; for this reason, they remain able to pay a higher cost as long as it meets their needs and presents good quality [227]. This chapter analyzes the literature and theories related to sustainable fashion and the practices that have been implemented in organizations.

2 Fast Fashion and Sustainable Fashion

As mentioned in previous lines, the fashion supply chain is prone to sustainability-related drawbacks due to intense competition. In the economic aspect, it invests in more production plants and different geographical areas, while, in the environmental aspect, a high amount of chemical products is used, for example, to transform the raw material into textiles, a minimum of 8000 is required. Finally, about the social dimension, unfair treatment is applied to employees on some occasions, granting them a low salary to produce as quickly as possible and with few expenses and then comply with the export times and manage to satisfy the tremendous market demand [111].

This rapid increase in the manufacture of clothing and footwear, fostered by the increase in consumption in many countries, gave rise to the term "fast fashion" [221]. According to Kim and Oh [161], several reports pointed out the negative impact on a social, environmental, and economic level. They experience continuous growth thanks to instant fashion or fast fashion, in other words, environmental pollution increases, labor exploitation, and promotes excessive consumption. The clothes that are not used much are discarded or stored in the closet, and new clothes continue to be bought that go with the new trends [221]. Consequently, the useful life of these garments becomes relatively short and below regular [67].

"Fast fashion" is considered a model of a garment supply chain that focuses on being up-to-date with the latest fashion trends, and designs are usually updated so that they are accessible in the market [90]. Ghemawat et al. [138] define it as an industrial practice implemented in fashion retail businesses, which supports obtaining the design most loved by consumers. On the other hand, Brooks [87] points to "fast fashion" as a term established by retailers in the fashion industry to explain that trends are constantly and rapidly moving from catwalks to stores. Barnes and Lea-Greenwood [72] relate "fast fashion" with mass production of garments at a low or fair cost and at the same time offer products that are haute couture and thus position themselves in the market while being competitive. Technological advances are taken advantage of from advances in production to accelerate further processes, such as 3D printing [84]. However, consumers are demonstrating to take a stance in favor of sustainable practices and consumption, which is why the instant fashion industry is being seen with different eyes due to bad practices that are irresponsible with the environment, exploitation labor, and the low quality of fashion items [190, 243].

Therefore, companies that abuse the "take-produce-dispose" model should consider taking a different path and implementing sustainable changes in their manufacturing processes [72, 222]. A proof of this revolution in consumer thinking is demonstrated by the Forbes ad on "responsible consumerism", which was exposed as one of the 6 global consumer trends in 2019 [108].

3 Slow Fashion and Sustainable Entrepreneurship

According to Diddi et al. [122], consumers purchased 60% more garments in 2014 but quickly stopped using them compared to 15 years ago. Broega et al. [86] consider that garments have a long useful life, but fast fashion causes them to dispose promptly. The useful life is measured by time, the number of times used, and the number of clients used [167]. Because of the impact experienced by the fashion sector and the repercussions that fast fashion has on the planet, its inhabitants, and the economy [195, 254]. Different authors have criticized current consumption and proposed a new way for entrepreneurs in the fashion sector to be sustainable [225]. Gurova and Morozova [141] mention that slow fashion originated because of the repercussions of fast fashion, which contains an agenda of slow production, equitable salary, and a longer duration of the useful life of the garments [240]. Likewise, it is related to quality products in labeling and durability [130], where the production process affects the environment and the consumer in lesser proportions and can be evaluated through traceability [143, 153].

Gardetti and Muthu [135] mention that sustainable fashion symbolizes that no harm was done to people or the planet during the development and use. The different enterprises of the fashion industry are leaning toward sustainable fashion since sustainability is an investment to stay in the market in the future [104]. So, sustainable ventures cover the environmental, social, and economic aspects [166]. The security in customer purchases and sustainable consumption [203]. In the same sense, Henninger et al. [146] point out that consumers have improved their notion about environmental and social problems for fashion, so collaboratively having fashion is a great opportunity. Consumers tend to buy second-hand clothing to avoid wasting it and have a longer useful life [62, 78, 167]. The exchange of garments allows them not to go to the landfill [168]. Therefore, consumers have a positive attitude toward sustainable fashion [150].

4 Theory of Sustainability in Business

Medcalfe and Miralles Miro [182] point out that sustainable practices positively affect the financial performance of the fashion industry. Wong and Ngai [245] improve the performance of startups internationally. However, new approaches are needed that can support the development of future research on sustainable entrepreneurship. Various theories related to sustainable companies are presented below [179].

4.1 Triple Income Statement Model

The model explains the performance of sustainable development in organizations from an economic, social, and environmental perspective [211]. There are several ways for entrepreneurs to interpret the theory, but the main one is to identify the problem in society or the economy [79, 189]. However, Friedrich [133] considers that the theory establishes that enterprises also have to consider costs about the pollution they generate to the environment. It should be noted that Freer Spreckley developed this model in 1981 [211]. The model helps identify organizations' actions to be responsible and sustainable. For Shen et al. [228], this theory is the basis for executing a Sustainable Business Model in all companies, particularly in the fashion industry. Yang et al. [247] require changes in translating business models involving customers, employees, and the planet, which is because, in different fashion ventures, they exploit employees or use polluting elements [56, 158]. However, fashion ventures seek to increase success with sustainability [238]. Companies are becoming more sustainable, manufacturers and distributors are beginning to spread the additional costs they incur through the supply process, but it depends on the partners' willingness [132, 149]. Large fashion companies are embracing sustainability due to customer demand, global rivals, and new policies [193]. However, they present different barriers [83].

4.2 Business Model Triangle

Business models have evolved over the years for companies to be competitive [237]. Each venture must have a business model to address the creation and delivery of value for customers (Evans et al. 2017; Jin et al. 2021). The models help to incorporate questions to identify the target audience and their present needs (Böhm et al. 2017). Gassmann et al. [136] developed a triangular model. The objective was to clarify its customer segments, value proposition, value chain, and benefits mechanism and make its business model understandable [237]. Companies must capture the forms of value [81]. The authors designed four questions related to value creation, as shown in Fig. 1.

4.3 The Theory of Social Practices (TPS)

Sustainable fashion needs significant changes regarding the consumer practices of the parties [60]. Therefore, the relevance of the Theory of Social Practices (TPS) and its directionality to the path of sustainability must be borne in mind [119]. It should be noted that there is an increase in purchases of sustainable materials [200]. Skjerven and Reitan [230] mentioned that garment design could become an essential

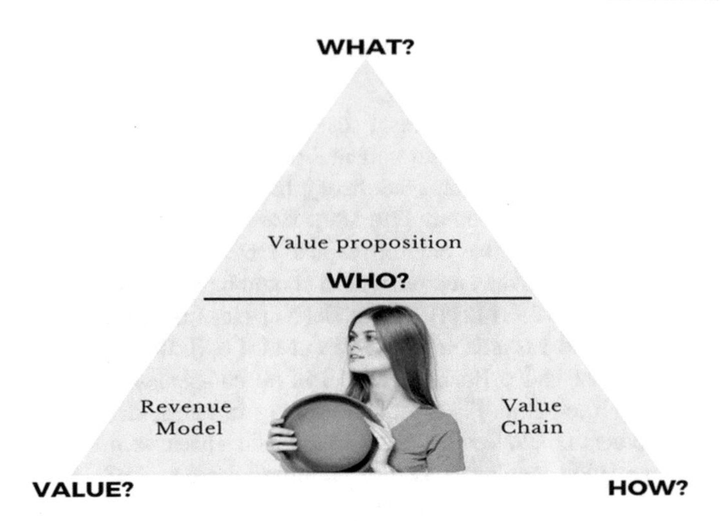

Fig. 1 Business model triangle. *Source* Adapted from Thorisdottir and Johannsdottir [237]

tool to develop a sustainable social and cultural environment. From the consumer perspective, Klepp and Bjerck [162] found that wearing and discarding garments has repercussions on the planet. Likewise, they communicate the identity and social environment to which consumers belong [61, 68, 130].

Customers are highly loyal brands that carry out sustainable practices, influenced by intrinsic and extrinsic singularities [8]. Such as culture and entrepreneurial orientation toward a more sustainable business [121, 236]. Through theory, enterprises can conduct rigorous manufacturing and consumer practices transformations to build a sustainable industry [23]. These must be related to the social aspect and can be included in marketing strategies to positively impact society [65, 101, 110].

5 Cases

The increase in population has caused the consumption and production of fashion products to increase in the same way [69], which has caused the emergence of a boom in creative entrepreneurs in the global clothing industry who are looking for ethics and sustainability to be applied in the sector. Therefore, the following are projects focused on providing alternatives responsible for the environment and their workers.

5.1 Pigments that Are Produced by Microorganisms

In the fashion industry, textile dyes are often used due to their low price; however, their effects on health and the environment are higher, as Lellis et al. [172] mentioned

Table 1 Microorganism used to dye fabrics

Microorganism	Color	From	References
Chromochloris zofingiensis	Pink–red	Microalgae	Chen et al. [99]
Pseudomonas aeruginosa	Blue–green	Soil	Alzahrani and Alqahtani [53]
Chryseobacterium rhizoplanae	Yellow	Soil and marine environments	Aruldass et al. [63]
Chromobacterium violaceum	Purple		Kanelli et al. [156]
Streptomyces glaucescens	Black		El-Naggar and El-Ewasy [123]

Source Mazotto et al. [181]

because they act as a toxic substance in human body and many living organisms. For that reason, entrepreneurs in the field can aim to offer green alternatives to mitigate this by using microorganisms as pigments. In dying fabrics, 15% and 50% of the textile dyes don't bind to the fabric on the first try, which would mean a higher number of wastewater which at the end, in developing countries, is used for agriculture according to Rehman et al. [218]. Consequently, the growing of the plants would be affected, the germination and the final product can cause many pathologies once the human consumes agricultural products because of all the toxins the plants were watered. Most of the time, this problem is considering how human health can be affected in the long term for workers because of cheap labor. Also, the toxicity of textile dyes can be transmitted to the workers who handled the reactive dyes by inhalation producing some allergic reactions such as conjunctivitis, asthma, dermatitis, and others [172]. Sustainable solution microorganisms can be used as an alternative for the dyes to reduce the worst effects. Some of them are shown in Table 1 and can be used to dye the fabrics in many colors. Entrepreneurs can provide a differentiated textile product using these innovative, sustainable alternatives in the process.

Other microorganisms can also be implemented in different parts of the textile life process, according to Mazotto et al. [181]. For example, in raw material extraction, industry entrepreneurs can focus there and implement sustainable textile practices. In Fig. 2, those alternatives for the process are shown.

5.2 Fish Leather

In the fashion industry, leather manufacturing generates a large carbon footprint due to the solid waste and water waste that it produces, the damage of fauna and flora, the number of toxic chemicals used like chromium, and that in the long-term the health

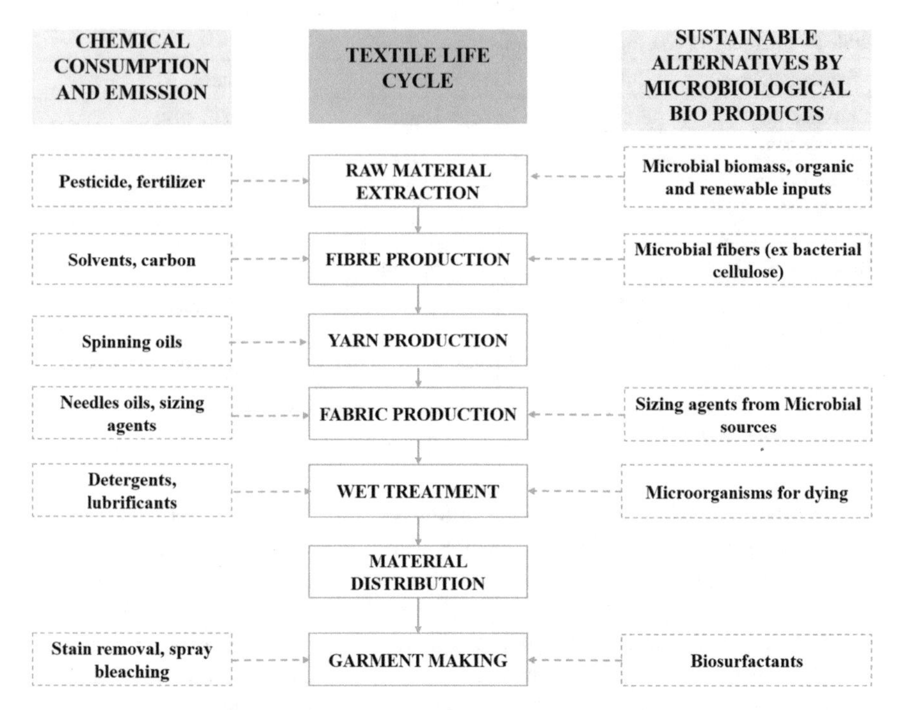

Fig. 2 Textile life cycle, traditional chemical consumption, and new alternatives offered by microbiological bioproducts. *Source* Mazotto et al. [181]

of workers who work directly in tanning can be damaged with an illness like cancer [24].

Over the years, this effect is becoming more and more noticeable; consumers are finding out where the cloth comes from, according to Palacios-Chavarro et al. [201]. This niche market is looking for sustainable alternatives to avoid excessive pollution and decent jobs for workers, as de Klerk et al. [112] mentioned. Entrepreneurs can provide them with the solution with the fish leather. An alternative to reduce environmental pollution generated by the fishing sector is to value fish skins. The skins must follow a process to be transformed into fish leather by using vegetable tanning to this reduction, according to Palomino [202]. In this way, the fish waste that could have ended up as waste is used to manufacture garments, handbags, wallets, shoes, and more.

To make this innovative and sustainable product, as shown in Fig. 3, some of the effects come with the use of this. For example, ocean acidification is reduced because no pesticides, no fertilizers, more water, or more soil are needed. Manufacturing this avoids the tons of fish skin waste that would end up in the ocean, thus negatively impacting the environment.

Also, entrepreneurs to decrease undocumented fisheries, entrepreneurs should have as suppliers fishermen registered to regulate harvesting to keep the business as sustainable as possible. Then there are the economic benefits, an additional income

Fig. 3 Positives activities that generate the use of fish skin for the environment, artisanal fisheries, and others. *Source* Palomino [202]

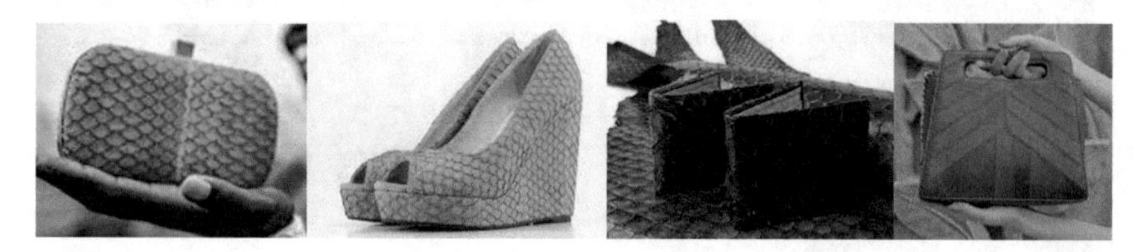

Fig. 4 Finals products with fish leather. *Source* Álvarez Acosta et al. [52]

for fishers, fish sellers, and many direct and indirect jobs that come with it. In addition, this field is further studied, and scientific knowledge allows the creation of new green technologies to produce fish leather [112]. Many entrepreneurs around the world in the fashion industry have seen fish skin as a sustainable alternative; countries located in Europe such as Finland, Iceland, Denmark, and Swede are developing these products based on fish skin to offer sustainable products to niche markets [202]. Also, entrepreneurs from South America in countries such as Perú and Ecuador, according to Álvarez Acosta et al. [52], are producing and selling those. Some of the products they sell are shown in Fig. 4.

6 Pineapple Leaf Fibers

Instead of waste being generated by improper treatment of pineapple leaf remains, Leão et al. [170] suggest using and processing them to manufacture textile products, which would be an alternative for entrepreneurs to venture into and manufacture sustainable products based on pineapple leaves.

For manual removal, first, scrapping the upper layer; second, unscrewing the clamp; third, inserting the leaves with the bottom laver facing up; fourth, screw the clamp; fifth, scrapping the bottom layer; sixth, pulling out the fibers; and finally, seventh get the scrapped pineapple leaf fiber. According to Yusof et al. [252], the process is shown in Fig. 5.

In addition, technology is a tool that these days can make it easier for the entrepreneur to manufacture pineapple fibers. For example, in the Asian continent,

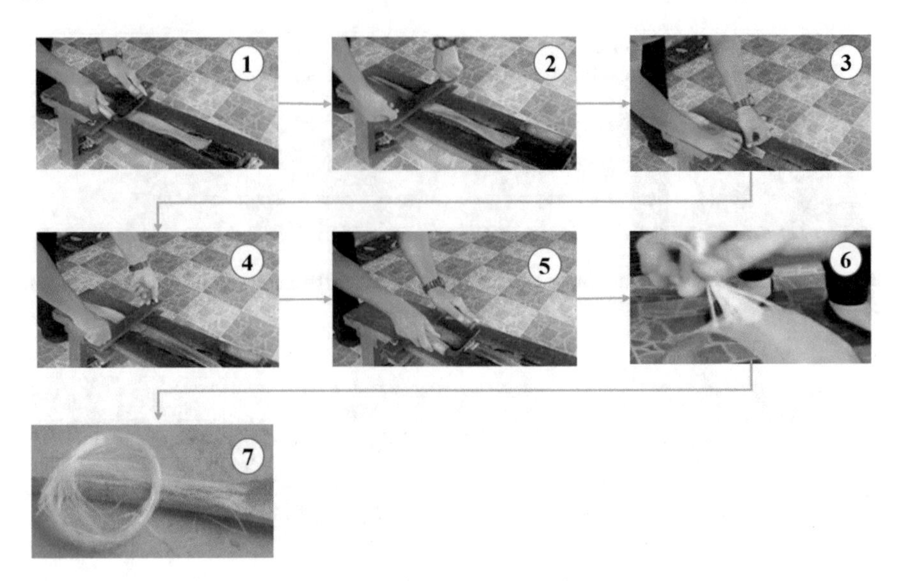

Fig. 5 Production of the pineapple leaf fiber by hand scrapping. *Source* Yusof et al. [252]

Yusof et al. [252], entrepreneurs use Pineapple Leaf Fiber Machine 1, a decortication machine to facilitate the work to produce the fibers.

Figure 6 shows how the machine works according to Yusof et al. [252]. First, insert the leaf into the feeder, start the extraction process, put out the leaf, and finally, get the extracted fiber. This machine allows a better amount of manufacturing in a shorter time. Figure 7 shows the final products with Pineapple Leaf Fibers Productions. A case that arose using this fiber is Piñatex, entrepreneurship based on fair economic principles, reducing waste, and saving raw materials and energy, according to Kowszyk and Maher [164]. These provide different brands, which generate fashion products with added and sustainable value, such as wallets, handbags, coats, shoes, and sandals.

7 Recycled Fibers

The amount of clothing that is worn and thrown away, according to Klepp and Bjerck [162], is high, and year by year, it is increasing. Because of that, an opportunity for businesses arrives to give garments a second chance by recycling so that they do not generate more waste. Currently, there is the recycling of fabrics such as cotton and polyester to make threads used later to produce polo shirts, pants, jackets, and sportswear, as Leonas [173] mentioned. For entrepreneurs to obtain textile waste, Payne [204] suggests forming alliances with regional or local governments, clothing manufacturers, among others. Instead of garments ending up in the trash and where

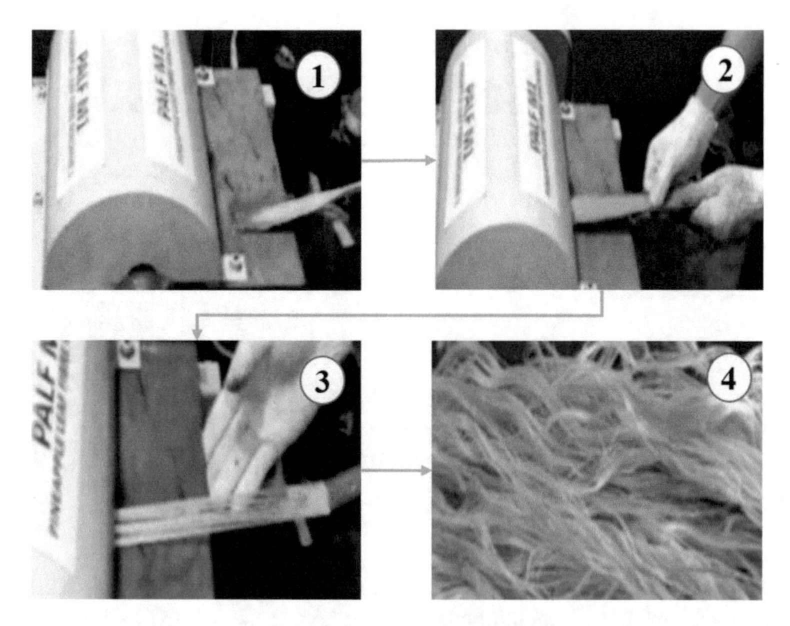

Fig. 6 Production of the pineapple leaf fiber by technology. *Source* Yusof et al. [252]

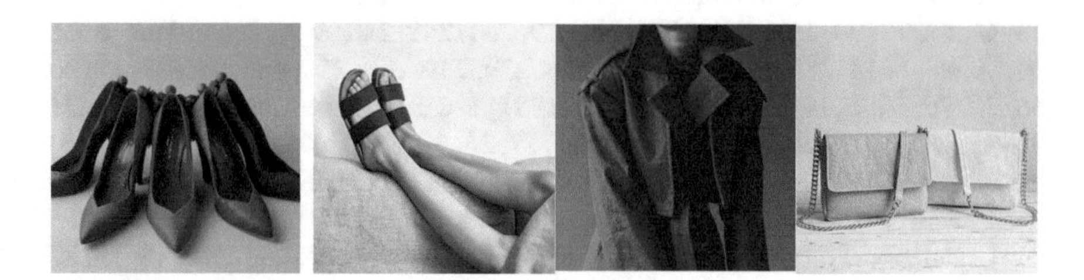

Fig. 7 Finals products with pineapple leaf fibers productions. *Source* Piñatex [210]

the garment's life cycle ends, as seen in Fig. 8, Payne [204] proposes "Cradle-to-cradle closed-loop recycling" which works as the garment's life cycle by using recycled fibers. Based on this, entrepreneurs can venture into the field.

8 Co-creation

For centuries, humans have needed garments to wear or accessories to take objects to other places, so they have had to develop techniques to make products that meet their needs because there was no textile industry and the facility that now exists to buy clothes instantly [131]. Whether of garments or accessories, these techniques for elaboration have been inherited from generation to generation [208]. As there were

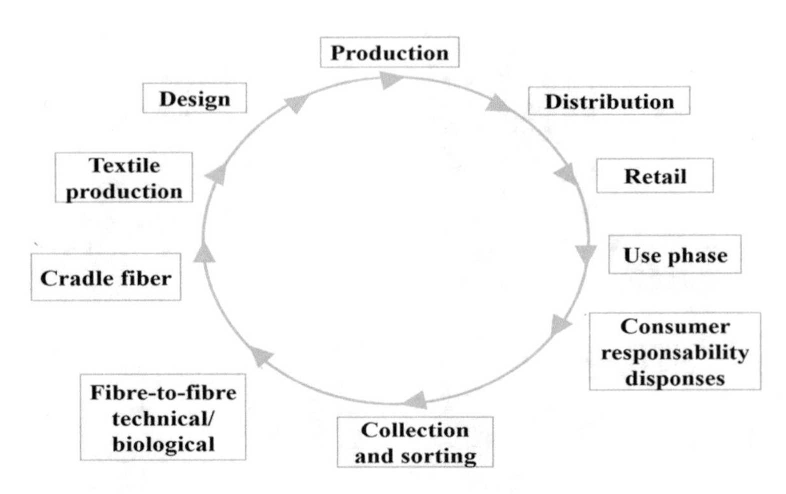

Fig. 8 Cradle-to-cradle closed-loop recycling. *Source* Payne [204]

few chemicals or synthetic textiles, the settlers carried out this work in an artisanal and sustainable way [117] mentioned that nowadays, these artisan products represent the culture and identity of the population and are well valued in niche markets.

The sustainable efforts include different actors as the citizens [13, 15, 17, 25, 37, 46–48, 58, 73, 96, 118, 134, 159, 178, 219, 234, 242, 248, 253], institutions [6, 26, 30, 42, 80, 82, 85, 88, 169, 175, 207, 224, 229, 239, 250, 251] and firms [7, 16, 20, 36, 70, 76, 105, 114, 177, 194, 209, 214, 231] and activities as tourism [11, 14, 21, 91–93, 129, 191, 206, 216], education [5, 17, 26, 35, 39, 43, 45, 95, 147, 171, 215], circular economy approach [3, 12, 28, 44, 75, 160, 174, 176, 241], prices [55, 57, 64, 66, 140, 142, 165, 171, 186, 226], hospitality [4, 22, 127, 137, 152, 180, 199, 235, 246], intellectual property [9, 29, 34, 54, 71, 74, 126, 184], health [2, 10, 18, 19, 27, 31–33, 38, 40, 49–51, 89, 100, 106, 116, 124, 183, 188, 213, 220, 249, 255–257], and research [102, 163, 196–198, 205, 212, 217, 223].

Entrepreneurs in the fashion industry can choose to co-create with communities that maintain these practices as a sustainable alternative for the environment, according to Diaz [120], which would mean working with the artisan to design fashion garments and accessories while maintaining inherited techniques [125]. These practices are developed more by women; in this sense, Strydom and Kempen [232] point out that working together, entrepreneurs working with artisans seek to improve their community well-being as their personal, since it seeks to generate income in the population, reduce poverty, fair wages, empowerment, and gender equality.

Working together benefits both parties; first, the entrepreneur is in charge of the administrative, logistics, sales, design staff, according to Williams [244], acts as a facilitator. Second, artisans obtain fair and dignified work since they provide their labor to manufacture products, transmitting their inherited knowledge through them, which are well valued in market niches [117], generating a sustainable end product for the fashion industry.

Recent suggested lectures about green approach

- Waste reduction and carbon footprint [139]
- Material selection for circularity and footprints [187]
- 3D Print, circularity, and footprints [109]
- Virtual tourism [109]
- Leadership for sustainability in crisis time [41]
- Virtual education and circularity [103]
- Circular Economy for packaging [94]
- Students oriented to circular learning [113]
- Food and circular economy [107]
- Water footprint and food supply chain management [154]
- Waste footprint [59]
- Measuring circular economy [115]
- Carbon footprint [128].

Closing Remarks

Prominent fashion brands are undergoing various transformations, such as using cheap materials from unknown sources and unleashing insecurity about customers' products [148]. Fashion factories produce large quantities of garments due to fast fashion, causing short shelf life and excess garments discarded. Sometimes, unfair treatment of employees occurs. To this is added that different authors design new perspectives on the practices that lead to new discoveries in relation to sustainable fashion [1]. New business ideas are emerging as sustainable alternatives that seek to generate a lower environmental impact than fast fashion. For this reason, sustainable fashion allows businesses to provide a quality product based on sustainable materials, fair payments to employees, preservation of the environment, and prolongation of the useful life of garments. In addition, it seeks to positively impact the lives of workers since they generate income through decent and responsible work.

Entrepreneurs have chosen to reject products that harm the environment and opt for sustainable alternatives. Pigmentation by microorganisms is one of them. In this, it seeks to avoid the use of chemical substances that harm the environment, human health, and workers. Using this alternative, the startups that emerge differ from the rest due to their added value. Another of the new options on the market for entrepreneurs is to enter the fish leather business, where the skin of this does not end up being a waste, but it is given a second life for manufacturing products such as wallets, shoes, handbags, and among others. Similarly, another of the alternatives that emerged was based on pineapple leaf fiber, which can be produced manually or through technology, such as the machine called PALF 1.

So that the garment does not end up as waste, it is proposed that it is a circular life cycle, where the entrepreneur collects these garments, it is suggested to talk to textile factories, governments, among others, to have the primary input. In addition to these, it is essential to point out the importance of co-creation between entrepreneurs and artisans to sustainably manufacture products for the fashion industry. Both are benefited, the craftsman offers his inherited knowledge over the years, and the entrepreneur acts as a facilitator.

References

1. Aakko M, Koskennurmi-Sivonen R (2013) Designing sustainable fashion: possibilities and challenges. Res J Text Appar 17(1):13–22. https://doi.org/10.1108/RJTA-17-01-2013-B002
2. Aaron L (2020) The oft forgotten part of nutrition: lessons from an integrated approach to understand consumer food safety behaviors. Curr Dev Nutr 4(Supplement_2):1322–1322. http://doi.org/10.1093/cdn/nzaa059_039
3. Abad-Segura E, Batlles-delaFuente A, González-Zamar M-D, Belmonte-Ureña LJ (2021) Implications for sustainability of the joint application of bioeconomy and circular economy: a worldwide trend study. Sustainability 13(13):7182. https://doi.org/10.3390/su13137182
4. Abad-Segura E, Cortés-García FJ, Belmonte-Ureña LJ (2019) The sustainable approach to corporate social responsibility: a global analysis and future trends. Sustainability 11(19):5382. https://doi.org/10.3390/su11195382
5. Abad-Segura E, González-Zamar M-D (2021) Sustainable economic development in higher education institutions: a global analysis within the SDGs framework. J Clean Prod 294:126133. http://doi.org/10.1016/j.jclepro.2021.126133
6. Abad-Segura E, González-Zamar M-D, Infante-Moro JC, Ruipérez García G (2020) Sustainable management of digital transformation in higher education: global research trends. Sustainability 12(5):2107. https://doi.org/10.3390/su12052107
7. Abad-Segura E, Morales ME, Cortés-García FJ, Belmonte-Ureña LJ (2020) Industrial processes management for a sustainable society: global research analysis. Processes 8(5). http://doi.org/10.3390/pr8050631
8. Abbes I, Hallem Y, Taga N (2020) Second-hand shopping and brand loyalty: the role of online collaborative redistribution platforms. J Retail Consum Serv 52:101885. https://doi.org/10.1016/j.jretconser.2019.101885
9. Abdaljaleel M, Singer EJ, Yong WH (2019) Sustainability in biobanking. In: Yong WH (ed) Biobanking: methods and protocols. Springer, New York, pp 1–6. http://doi.org/10.1007/978-1-4939-8935-5_1
10. Aberilla JM, Gallego-Schmid A, Stamford L, Azapagic A (2020) Environmental sustainability of cooking fuels in remote communities: life cycle and local impacts. Sci Total Environ 713:136445. https://doi.org/10.1016/j.scitotenv.2019.136445
11. Acevedo-Duque Á, Vega-Muñoz A, Salazar-Sepúlveda G (2020) Analysis of hospitality, leisure, and tourism studies in Chile. Sustainability 12(18):7238. https://doi.org/10.3390/su12187238
12. Ada E, Sagnak M, Mangla SK, Kazancoglu Y (2021) A circular business cluster model for sustainable operations management. Int J Logistics Res Appl 1–19. https://doi.org/10.1080/13675567.2021.2008335
13. Adam A (2015) Challenges of public finance sustainability in the European Union. Procedia Econ Financ 23:298–302. https://doi.org/10.1016/S2212-5671(15)00507-9
14. Adam I, Agyeiwaah E, Dayour F (2021) Understanding the social identity, motivations, and sustainable behaviour among backpackers: a clustering approach. J Travel Tour Mark 38(2):139–154. https://doi.org/10.1080/10548408.2021.1887053
15. Adam JN, Adams T, Gerber J-D, Haller T (2021) Decentralization for increased sustainability in natural resource management? Two cautionary cases from Ghana. Sustainability 13(12). http://doi.org/10.3390/su13126885
16. Adam M (2018) The role of human resource management (HRM) for the implementation of sustainable product-service systems (PSS)—an analysis of fashion retailers. Sustainability 10(7). http://doi.org/10.3390/su10072518

17. Adams R, Martin S, Boom K (2018) University culture and sustainability: designing and implementing an enabling framework. J Clean Prod 171:434–445. https://doi.org/10.1016/j.jclepro.2017.10.032
18. Adesogan AT, Havelaar AH, McKune SL, Eilittä M, Dahl GE (2020) Animal source foods: sustainability problem or malnutrition and sustainability solution? Perspective matters. Glob Food Secur 25:100325. https://doi.org/10.1016/j.gfs.2019.100325
19. Adomah-Afari A, Chandler JA (2018) The role of government and community in the scaling up and sustainability of mutual health organisations: an exploratory study in Ghana. Soc Sci Med 207:25–37. https://doi.org/10.1016/j.socscimed.2018.04.044
20. Adomako S, Amankwah-Amoah J, Danso A, Konadu R, Owusu-Agyei S (2019) Environmental sustainability orientation and performance of family and nonfamily firms. Bus Strateg Environ 28(6):1250–1259. https://doi.org/10.1002/bse.2314
21. Adongo CA, Taale F, Adam I (2018) Tourists' values and empathic attitude toward sustainable development in tourism. Ecol Econ 150:251–263. https://doi.org/10.1016/j.ecolecon.2018.04.013
22. Agyeiwaah E (2019) Exploring the relevance of sustainability to micro tourism and hospitality accommodation enterprises (MTHAEs): evidence from home-stay owners. J Clean Prod 226:159–171. https://doi.org/10.1016/j.jclepro.2019.04.089
23. Aitken R, Watkins L, Kemp S (2019) Envisioning a sustainable consumption future. Young Consumers 20(4):299–313. https://doi.org/10.1108/YC-12-2018-0905
24. Al-Jabari M, Sawalha H, Pugazhendhi A, Rene ER (2021) Cleaner production and resource recovery opportunities in leather tanneries: technological applications and perspectives. Bioresour Technol Rep 16:100815. https://doi.org/10.1016/j.biteb.2021.100815
25. Al-Naqbi AK, Alshannag Q (2018) The status of education for sustainable development and sustainability knowledge, attitudes, and behaviors of UAE University students. Int J Sustain High Educ 19(3):566–588. https://doi.org/10.1108/IJSHE-06-2017-0091
26. Aleixo AM, Leal S, Azeiteiro UM (2018) Conceptualization of sustainable higher education institutions, roles, barriers, and challenges for sustainability: an exploratory study in Portugal. J Clean Prod 172:1664–1673. https://doi.org/10.1016/j.jclepro.2016.11.010
27. Ali M, de Azevedo ARG, Marvila MT, Khan MI, Memon AM, Masood F et al (2021) The influence of COVID-19-induced daily activities on health parameters—a case study in Malaysia. Sustainability 13(13):7465. https://doi.org/10.3390/su13137465
28. Alkhuzaim L, Zhu Q, Sarkis J (2021) Evaluating emergy analysis at the nexus of circular economy and sustainable supply chain management. Sustain Prod Consumption 25:413–424. https://doi.org/10.1016/j.spc.2020.11.022
29. Alonso-Fradejas A (2021) The resource property question in climate stewardship and sustainability transitions. Land Use Policy 108:105529. https://doi.org/10.1016/j.landusepol.2021.105529
30. Alshuwaikhat HM, Adenle YA, Saghir B (2016) Sustainability assessment of higher education institutions in Saudi Arabia. Sustainability 8(8). http://doi.org/10.3390/su8080750
31. Álvarez-Risco A, Arellano EZ, Valerio EM, Acosta NM, Tarazona ZS (2013) Pharmaceutical care campaign as a strategy for implementation of pharmaceutical services: experience Peru. Pharm Care Esp 15(1):35–37
32. Alvarez-Risco A, Dawson J, Johnson W, Conteh-Barrat M, Aslani P, Del-Aguila-Arcentales S, Diaz-Risco S (2020) Ebola virus disease outbreak: a global approach for health systems. Rev Cubana de Farmacia 53(4):1–13, Article e491
33. Alvarez-Risco A, Del-Aguila-Arcentales S (2015) Prescription errors as a barrier to pharmaceutical care in public health facilities: experience Peru. Pharm Care Esp 17(6):725–731
34. Alvarez-Risco A, Del-Aguila-Arcentales S (2021a) A note on changing regulation in international business: the world intellectual property organization (wipo) and artificial intelligence. In: Progress in international business research, vol 15, pp 363–371. http://doi.org/10.1108/S1745-886220210000015020
35. Alvarez-Risco A, Del-Aguila-Arcentales S (2021b) Public policies and private efforts to increase women entrepreneurship based on STEM background. In: Contributions to management science, pp 75–87. http://doi.org/10.1007/978-3-030-83792-1_5

36. Alvarez-Risco A, Del-Aguila-Arcentales S (2022) Sustainable initiatives in international markets. In: Contributions to management science, pp 181–191. http://doi.org/10.1007/978-3-030-85950-3_10
37. Alvarez-Risco A, Del-Aguila-Arcentales S, Delgado-Zegarra J, Yáñez JA, Diaz-Risco S (2019) Doping in sports: findings of the analytical test and its interpretation by the public. Sport Sci Health 15(1):255–257. https://doi.org/10.1007/s11332-018-0484-8
38. Alvarez-Risco A, Del-Aguila-Arcentales S, Diaz-Risco S (2018) Pharmacovigilance as a tool for sustainable development of healthcare in Peru. Pharmacovigil Rev 10(2):4–6
39. Alvarez-Risco A, Del-Aguila-Arcentales S, Rosen MA, García-Ibarra V, Maycotte-Felkel S, Martínez-Toro GM (2021) Expectations and interests of university students in covid-19 times about sustainable development goals: evidence from Colombia, Ecuador, Mexico, and Peru. Sustainability (Switzerland) 13(6), Article 3306. http://doi.org/10.3390/su13063306
40. Alvarez-Risco A, Del-Aguila-Arcentales S, Stevenson JG (2015) Pharmacists and mass communication for implementing pharmaceutical care. Am J Pharm Benefits 7(3):e125–e126
41. Alvarez-Risco A, Del-Aguila-Arcentales S, Villalobos-Alvarez D, Diaz-Risco S (2022) Leadership for sustainability in crisis time. In: Alvarez-Risco A, Muthu SS, Del-Aguila-Arcentales S (eds) Circular economy: impact on carbon and water footprint. Springer, Singapore, pp 41–64. http://doi.org/10.1007/978-981-19-0549-0_3
42. Alvarez-Risco A, Del-Aguila-Arcentales S, Yáñez JA, Alvarez-Risco A (2021) Telemedicine in Peru as a result of the COVID-19 pandemic: perspective from a country with limited internet access. Am J Trop Med Hyg 105(1):6–11. https://doi.org/10.4269/ajtmh.21-0255
43. Alvarez-Risco A, Del-Aguila-Arcentales S, Yáñez JA, Rosen MA, Mejia CR (2021) Influence of technostress on academic performance of university medicine students in peru during the covid-19 pandemic. Sustainability (Switzerland) 13(16), Article 8949. http://doi.org/10.3390/su13168949
44. Alvarez-Risco A, Delgado-Zegarra J, Yáñez JA, Diaz-Risco S, Del-Aguila-Arcentales S (2018) Predation risk by gastronomic boom—case Peru. J Landscape Ecol (Czech Repub) 11(1):100–103. http://doi.org/10.2478/jlecol-2018-0003
45. Alvarez-Risco A, Estrada-Merino A, Anderson-Seminario MM, Mlodzianowska S, García-Ibarra V, Villagomez-Buele C, Carvache-Franco M (2021) Multitasking behavior in online classrooms and academic performance: case of university students in Ecuador during COVID-19 outbreak. Interact Technol Smart Educ 18(3):422–434. https://doi.org/10.1108/ITSE-08-2020-0160
46. Alvarez-Risco A, Mejia CR, Delgado-Zegarra J, Del-Aguila-Arcentales S, Arce-Esquivel AA, Valladares-Garrido MJ et al (2020) The Peru approach against the COVID-19 infodemic: insights and strategies. Am J Trop Med Hyg 103(2):583–586. https://doi.org/10.4269/ajtmh.20-0536
47. Alvarez-Risco A, Mlodzianowska S, García-Ibarra V, Rosen MA, Del-Aguila-Arcentales S (2021) Factors affecting green entrepreneurship intentions in business university students in covid-19 pandemic times: case of Ecuador. Sustainability (Switzerland) 13(11), Article 6447. http://doi.org/10.3390/su13116447
48. Alvarez-Risco A, Mlodzianowska S, Zamora-Ramos U, Del-Aguila-Arcentales S (2021) Green entrepreneurship intention in university students: the case of Peru. Entrepreneurial Bus Econ Rev 9(4):85–100. http://doi.org/10.15678/EBER.2021.090406
49. Alvarez-Risco A, Quiroz-Delgado D, Del-Aguila-Arcentales S (2016) Pharmaceutical care in hypertension patients in a Peruvian hospital [Article]. Indian J Public Health Res Dev 7(3):183–188. https://doi.org/10.5958/0976-5506.2016.00153.4
50. Alvarez-Risco A, Turpo-Cama A, Ortiz-Palomino L, Gongora-Amaut N, Del-Aguila-Arcentales S (2016) Barriers to the implementation of pharmaceutical care in pharmacies in Cusco, Peru. Pharm Care Esp 18(5):194–205
51. Alvarez-Risco A, Van Mil JWF (2007) Pharmaceutical care in community pharmacies: practice and research in Peru. Ann Pharmacother 41(12):2032–2037. https://doi.org/10.1345/aph.1K117

52. Álvarez Acosta R, Núñez Guale L, Calderón Pineda F, Mendoza Tarabó AE (2020) Production and commercialization of fish skin tannery products, Santa Elena-Ecuador [Producción y comercialización de productos de curtiembre de piel de pescado, Santa Elena–Ecuador]. Revista de ciencias sociales 26(4):353–367
53. Alzahrani S, Alqahtani F (2016) Pyocyanin pigment extracted from Pseudomonas aeruginosa isolate as antimicrobial agent and textile colorant. Science 5(9):467–470
54. Ang KL, Saw ET, He W, Dong X, Ramakrishna S (2021) Sustainability framework for pharmaceutical manufacturing (PM): a review of research landscape and implementation barriers for circular economy transition. J Clean Prod 280:124264. https://doi.org/10.1016/j.jclepro.2020.124264
55. Angulo-Mosquera LS, Alvarado-Alvarado AA, Rivas-Arrieta MJ, Cattaneo CR, Rene ER, García-Depraect O (2021) Production of solid biofuels from organic waste in developing countries: a review from sustainability and economic feasibility perspectives. Sci Total Environ 795:148816. http://doi.org/10.1016/j.scitotenv.2021.148816
56. Anner M (2019) Predatory purchasing practices in global apparel supply chains and the employment relations squeeze in the Indian garment export industry. Int Labour Rev 158(4):705–727. https://doi.org/10.1111/ilr.12149
57. Apcho-Ccencho LV, Cuya-Velásquez BB, Rodríguez DA, Anderson-Seminario MLM, Alvarez-Risco A, Estrada-Merino A, Mlodzianowska S (2021) The impact of international price on the technological industry in the United States and China during times of crisis: commercial war and covid-19. In: Advances in business and management forecasting, vol 14, pp 149–160
58. Aragon-Correa JA, Marcus AA, Rivera JE, Kenworthy AL (2017) Sustainability management teaching resources and the challenge of balancing planet, people, and profits. Acad Manag Learn Educ 16(3):469–483. https://doi.org/10.5465/amle.2017.0180
59. Arias-Meza M, Alvarez-Risco A, Cuya-Velásquez BB, de las Mercedes Anderson-Seminario M, Del-Aguila-Arcentales S (2022) Fashion and textile circularity and waste footprint. In: Alvarez-Risco A, Muthu SS, Del-Aguila-Arcentales S (eds) Circular economy: impact on carbon and water footprint. Springer, Singapore, pp 181–204. http://doi.org/10.1007/978-981-19-0549-0_9
60. Ariztía T (2017) The theory of social practices: particularities, possibilities and limits [La teoría de las prácticas sociales: particularidades, posibilidades y límites]. Cinta de moebio 221–234. http://www.scielo.cl/scielo.php?script=sci_arttext&pid=S0717-554X2017000200221&nrm=iso
61. Armstrong CM, Niinimäki K, Kujala S, Karell E, Lang C (2015) Sustainable product-service systems for clothing: exploring consumer perceptions of consumption alternatives in Finland. J Clean Prod 97:30–39. https://doi.org/10.1016/j.jclepro.2014.01.046
62. Arrigo E (2021) Digital platforms in fashion rental: a business model analysis. J Fashion Mark Manag Int J (ahead-of-print). http://doi.org/10.1108/JFMM-03-2020-0044
63. Aruldass CA, Dufossé L, Ahmad WA (2018) Current perspective of yellowish-orange pigments from microorganisms—a review. J Clean Prod 180:168–182. https://doi.org/10.1016/j.jclepro.2018.01.093
64. Aschemann-Witzel J, Giménez A, Ares G (2018) Convenience or price orientation? Consumer characteristics influencing food waste behaviour in the context of an emerging country and the impact on future sustainability of the global food sector. Glob Environ Chang 49:85–94. https://doi.org/10.1016/j.gloenvcha.2018.02.002
65. Athwal N, Wells VK, Carrigan M, Henninger CE (2019) Sustainable luxury marketing: a synthesis and research agenda. Int J Manag Rev 21(4):405–426. https://doi.org/10.1111/ijmr.12195
66. Aydın B, Alvarez MD (2020) Understanding the tourists' perspective of sustainability in cultural tourist destinations. Sustainability 12(21):8846. https://doi.org/10.3390/su12218846
67. Azeiteiro UM, Bacelar-Nicolau P, Caetano FJP, Caeiro S (2015) Education for sustainable development through e-learning in higher education: experiences from Portugal. J Clean Prod 106:308–319. https://doi.org/10.1016/j.jclepro.2014.11.056

68. Balderjahn I, Peyer M, Seegebarth B, Wiedmann K-P, Weber A (2018) The many faces of sustainability-conscious consumers: a category-independent typology. J Bus Res 91:83–93. https://doi.org/10.1016/j.jbusres.2018.05.022
69. Bali Swain R, Sweet S (2021) Sustainable consumption and production: introduction to circular economy and beyond. In: Bali Swain R, Sweet S (eds) Sustainable consumption and production. Volume II: circular economy and beyond. Springer International Publishing, Cham, pp 1–16. http://doi.org/10.1007/978-3-030-55285-5_1
70. Bamgbade JA, Kamaruddeen AM, Nawi MNM, Adeleke AQ, Salimon MG, Ajibike WA (2019) Analysis of some factors driving ecological sustainability in construction firms. J Clean Prod 208:1537–1545. https://doi.org/10.1016/j.jclepro.2018.10.229
71. Bannerman S (2020) The World Intellectual Property Organization and the sustainable development agenda. Futures 122:102586. https://doi.org/10.1016/j.futures.2020.102586
72. Barnes L, Lea-Greenwood G (2010) Fast fashion in the retail store environment. Int J Retail Distrib Manag 38(10):760–772. https://doi.org/10.1108/09590551011076533
73. Barr S, Gilg A, Shaw G (2011) 'Helping people make better choices': exploring the behaviour change agenda for environmental sustainability. Appl Geogr 31(2):712–720. https://doi.org/10.1016/j.apgeog.2010.12.003
74. Barragán-Ocaña A, Silva-Borjas P, Olmos-Peña S, Polanco-Olguín M (2020) Biotechnology and bioprocesses: their contribution to sustainability. Processes 8(4). http://doi.org/10.3390/pr8040436
75. Barros MV, Salvador R, do Prado GF, de Francisco AC, Piekarski CM (2021) Circular economy as a driver to sustainable businesses. Cleaner Environ Syst 2:100006. https://doi.org/10.1016/j.cesys.2020.100006
76. Batista AA, Francisco AC (2018) Organizational sustainability practices: a study of the firms listed by the corporate sustainability index. Sustainability 10(1). http://doi.org/10.3390/su10010226
77. Battaglia M, Testa F, Bianchi L, Iraldo F, Frey M (2014) Corporate social responsibility and competitiveness within SMEs of the fashion industry: evidence from Italy and France. Sustainability 6(2). http://doi.org/10.3390/su6020872
78. Becker-Leifhold CV (2018) The role of values in collaborative fashion consumption—a critical investigation through the lenses of the theory of planned behavior. J Clean Prod 199:781–791. http://doi.org/10.1016/j.jclepro.2018.06.296
79. Belz FM, Binder JK (2017) Sustainable entrepreneurship: a convergent process model. Bus Strateg Environ 26(1):1–17. https://doi.org/10.1002/bse.1887
80. Ben Youssef A, Boubaker S, Omri A (2018) Entrepreneurship and sustainability: the need for innovative and institutional solutions. Technol Forecast Soc Chang 129:232–241. https://doi.org/10.1016/j.techfore.2017.11.003
81. Birkie SE (2018) Exploring business model innovation for sustainable production: lessons from Swedish manufacturers. Procedia Manuf 25:247–254. https://doi.org/10.1016/j.promfg.2018.06.080
82. Bokpin GA (2017) Foreign direct investment and environmental sustainability in Africa: the role of institutions and governance. Res Int Bus Financ 39:239–247. http://doi.org/10.1016/j.ribaf.2016.07.038
83. Brandão A, da Costa AG (2021) Extending the theory of planned behaviour to understand the effects of barriers towards sustainable fashion consumption. Eur Bus Rev 33(5):742–774. https://doi.org/10.1108/EBR-11-2020-0306
84. Brewer MK (2019) Slow fashion in a fast fashion world: promoting sustainability and responsibility. Laws 8(4). http://doi.org/10.3390/laws8040024
85. Brito RM, Rodríguez C, Aparicio JL (2018) Sustainability in teaching: an evaluation of university teachers and students. Sustainability 10(2):439. https://doi.org/10.3390/su10020439
86. Broega AC, Jordão C, Martins SB (2017) Textile sustainability: reuse of clean waste from the textile and apparel industry. IOP Conf Ser Mater Sci Eng 254:192006. https://doi.org/10.1088/1757-899x/254/19/192006

87. Brooks A (2015) Systems of provision: fast fashion and jeans. Geoforum 63:36–39. https://doi.org/10.1016/j.geoforum.2015.05.018
88. Brown HS, de Jong M, Levy DL (2009) Building institutions based on information disclosure: lessons from GRI's sustainability reporting. J Clean Prod 17(6):571–580. https://doi.org/10.1016/j.jclepro.2008.12.009
89. Brown KA, Harris F, Potter C, Knai C (2020) The future of environmental sustainability labelling on food products. Lancet Planet Health 4(4):e137–e138. https://doi.org/10.1016/S2542-5196(20)30074-7
90. Byun S-E, Sternquist B (2011) Fast fashion and in-store hoarding: the drivers, moderator, and consequences. Cloth Text Res J 29(3):187–201. https://doi.org/10.1177/0887302X11411709
91. Carvache-Franco M, Alvarez-Risco A, Carvache-Franco O, Carvache-Franco W, Estrada-Merino A, Villalobos-Alvarez D (2021) Perceived value and its influence on satisfaction and loyalty in a coastal city: a study from Lima, Peru. J Policy Res Tourism Leisure Events. http://doi.org/10.1080/19407963.2021.1883634
92. Carvache-Franco M, Alvarez-Risco A, Carvache-Franco W, Carvache-Franco O, Estrada-Merino A, Rosen MA (2021) Coastal cities seen from loyalty and their tourist motivations: a study in Lima, Peru. Sustainability (Switzerland) 13(21), Article 11575. http://doi.org/10.3390/su132111575
93. Carvache-Franco M, Carvache-Franco O, Carvache-Franco W, Alvarez-Risco A, Estrada-Merino A (2021) Motivations and segmentation of the demand for coastal cities: a study in Lima, Peru. Int J Tourism Res 23(4):517–531. https://doi.org/10.1002/jtr.2423
94. Castillo-Benancio S, Alvarez-Risco A, Esquerre-Botton S, Leclercq-Machado L, Calle-Nole M, Morales-Ríos F et al (2022) Circular economy for packaging and carbon footprint. In: Alvarez-Risco A, Muthu SS, Del-Aguila-Arcentales S (eds) Circular economy: impact on carbon and water footprint. Springer, Singapore, pp 115–138. http://doi.org/10.1007/978-981-19-0549-0_6
95. Chafloque-Cespedes R, Alvarez-Risco A, Robayo-Acuña PV, Gamarra-Chavez CA, Martinez-Toro GM, Vicente-Ramos W (2021) Effect of sociodemographic factors in entrepreneurial orientation and entrepreneurial intention in university students of Latin American business schools. In: Contemporary issues in entrepreneurship research, vol 11, pp 151–165. http://doi.org/10.1108/S2040-724620210000011010
96. Chafloque-Céspedes R, Vara-Horna A, Asencios-Gonzales Z, López-Odar D, Alvarez-Risco A, Quipuzco-Chicata L et al (2020) Academic presenteeism and violence against women in schools of business and engineering in Peruvian universities. Lecturas de Economia 93:127–153. http://doi.org/10.17533/udea.le.n93a340726
97. Chandran C, Bhattacharya P (2019) Hotel's best practices as strategic drivers for environmental sustainability and green marketing. J Glob Scholars Market Sci 29(2):218–233. https://doi.org/10.1080/21639159.2019.1577156
98. Chen H-L, Burns LD (2006) Environmental analysis of textile products. Cloth Text Res J 24(3):248–261. https://doi.org/10.1177/0887302X06293065
99. Chen J-H, Liu L, Wei D (2017) Enhanced production of astaxanthin by Chromochloris zofingiensis in a microplate-based culture system under high light irradiation. Biores Technol 245:518–529. https://doi.org/10.1016/j.biortech.2017.08.102
100. Chen X, Zhang SX, Jahanshahi AA, Alvarez-Risco A, Dai H, Li J, Ibarra VG (2020) Belief in a COVID-19 conspiracy theory as a predictor of mental health and well-being of health care workers in Ecuador: cross-sectional survey study. JMIR Public Health Surveill 6(3), Article e20737. http://doi.org/10.2196/20737
101. Choi Y-H, Yoon S, Xuan B, Lee S-YT, Lee K-H (2021) Fashion informatics of the Big 4 Fashion Weeks using topic modeling and sentiment analysis. Fashion and Textiles 8(1):33. https://doi.org/10.1186/s40691-021-00265-6
102. Chung SA, Olivera S, Román BR, Alanoca E, Moscoso S, Terceros BL et al (2021) Themes of scientific production of the cuban journal of pharmacy indexed in scopus (1967–2020). Revista Cubana de Farmacia 54(1), Article e511

103. Contreras-Taica A, Alvarez-Risco A, Arias-Meza M, Campos-Dávalos N, Calle-Nole M, Almanza-Cruz C et al (2022) Virtual education: carbon footprint and circularity. In: Alvarez-Risco A, Muthu SS, Del-Aguila-Arcentales S (eds) Circular economy: impact on carbon and water footprint. Springer, Singapore, pp 265–285. http://doi.org/10.1007/978-981-19-0549-0_13
104. Criado-Gomis A, Cervera-Taulet A, Iniesta-Bonillo M-A (2017) Sustainable entrepreneurial orientation: a business strategic approach for sustainable development. Sustainability 9(9):1667. https://doi.org/10.3390/su9091667
105. Cruz-Torres W, Alvarez-Risco A, Del-Aguila-Arcentales S (2021) Impact of Enterprise Resource Planning (ERP) implementation on performance of an education enterprise: a Structural Equation Modeling (SEM) [Article]. Stud Bus Econ 16(2):37–52. https://doi.org/10.2478/sbe-2021-0023
106. Cruz JP, Alshammari F, Felicilda-Reynaldo RFD (2018) Predictors of Saudi nursing students' attitudes towards environment and sustainability in health care. Int Nurs Rev 65(3):408–416. https://doi.org/10.1111/inr.12432
107. Cuya-Velásquez BB, Alvarez-Risco A, Gomez-Prado R, Juarez-Rojas L, Contreras-Taica A, Ortiz-Guerra A et al (2022) Circular economy for food loss reduction and water footprint. In: Alvarez-Risco A, Muthu SS, Del-Aguila-Arcentales S (eds) Circular economy: impact on carbon and water footprint. Springer, Singapore, pp 65–91. http://doi.org/10.1007/978-981-19-0549-0_4
108. Danziger P (2019) 6 global consumer trends for 2019, and the brands that are out in front of them. Retrieved 12/02/2021 from https://www.forbes.com/sites/pamdanziger/2019/01/13/6-global-consumer-trends-and-brands-that-are-out-in-front-of-them-in-2019/?sh=55ab1154fe4c
109. De-la-Cruz-Diaz M, Alvarez-Risco A, Jaramillo-Arévalo M, de las Mercedes Anderson-Seminario M, Del-Aguila-Arcentales S (2022) 3D print, circularity, and footprints. In: Alvarez-Risco A, Muthu SS, Del-Aguila-Arcentales S (eds) Circular economy: impact on carbon and water footprint. Springer, Singapore, pp 93–112. http://doi.org/10.1007/978-981-19-0549-0_5
110. de Aguiar Hugo A, de Nadae J, da Silva Lima R (2021) Can fashion be circular? A literature review on circular economy barriers, drivers, and practices in the fashion industry's productive chain. Sustainability 13(21):12246. https://doi.org/10.3390/su132112246
111. de Brito MP, Carbone V, Blanquart CM (2008) Towards a sustainable fashion retail supply chain in Europe: organisation and performance. Int J Prod Econ 114(2):534–553. https://doi.org/10.1016/j.ijpe.2007.06.012
112. de Klerk HM, Kearns M, Redwood M (2019) Controversial fashion, ethical concerns and environmentally significant behaviour. Int J Retail Distrib Manag 47(1):19–38. https://doi.org/10.1108/IJRDM-05-2017-0106
113. de las Mercedes Anderson-Seminario M, Alvarez-Risco A (2022) Better students, better companies, better life: circular learning. In: Alvarez-Risco A, Muthu SS, Del-Aguila-Arcentales S (eds) Circular economy: impact on carbon and water footprint. Springer, Singapore, pp 19–40. http://doi.org/10.1007/978-981-19-0549-0_2
114. Del-Aguila-Arcentales S, Alvarez-Risco A (2013) Human error or burnout as explanation for mistakes in pharmaceutical laboratories. Accred Qual Assur 18(5):447–448. https://doi.org/10.1007/s00769-013-1000-0
115. Del-Aguila-Arcentales S, Alvarez-Risco A, Muthu SS (2022) Measuring circular economy. In: Alvarez-Risco A, Muthu SS, Del-Aguila-Arcentales S (eds) Circular economy: impact on carbon and water footprint. Springer, Singapore, pp 3–17. http://doi.org/10.1007/978-981-19-0549-0_1
116. Delgado-Zegarra J, Alvarez-Risco A, Yáñez JA (2018) Indiscriminate use of pesticides and lack of sanitary control in the domestic market in Peru. Revista Panamericana de Salud Publica/Pan Am J Public Health 42, Article e3. http://doi.org/10.26633/RPSP.2018.3
117. Delgado MJBL, Albuquerque MHF (2015) The contribution of regional costume in fashion. Procedia Manuf 3:6380–6387. https://doi.org/10.1016/j.promfg.2015.07.966

118. Deslatte A, Swann WL (2019) Elucidating the linkages between entrepreneurial orientation and local government sustainability performance. Am Rev Public Adm 50(1):92–109. https://doi.org/10.1177/0275074019869376
119. Di Benedetto CA (2017) Corporate social responsibility as an emerging business model in fashion marketing. J Glob Fash Market 8(4):251–265. https://doi.org/10.1080/20932685.2017.1329023
120. Diaz VC (2018) A responsible fashion? Clothing design enterprises with handmade production of native and rural peoples from the perspective of corporate social responsibility [¿Una moda responsable? Emprendimientos de diseño de indumentaria con producción artesanal de pueblos originarios y rurales desde la perspectiva de la responsabilidad social empresarial]. http://doi.org/10.14409/rce.v1i0.7737
121. Dicuonzo G, Galeone G, Ranaldo S, Turco M (2020) The key drivers of born-sustainable businesses: evidence from the Italian fashion industry. Sustainability 12(24):10237. https://doi.org/10.3390/su122410237
122. Diddi S, Yan R-N, Bloodhart B, Bajtelsmit V, McShane K (2019) Exploring young adult consumers' sustainable clothing consumption intention-behavior gap: a behavioral reasoning theory perspective. Sustain Prod Consumption 18:200–209. https://doi.org/10.1016/j.spc.2019.02.009
123. El-Naggar NE-A, El-Ewasy SM (2017) Bioproduction, characterization, anticancer and antioxidant activities of extracellular melanin pigment produced by newly isolated microbial cell factories Streptomyces glaucescens NEAE-H. Sci Rep 7(1):42129. https://doi.org/10.1038/srep42129
124. Enciso-Zarate A, Guzmán-Oviedo J, Sánchez-Cardona F, Martínez-Rohenes D, Rodríguez-Palomino JC, Alvarez-Risco A et al (2016) Evaluation of contamination by cytotoxic agents in Colombian hospitals. Pharm Care Esp 18(6):241–250
125. England L, Ikpe E, Comunian R, Kabir AJ (2021) Africa fashion futures: creative economies, global networks and local development. Geogr Compass 15(9):e12589. https://doi.org/10.1111/gec3.12589
126. Eppinger E, Jain A, Vimalnath P, Gurtoo A, Tietze F, Hernandez Chea R (2021) Sustainability transitions in manufacturing: the role of intellectual property. Curr Opin Environ Sustain 49:118–126. https://doi.org/10.1016/j.cosust.2021.03.018
127. Ertuna B, Karatas-Ozkan M, Yamak S (2019) Diffusion of sustainability and CSR discourse in hospitality industry. Int J Contemp Hosp Manag 31(6):2564–2581. https://doi.org/10.1108/IJCHM-06-2018-0464
128. Esquerre-Botton S, Alvarez-Risco A, Leclercq-Machado L, de las Mercedes Anderson-Seminario M, Del-Aguila-Arcentales S (2022) Food loss reduction and carbon footprint practices worldwide: a benchmarking approach of circular economy. In: Alvarez-Risco A, Muthu SS, Del-Aguila-Arcentales S (eds) Circular economy: impact on carbon and water footprint. Springer, Singapore, pp 161–179. http://doi.org/10.1007/978-981-19-0549-0_8
129. Figueroa-Domecq C, Kimbu A, de Jong A, Williams AM (2020) Sustainability through the tourism entrepreneurship journey: a gender perspective. J Sustain Tourism 1–24. https://doi.org/10.1080/09669582.2020.1831001
130. Fletcher K (2018) The fashion land ethic: localism, clothing activity, and Macclesfield. Fash Pract 10(2):139–159. https://doi.org/10.1080/17569370.2018.1458495
131. Flores-Montes J (2020) Unique commodities. Ideological fantasy of craft textiles production. LiminaR 18(1):49–60
132. Friedrich D (2021) Market and business-related key factors supporting the use of compostable bioplastics in the apparel industry: a cross-sector analysis. J Clean Prod 297:126716. https://doi.org/10.1016/j.jclepro.2021.126716
133. Friedrich D (2021) What makes bioplastics innovative for fashion retailers? An in-depth analysis according to the Triple Bottom Line Principle. J Clean Prod 316:128257. https://doi.org/10.1016/j.jclepro.2021.128257
134. Galleli B, Teles NEB, dos Santos JAR, Freitas-Martins MS, Hourneaux Junior F (2021) Sustainability university rankings: a comparative analysis of UI green metric and the times

higher education world university rankings. Int J Sustain High Educ (ahead-of-print). http://doi.org/10.1108/IJSHE-12-2020-0475
135. Gardetti MA, Muthu SS (2018) Sustainable luxury: cases on circular economy and entrepreneurship. Springer, Berlin
136. Gassmann O, Frankenberger K, Csik M (2014) Revolutionizing the business model. In: Gassmann O, Schweitzer F (eds) Management of the fuzzy front end of innovation. Springer International Publishing, pp 89–97. http://doi.org/10.1007/978-3-319-01056-4_7
137. Gerdt S-O, Wagner E, Schewe G (2019) The relationship between sustainability and customer satisfaction in hospitality: an explorative investigation using eWOM as a data source. Tour Manage 74:155–172. https://doi.org/10.1016/j.tourman.2019.02.010
138. Ghemawat P, Nueno JL, Dailey M (2003) ZARA: fast fashion, vol 1. Harvard Business School Boston, MA
139. Gómez-Prado R, Alvarez-Risco A, Sánchez-Palomino J, de las Mercedes Anderson-Seminario M, Del-Aguila-Arcentales S (2022) Circular economy for waste reduction and carbon footprint. In: Alvarez-Risco A, Muthu SS, Del-Aguila-Arcentales S (eds) Circular economy: impact on carbon and water footprint. Springer, Singapore, pp 139–159. http://doi.org/10.1007/978-981-19-0549-0_7
140. Grewal J, Hauptmann C, Serafeim G (2021) Material sustainability information and stock price informativeness. J Bus Ethics 171(3):513–544. https://doi.org/10.1007/s10551-020-04451-2
141. Gurova O, Morozova D (2018) A critical approach to sustainable fashion: practices of clothing designers in the Kallio neighborhood of Helsinki. J Consum Cult 18(3):397–413. https://doi.org/10.1177/1469540516668227
142. Hall MR (2019) The sustainability price: expanding environmental life cycle costing to include the costs of poverty and climate change. Int J Life Cycle Assess 24(2):223–236. https://doi.org/10.1007/s11367-018-1520-2
143. Han SL-C, Henninger CE, Apeagyei P, Tyler D (2017) Determining effective sustainable fashion communication strategies. In: Henninger CE, Alevizou PJ, Goworek H, Ryding D (eds) Sustainability in fashion: a cradle to upcycle approach. Springer International Publishing, Cham, pp 127–149. http://doi.org/10.1007/978-3-319-51253-2_7
144. Hansmann R, Mieg HA, Frischknecht P (2012) Principal sustainability components: empirical analysis of synergies between the three pillars of sustainability. Int J Sust Dev World 19(5):451–459. https://doi.org/10.1080/13504509.2012.696220
145. Henninger CE (2015) Traceability the new eco-label in the slow-fashion industry? Consumer perceptions and micro-organisations responses. Sustainability 7(5):6011–6032. https://doi.org/10.3390/su7056011
146. Henninger CE, Brydges T, Iran S, Vladimirova K (2021) Collaborative fashion consumption—a synthesis and future research agenda. J Clean Prod 319:128648. https://doi.org/10.1016/j.jclepro.2021.128648
147. Hermann RR, Bossle MB (2020) Bringing an entrepreneurial focus to sustainability education: a teaching framework based on content analysis. J Clean Prod 246:119038. http://doi.org/10.1016/j.jclepro.2019.119038
148. Hethorn J (2008) Consideration of consumer desire. In: Hethorn J, Ulasewicz C (eds) Sustainable fashion: why now? A conversation about issues, practices, and possibilities, pp 53–76
149. Hong Z, Guo X (2019) Green product supply chain contracts considering environmental responsibilities. Omega 83:155–166. https://doi.org/10.1016/j.omega.2018.02.010
150. Jacobs K, Petersen L, Hörisch J, Battenfeld D (2018) Green thinking but thoughtless buying? An empirical extension of the value-attitude-behaviour hierarchy in sustainable clothing. J Clean Prod 203:1155–1169. https://doi.org/10.1016/j.jclepro.2018.07.320
151. Joergens C (2006) Ethical fashion: myth or future trend? J Fashion Mark Manag Int J 10(3):360–371. https://doi.org/10.1108/13612020610679321
152. Jones P, Comfort D (2020) The COVID-19 crisis and sustainability in the hospitality industry. Int J Contemp Hosp Manag 32(10):3037–3050. https://doi.org/10.1108/IJCHM-04-2020-0357

153. Joy A, Peña C (2017) Sustainability and the fashion industry: conceptualizing nature and traceability. In: Henninger CE, Alevizou PJ, Goworek H, Ryding D (eds) Sustainability in fashion: a cradle to upcycle approach. Springer International Publishing, Cham, pp 31–54. http://doi.org/10.1007/978-3-319-51253-2_3
154. Juarez-Rojas L, Alvarez-Risco A, Campos-Dávalos N, de las Mercedes Anderson-Seminario M, Del-Aguila-Arcentales S (2022) Water footprint in the textile and food supply chain management: trends to become circular and sustainable. In: Alvarez-Risco A, Muthu SS, Del-Aguila-Arcentales S (eds) Circular economy: impact on carbon and water footprint. Springer, Singapore, pp 225–243. http://doi.org/10.1007/978-981-19-0549-0_11
155. Jung S, Jin B (2014) A theoretical investigation of slow fashion: sustainable future of the apparel industry. Int J Consum Stud 38(5):510–519. https://doi.org/10.1111/ijcs.12127
156. Kanelli M, Mandic M, Kalakona M, Vasilakos S, Kekos D, Nikodinovic-Runic J, Topakas E (2018) Microbial production of violacein and process optimization for dyeing polyamide fabrics with acquired antimicrobial properties. Front Microbiol 9(1495). http://doi.org/10.3389/fmicb.2018.01495
157. Keeble BR (1988) The Brundtland report: 'our common future.' Med War 4(1):17–25. https://doi.org/10.1080/07488008808408783
158. Khajavi SH (2021) Additive manufacturing in the clothing industry: towards sustainable new business models. Appl Sci 11(19). http://doi.org/10.3390/app11198994
159. Khan S, Henderson C (2020) How Western Michigan University is approaching its commitment to sustainability through sustainability-focused courses. J Clean Prod 253:119741. https://doi.org/10.1016/j.jclepro.2019.119741
160. Khan SAR, Razzaq A, Yu Z, Miller S (2021) Industry 4.0 and circular economy practices: a new era business strategies for environmental sustainability. Bus Strategy Environ 30(8):4001–4014. http://doi.org/10.1002/bse.2853
161. Kim Y, Oh KW (2020) Which consumer associations can build a sustainable fashion brand image? Evidence from fast fashion brands. Sustainability 12(5):1703. https://doi.org/10.3390/su12051703
162. Klepp IG, Bjerck M (2014) A methodological approach to the materiality of clothing: wardrobe studies. Int J Soc Res Methodol 17(4):373–386. https://doi.org/10.1080/13645579.2012.737148
163. Kong L, Liu Z, Wu J (2020) A systematic review of big data-based urban sustainability research: state-of-the-science and future directions. J Clean Prod 273:123142. https://doi.org/10.1016/j.jclepro.2020.123142
164. Kowszyk Y, Maher R (2018) Case studies on circular economy models and integration of the sustainable development goals into business strategies in the EU and LAC [Estudios de caso sobre modelos de Economía Circular e integración de los Objetivos de Desarrollo Sostenible en estrategias empresariales en la UE y ALC]. Federal Foering Oficce, Hamburgo-Alemania
165. Kozlowski A, Searcy C, Bardecki M (2018) The reDesign canvas: fashion design as a tool for sustainability. J Clean Prod 183:194–207. http://doi.org/10.1016/j.jclepro.2018.02.014
166. Kraus S, Burtscher J, Vallaster C, Angerer M (2018) Sustainable entrepreneurship orientation: a reflection on status-quo research on factors facilitating responsible managerial practices. Sustainability 10(2):444. https://doi.org/10.3390/su10020444
167. Laitala K, Klepp IG (2020) What affects garment lifespans? International clothing practices based on a wardrobe survey in China, Germany, Japan, the UK, and the USA. Sustainability 12(21):9151. https://doi.org/10.3390/su12219151
168. Lang C, Zhang R (2019) Second-hand clothing acquisition: the motivations and barriers to clothing swaps for Chinese consumers. Sustain Prod Consumption 18:156–164. https://doi.org/10.1016/j.spc.2019.02.002
169. Larrán Jorge M, Andrades Peña FJ, Herrera Madueño J (2019) An analysis of university sustainability reports from the GRI database: an examination of influential variables. J Environ Planning Manage 62(6):1019–1044. https://doi.org/10.1080/09640568.2018.1457952
170. Leão AL, Cherian BM, Narine S, Souza SF, Sain M, Thomas S (2015) 7—the use of pineapple leaf fibers (PALFs) as reinforcements in composites. In: Faruk O, Sain M (eds) Biofiber

reinforcements in composite materials. Woodhead Publishing, pp 211–235. http://doi.org/10.1533/9781782421276.2.211

171. Leiva-Martinez MA, Anderson-Seminario MLM, Alvarez-Risco A, Estrada-Merino A, Mlodzianowska S (2021) Price variation in lower goods as of previous economic crisis and the contrast of the current price situation in the context of covid-19 in Peru. In: Advances in business and management forecasting, vol 14, pp 161–166
172. Lellis B, Fávaro-Polonio CZ, Pamphile JA, Polonio JC (2019) Effects of textile dyes on health and the environment and bioremediation potential of living organisms. Biotechnol Res Innov 3(2):275–290. https://doi.org/10.1016/j.biori.2019.09.001
173. Leonas KK (2017) The use of recycled fibers in fashion and home products. In: Muthu SS (ed) Textiles and clothing sustainability: recycled and upcycled textiles and fashion. Springer, Singapore, pp 55–77. http://doi.org/10.1007/978-981-10-2146-6_2
174. Lopez-Odar D, Alvarez-Risco A, Vara-Horna A, Chafloque-Cespedes R, Sekar MC (2020) Validity and reliability of the questionnaire that evaluates factors associated with perceived environmental behavior and perceived ecological purchasing behavior in Peruvian consumers [Article]. Soc Responsib J 16(3):403–417. https://doi.org/10.1108/SRJ-08-2018-0201
175. Lozano R, Barreiro-Gen M, Lozano FJ, Sammalisto K (2019) Teaching sustainability in European higher education institutions: assessing the connections between competences and pedagogical approaches. Sustainability 11(6):1602. https://doi.org/10.3390/su11061602
176. Manavalan E, Jayakrishna K (2019) An analysis on sustainable supply chain for circular economy. Procedia Manuf 33:477–484. https://doi.org/10.1016/j.promfg.2019.04.059
177. Mani V, Gunasekaran A, Delgado C (2018) Supply chain social sustainability: standard adoption practices in Portuguese manufacturing firms. Int J Prod Econ 198:149–164. https://doi.org/10.1016/j.ijpe.2018.01.032
178. Manzoor SR, Ho JSY, Al Mahmud A (2021) Revisiting the 'university image model' for higher education institutions' sustainability. J Mark High Educ 31(2):220–239. https://doi.org/10.1080/08841241.2020.1781736
179. Martínez-Barreiro A (2020) Sustainable fashion: beyond scientific prejudice, a research field of social practices [Moda sostenible: más allá del prejuicio científico, un campo de investigación de prácticas sociales]. Sociedad y economía (40):51–68
180. Martinez-Martinez A, Cegarra-Navarro J-G, Garcia-Perez A, Wensley A (2019) Knowledge agents as drivers of environmental sustainability and business performance in the hospitality sector. Tour Manage 70:381–389. https://doi.org/10.1016/j.tourman.2018.08.030
181. Mazotto AM, de Ramos Silva J, de Brito LAA, Rocha NU, de Souza Soares A (2021) How can microbiology help to improve sustainability in the fashion industry? Environ Technol Innov 23:101760. https://doi.org/10.1016/j.eti.2021.101760
182. Medcalfe S, Miralles Miro E (2021) Sustainable practices and financial performance in fashion firms. J Fashion Mark Manag Int J (ahead-of-print). http://doi.org/10.1108/JFMM-10-2020-0217
183. Mejía-Acosta N, Alvarez-Risco A, Solís-Tarazona Z, Matos-Valerio E, Zegarra-Arellano E, Del-Aguila-Arcentales S (2016) Adverse drug reactions reported as a result of the implementation of pharmaceutical care in the Institutional Pharmacy DIGEMID—Ministry of Health. Pharm Care Esp 18(2):67–74
184. Michelino F, Cammarano A, Celone A, Caputo M (2019) The linkage between sustainability and innovation performance in IT hardware sector. Sustainability 11(16):4275. https://doi.org/10.3390/su11164275
185. Min Kong H, Ko E (2017) Why do consumers choose sustainable fashion? A cross-cultural study of South Korean, Chinese, and Japanese consumers. J Glob Fash Market 8(3):220–234. https://doi.org/10.1080/20932685.2017.1336458
186. Mohamued EA, Ahmed M, Pypłacz P, Liczmańska-Kopcewicz K, Khan MA (2021) Global oil price and innovation for sustainability: the impact of R&D spending, oil price and oil price volatility on GHG emissions. Energies 14(6). http://doi.org/10.3390/en14061757
187. Morales-Ríos F, Alvarez-Risco A, Castillo-Benancio S, de las Mercedes Anderson-Seminario M, Del-Aguila-Arcentales S (2022) Material selection for circularity and footprints. In:

Alvarez-Risco A, Muthu SS, Del-Aguila-Arcentales S (eds) Circular economy: impact on carbon and water footprint. Springer, Singapore, pp 205–221. http://doi.org/10.1007/978-981-19-0549-0_10
188. Mousa SK, Othman M (2020) The impact of green human resource management practices on sustainable performance in healthcare organisations: a conceptual framework. J Clean Prod 243:118595. https://doi.org/10.1016/j.jclepro.2019.118595
189. Mukendi A, Davies I, Glozer S, McDonagh P (2020) Sustainable fashion: current and future research directions. Eur J Mark 54(11):2873–2909. https://doi.org/10.1108/EJM-02-2019-0132
190. Murphy PE, Schlegelmilch BB (2013) Corporate social responsibility and corporate social irresponsibility: introduction to a special topic section. J Bus Res 66(10):1807–1813. https://doi.org/10.1016/j.jbusres.2013.02.001
191. Mzembe AN, Lindgreen A, Idemudia U, Melissen F (2020) A club perspective of sustainability certification schemes in the tourism and hospitality industry. J Sustain Tour 28(9):1332–1350. https://doi.org/10.1080/09669582.2020.1737092
192. Nagurney A, Yu M (2012) Sustainable fashion supply chain management under oligopolistic competition and brand differentiation. Int J Prod Econ 135(2):532–540. https://doi.org/10.1016/j.ijpe.2011.02.015
193. Nayak R, Panwar T, Nguyen LVT (2020) 1—Sustainability in fashion and textiles: a survey from developing country. In: Nayak R (ed) Sustainable technologies for fashion and textiles. Woodhead Publishing, pp 3–30. http://doi.org/10.1016/B978-0-08-102867-4.00001-3
194. Niemann CC, Dickel P, Eckardt G (2020) The interplay of corporate entrepreneurship, environmental orientation, and performance in clean-tech firms—a double-edged sword. Bus Strategy Environ 29(1):180–196. http://doi.org/10.1002/bse.2357
195. Niinimäki K, Peters G, Dahlbo H, Perry P, Rissanen T, Gwilt A (2020) The environmental price of fast fashion. Nat Rev Earth Environ 1(4):189–200. https://doi.org/10.1038/s43017-020-0039-9
196. Norström AV, Cvitanovic C, Löf MF, West S, Wyborn C, Balvanera P et al (2020) Principles for knowledge co-production in sustainability research. Nat Sustain 3(3):182–190. http://doi.org/10.1038/s41893-019-0448-2
197. Olawumi TO, Chan DWM (2018) A scientometric review of global research on sustainability and sustainable development. J Clean Prod 183:231–250. http://doi.org/10.1016/j.jclepro.2018.02.162
198. Omoloso O, Mortimer K, Wise WR, Jraisat L (2021) Sustainability research in the leather industry: a critical review of progress and opportunities for future research. J Clean Prod 285:125441. http://doi.org/10.1016/j.jclepro.2020.125441
199. Ozturkoglu Y, Sari FO, Saygili E (2021) A new holistic conceptual framework for sustainability oriented hospitality innovation with triple bottom line perspective. J Hosp Tour Technol 12(1):39–57. https://doi.org/10.1108/JHTT-02-2019-0022
200. Pal R, Gander J (2018) Modelling environmental value: an examination of sustainable business models within the fashion industry. J Clean Prod 184:251–263. https://doi.org/10.1016/j.jclepro.2018.02.001
201. Palacios-Chavarro J-A, Marroquín-Ciendúa F, Bohórquez-Lazdhaluz R (2021) Social campaigns to encourage responsible fashion consumption: qualitative study with university students. Commun Soc 34(3):153–169
202. Palomino E (2020) SDG 14 life below water. In: Franco IB, Chatterji T, Derbyshire E, Tracey J (eds) Actioning the global goals for local impact: towards sustainability science, policy, education and practice. Springer, Singapore, pp 229–246. http://doi.org/10.1007/978-981-32-9927-6_15
203. Palomo-Lovinski N (2021) Generational cohorts' views on local sustainable practices in fashion. Fashion Style Popular Cult 8(4):437–454. https://doi.org/10.1386/fspc_00032_1
204. Payne A (2015) 6—Open- and closed-loop recycling of textile and apparel products. In: Muthu SS (ed) Handbook of life cycle assessment (LCA) of textiles and clothing. Woodhead Publishing, pp 103–123. http://doi.org/10.1016/B978-0-08-100169-1.00006-X

205. Peçanha Enqvist J, West S, Masterson VA, Haider LJ, Svedin U, Tengö M (2018) Stewardship as a boundary object for sustainability research: linking care, knowledge and agency. Landsc Urban Plan 179:17–37. https://doi.org/10.1016/j.landurbplan.2018.07.005
206. Peeters P (2018) Why space tourism will not be part of sustainable tourism. Tour Recreat Res 43(4):540–543. https://doi.org/10.1080/02508281.2018.1511942
207. Peña Miguel N, Corral Lage J, Mata Galindez A (2020) Assessment of the development of professional skills in university students: sustainability and serious games. Sustainability 12(3). http://doi.org/10.3390/su12031014
208. Pérez Hernández DM, Neme Calacich S (2021) The value chain in textile handicrafts: the case of the embroidered strip from Tabasco [La cadena de valor en la artesanía textil: el caso de la tira bordada tabasqueña]. Nova scientia 13(26)
209. Pham H, Kim S-Y (2019) The effects of sustainable practices and managers' leadership competences on sustainability performance of construction firms. Sustain Prod Consumption 20:1–14. https://doi.org/10.1016/j.spc.2019.05.003
210. Piñatex (n.d.) Products. https://www.ananas-anam.com/products-2
211. Plasencia-Soler JA, Marrero-Delgado F, Bajo-Sanjuán AM, Nicado-García M (2018) Models for assessing the sustainability of organizations [Modelos para evaluar la sostenibilidad de las organizaciones]. Estudios Gerenciales 34(146):63–73
212. Plewnia F, Guenther E (2018) Mapping the sharing economy for sustainability research. Manag Decis 56(3):570–583. https://doi.org/10.1108/MD-11-2016-0766
213. Quispe-Cañari JF, Fidel-Rosales E, Manrique D, Mascaró-Zan J, Huamán-Castillón KM, Chamorro–Espinoza SE et al (2021) Self-medication practices during the COVID-19 pandemic among the adult population in Peru: a cross-sectional survey. Saudi Pharm J 29(1):1–11. http://doi.org/10.1016/j.jsps.2020.12.001
214. Rajesh R (2020) Exploring the sustainability performances of firms using environmental, social, and governance scores. J Clean Prod 247:119600. https://doi.org/10.1016/j.jclepro.2019.119600
215. Ramísio PJ, Pinto LMC, Gouveia N, Costa H, Arezes D (2019) Sustainability strategy in higher education institutions: lessons learned from a nine-year case study. J Clean Prod 222:300–309. https://doi.org/10.1016/j.jclepro.2019.02.257
216. Ransfield AK, Reichenberger I (2021) Māori indigenous values and tourism business sustainability. Altern Int J Indigenous Peoples 17(1):49–60. http://doi.org/10.1177/1177180121994680
217. Rau H, Goggins G, Fahy F (2018) From invisibility to impact: recognising the scientific and societal relevance of interdisciplinary sustainability research. Res Policy 47(1):266–276. https://doi.org/10.1016/j.respol.2017.11.005
218. Rehman K, Shahzad T, Sahar A, Hussain S, Mahmood F, Siddique MH et al (2018) Effect of Reactive Black 5 azo dye on soil processes related to C and N cycling. PeerJ 6:e4802
219. Rojas-Osorio M, Alvarez-Risco A (2019) Intention to use smartphones among Peruvian university students. Int J Interact Mobile Technol 13(3):40–52. https://doi.org/10.3991/ijim.v13i03.9356
220. Román BR, Moscoso S, Chung SA, Terceros BL, Álvarez-Risco A, Yáñez JA (2020) Treatment of COVID-19 in Peru and Bolivia, and self-medication risks. Revista Cubana de Farmacia 53(2):1–20, Article e435
221. Roos S, Zamani B, Sandin G, Peters GM, Svanström M (2016) A life cycle assessment (LCA)-based approach to guiding an industry sector towards sustainability: the case of the Swedish apparel sector. J Clean Prod 133:691–700. https://doi.org/10.1016/j.jclepro.2016.05.146
222. Runfola A, Guercini S (2013) Fast fashion companies coping with internationalization: driving the change or changing the model? J Fashion Mark Manag Int J 17(2):190–205. https://doi.org/10.1108/JFMM-10-2011-0075
223. Sakao T, Brambila-Macias SA (2018) Do we share an understanding of transdisciplinarity in environmental sustainability research? J Clean Prod 170:1399–1403. http://doi.org/10.1016/j.jclepro.2017.09.226

224. Salmerón-Manzano E, Manzano-Agugliaro F (2018) The higher education sustainability through virtual laboratories: the Spanish University as case of study. Sustainability 10(11). http://doi.org/10.3390/su10114040
225. Sánchez Vázquez P, Gago-Cortés C, Alló M (2020) Moda sostenible y preferencias del consumidor. 3C Empresa. Investigación y pensamiento crítico 9(3):39–57
226. Serafeim G (2020) Public sentiment and the price of corporate sustainability. Financ Anal J 76(2):26–46. https://doi.org/10.1080/0015198X.2020.1723390
227. Shen B (2014) Sustainable fashion supply chain: lessons from H&M. Sustainability 6(9):6236–6249. https://doi.org/10.3390/su6096236
228. Shen S-F, Emlen ST, Koenig WD, Rubenstein DR (2017) The ecology of cooperative breeding behaviour. Ecol Lett 20(6):708–720. https://doi.org/10.1111/ele.12774
229. Shuqin C, Minyan L, Hongwei T, Xiaoyu L, Jian G (2019) Assessing sustainability on Chinese university campuses: development of a campus sustainability evaluation system and its application with a case study. J Build Eng 24:100747. https://doi.org/10.1016/j.jobe.2019.100747
230. Skjerven A, Reitan J (2017) Design for a sustainable culture: perspectives, practices and education. Taylor & Francis
231. Sroufe R, Gopalakrishna-Remani V (2018) Management, social sustainability, reputation, and financial performance relationships: an empirical examination of U.S. firms. Organ Environ 32(3):331–362. http://doi.org/10.1177/1086026618756611
232. Strydom M, Kempen E (2021) Towards economic sustainability: how higher education can support the business operations of emerging clothing manufacturing micro enterprises. Int J Sustain High Educ 22(7):1469–1486. https://doi.org/10.1108/IJSHE-05-2020-0152
233. Sun Y, Ko E (2016) Influence of sustainable marketing activities on customer equity. J Glob Scholars Market Sci 26(3):270–283. https://doi.org/10.1080/21639159.2016.1174537
234. Sung E, Kim H, Lee D (2018) Why do people consume and provide sharing economy accommodation? A sustainability perspective. Sustainability 10(6):2072. https://doi.org/10.3390/su10062072
235. Thirumalesh Madanaguli A, Kaur P, Bresciani S, Dhir A (2021) Entrepreneurship in rural hospitality and tourism. A systematic literature review of past achievements and future promises. Int J Contemp Hospitality Manag 33(8):2521–2558. http://doi.org/10.1108/IJCHM-09-2020-1121
236. Thomas K (2020) Cultures of sustainability in the fashion industry. Fash Theory 24(5):715–742. https://doi.org/10.1080/1362704X.2018.1532737
237. Thorisdottir TS, Johannsdottir L (2019) Sustainability within fashion business models: a systematic literature review. Sustainability 11(8):2233. https://doi.org/10.3390/su11082233
238. Todeschini BV, Cortimiglia MN, Callegaro-de-Menezes D, Ghezzi A (2017) Innovative and sustainable business models in the fashion industry: entrepreneurial drivers, opportunities, and challenges. Bus Horiz 60(6):759–770. https://doi.org/10.1016/j.bushor.2017.07.003
239. Trosper RL (2002) Northwest coast indigenous institutions that supported resilience and sustainability. Ecol Econ 41(2):329–344. https://doi.org/10.1016/S0921-8009(02)00041-1
240. Vehmas K, Raudaskoski A, Heikkilä P, Harlin A, Mensonen A (2018) Consumer attitudes and communication in circular fashion. J Fashion Mark Manag Int J 22(3):286–300. https://doi.org/10.1108/JFMM-08-2017-0079
241. Veleva V, Bodkin G (2018) Corporate-entrepreneur collaborations to advance a circular economy. J Clean Prod 188:20–37. http://doi.org/10.1016/j.jclepro.2018.03.196
242. Vizcardo D, Salvador LF, Nole-Vara A, Dávila KP, Alvarez-Risco A, Yáñez JA, Mejia CR (2022) Sociodemographic predictors associated with the willingness to get vaccinated against COVID-19 in Peru: a cross-sectional survey. Vaccines 10(1), Article 48. http://doi.org/10.3390/vaccines10010048
243. Wang H, Liu H, Kim SJ, Kim KH (2019) Sustainable fashion index model and its implication. J Bus Res 99:430–437. https://doi.org/10.1016/j.jbusres.2017.12.027
244. Williams D (2015) 6—fashion design and sustainability. In: Blackburn R (ed) Sustainable apparel. Woodhead Publishing, pp 163–185. http://doi.org/10.1016/B978-1-78242-339-3.00006-6

245. Wong DTW, Ngai EWT (2021) Economic, organizational, and environmental capabilities for business sustainability competence: findings from case studies in the fashion business. J Bus Res 126:440–471. https://doi.org/10.1016/j.jbusres.2020.12.060
246. Yan J, Kim S, Zhang SX, Foo MD, Alvarez-Risco A, Del-Aguila-Arcentales S, Yáñez JA (2021) Hospitality workers' COVID-19 risk perception and depression: a contingent model based on transactional theory of stress model. Int J Hospitality Manag 95, Article 102935. http://doi.org/10.1016/j.ijhm.2021.102935
247. Yang S, Song Y, Tong S (2017) Sustainable retailing in the fashion industry: a systematic literature review. Sustainability 9(7):1266. https://doi.org/10.3390/su9071266
248. Yáñez JA, Alvarez-Risco A, Delgado-Zegarra J (2020) Covid-19 in Peru: from supervised walks for children to the first case of Kawasaki-like syndrome. The BMJ 369, Article m2418. http://doi.org/10.1136/bmj.m2418
249. Yáñez JA, Jahanshahi AA, Alvarez-Risco A, Li J, Zhang SX (2020) Anxiety, distress, and turnover intention of healthcare workers in Peru by their distance to the epicenter during the COVID-19 crisis. Am J Trop Med Hyg 103(4):1614–1620. https://doi.org/10.4269/ajtmh.20-0800
250. Yáñez S, Uruburu Á, Moreno A, Lumbreras J (2019) The sustainability report as an essential tool for the holistic and strategic vision of higher education institutions. J Clean Prod 207:57–66. https://doi.org/10.1016/j.jclepro.2018.09.171
251. Yarime M, Tanaka Y (2012) The issues and methodologies in sustainability assessment tools for higher education institutions: a review of recent trends and future challenges. J Educ Sustain Dev 6(1):63–77. https://doi.org/10.1177/097340821100600113
252. Yusof Y, Yahya SA, Adam A (2015) Novel technology for sustainable pineapple leaf fibers productions. Procedia CIRP 26:756–760. https://doi.org/10.1016/j.procir.2014.07.160
253. Zamora-Polo F, Sánchez-Martín J (2019) Teaching for a better world. Sustainability and sustainable development goals in the construction of a change-maker university. Sustainability 11(15). http://doi.org/10.3390/su11154224
254. Zhang B, Zhang Y, Zhou P (2021) Consumer attitude towards sustainability of fast fashion products in the UK. Sustainability 13(4):1646. https://doi.org/10.3390/su13041646
255. Zhang SX, Chen J, Afshar Jahanshahi A, Alvarez-Risco A, Dai H, Li J, Patty-Tito RM (2021) Succumbing to the COVID-19 pandemic—healthcare workers not satisfied and intend to leave their jobs. Int J Ment Heal Addict. https://doi.org/10.1007/s11469-020-00418-6
256. Zhang SX, Chen J, Jahanshahi AA, Alvarez-Risco A, Dai H, Li J, Patty-Tito RM (2021) Correction to: succumbing to the COVID-19 pandemic—healthcare workers not satisfied and intend to leave their jobs. Int J Ment Heal Addict. https://doi.org/10.1007/s11469-021-005 02-5
257. Zhang SX, Sun S, Afshar Jahanshahi A, Alvarez-Risco A, Ibarra VG, Li J, Patty-Tito RM (2020) Developing and testing a measure of COVID-19 organizational support of healthcare workers—results from Peru, Ecuador, and Bolivia. Psychiatry Res 291. http://doi.org/10.1016/j.psychres.2020.113174

Strategies in Small Businesses to Combat Plastic Overproduction

Myreya De-La-Cruz-Diaz, Aldo Alvarez-Risco, Micaela Jaramillo-Arévalo, María de las Mercedes Anderson-Seminario, and Shyla Del-Aguila-Arcentales

Abstract Modern production processes have been increasingly responsible for plastic overproduction for various reasons, not limited to containers and packaging. Said situation has reflected its consequences on human and animal health, especially with the COVID-19 pandemic. Because of this, it is imperative to consider the measures taken to reduce plastic contamination and the role of sustainability in plastic production processes. Previous studies and literature about the effects of plastics, the provisional standards, and programs to reduce contamination caused by this material have been analyzed in order to highlight its importance. Green businesses are also proposed as a more efficient solution to the problem of plastic overproduction. Some examples of small green businesses are presented to demonstrate the adoption of sustainable, eco-friendly measures in their business model and that even after some years, said models have proved to be effective by providing revenues and still following the ecological practices such as recycling different kinds of plastics and wastes.

Keywords Single-use plastic · Microplastics · Marine litter · Entrepreneurship · Sustainability · Green business · Plastic · Contamination · Sustainable development goal · Circularity · Circular economy

1 Introduction

Contamination has been a topic of discussion even before the pandemic started. However, now it has become one of the most important problems to take care of, which is why many companies have started changing parts of their production process and even their entire business model to meet the environmental goals and expectations

M. De-La-Cruz-Diaz · A. Alvarez-Risco · M. Jaramillo-Arévalo · M. de las Mercedes Anderson-Seminario
Universidad de Lima, Lima, Peru

S. Del-Aguila-Arcentales (✉)
Escuela de Posgrado, Universidad San Ignacio de Loyola, Lima, Peru
e-mail: sdelaguila@usil.edu.pe

A. Alvarez-Risco et al. (eds.), *Footprint and Entrepreneurship*, Environmental Footprints and Eco-design of Products and Processes,
https://doi.org/10.1007/978-981-19-8895-0_5

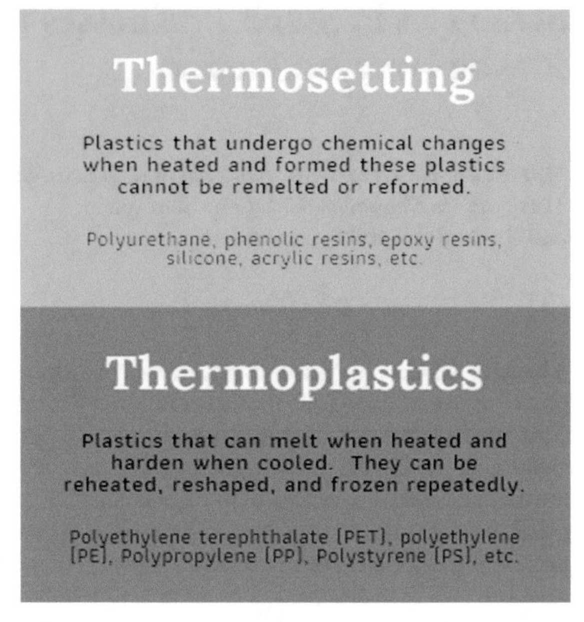

Fig. 1 Categories of plastic. *Source* Meneses [147]

of the future. Recycling was one of the several measures governments and people proposed to help reduce contamination and preserve the environment. According to Piattu and Ferreira, plastic is defined as a material whose prime component is a polymer, organic and synthetic, solid in its final state. It was previously transformed into the fluid for molding by heat and pressure (as aforementioned in Meneses [147]). This material can be classified into two main categories depending on its capacity to be reshaped: thermosetting and thermoplastics (Fig. 1).

A material such as plastic seems to be a natural part of our everyday lives; however, its appearance as we usually use it goes back to the end of World War II. Before that, according to Geyer et al. [120], plastic was only used by the military and in small quantities. Now, it is rapidly surpassing most of the other materials produced by man, mainly because of its massive use in the creation of single-use packaging and other first-need products such as toothbrushes and Tupperware [190], which can create different perceptions in consumers [205]. Plastics, although stable physically, over time tend to degrade into smaller fragments [72], creating microplastics. As stated by Frias and Nash [116], microplastics can be described as solid synthetic particles or polymeric matrices with regular or irregular shapes and between 1 and 5 mm in size. These particles can be from primary or secondary manufacturing origin that is insoluble. The primary microplastics are those made for particular applications, meaning that this kind of microplastic is created to substitute some organic ingredients in the making of cosmetics, cleaning products, and paints. These primary microplastics are used to create larger plastics, like pellets. Secondary microplastics are created due to the degradation of plastic previously exposed to the environment caused by chemical or physical phenomena such as photolytic degradation [82] (Fig. 2).

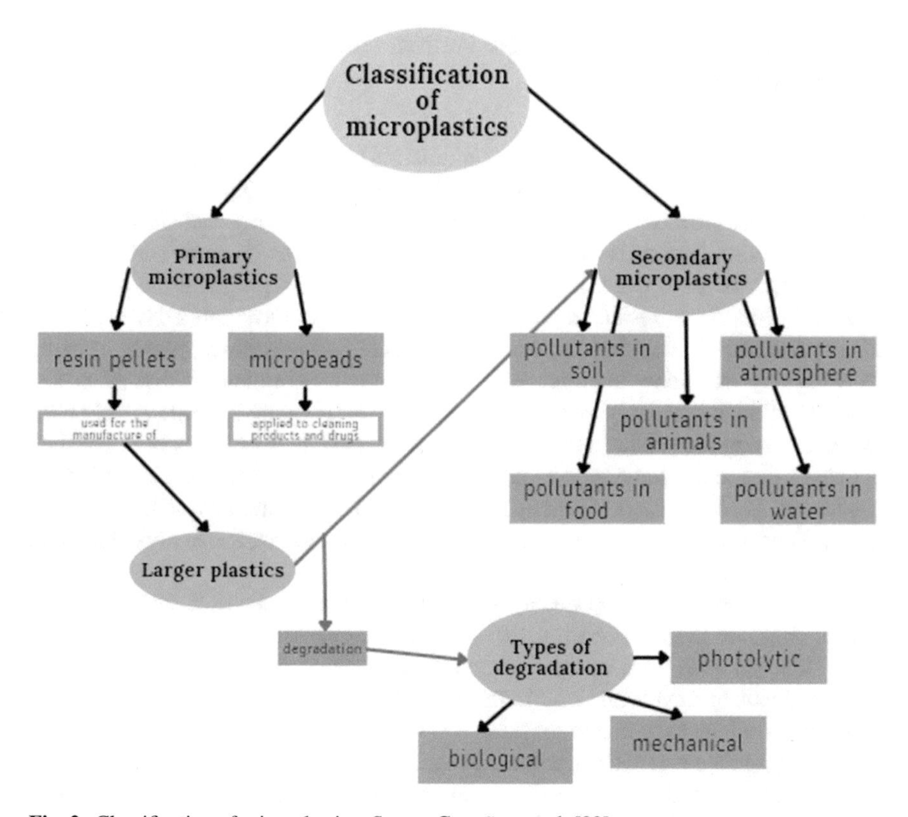

Fig. 2 Classification of microplastics. *Source* Castañeta et al. [82]

The excessive production of such material has resulted in vast amounts of waste and contamination, difficult to reduce and eradicate. Since derivatives make commonly-used plastic of fossil hydrocarbons, which are not biodegradable, it is almost impossible to eliminate; thus, it ends in landfills, dumps, and even the oceans, damaging the environment [120]. Unfortunately, the constant production of plastic and the waste that it generates is not the only problem for us; these last years, the major problem was how to dispose of this waste and how inefficient it has been with the latest measures that have been applied. The consequences of the disposal in aquatic and terrestrial ecosystems worldwide are tremendous and cause a large amount of damage [70].

2 Current Problem of Plastic Before the COVID-19 Pandemic

Before the COVID-19 outbreak, the contamination of plastic in marine and terrestrial areas was an issue studied at different levels, but during 2019 governments started taking measures to reduce their waste. A topic that got more relevant was the plastic of only one use. The definition of single-use plastics is, as its name states, plastic that can only be used once, like the plastic used for packaging and the ones used to contain food from restaurants. Although governments have applied measures to restrict the use of shopping bags in supermarkets and encourage people to bring their bags by increasing the price per bag, it is essential to mention that the whole lifecycle of the bag must be considered. In this case, shopping bags are mainly used for transporting the products from the store to the customer's homes, but after that, people often use the same bags to take out the trash. In addition, the United States' production of plastic bags is mainly made from recycled materials, so the term and the applications on measures based on that are tricky and must be analyzed for it to work [147].

For the classification and management of solid plastic waste, a system is used where a number is assigned from 1 to 7 to a piece of plastic that indicates its kind [58]. This system is used worldwide, and its objective is to identify the most challenging plastic to recycle and which elements it can find. Table 1 describes the plastic depending on how easy it is to recycle.

According to the UN Environment Program [197], about 8.300 million tons of plastic has been produced since 1950, and around 60% of that plastic ended up in a landfill or the natural environment like oceans and had devastating effects on marine life [200]. As reported by "Plastics Europe the Facts 2018", the production of plastic reached 348 million tons, an increase of 3.8% compared to the previous years. Adding the lousy management of waste, the result of such an increase in production is finding waste in illegal landfills or more contamination in the environment even far away from the place of origin, that is why we even found it in areas like Arctic Sea ice [161], among other places. Plastic degradation can take hundreds of years, depending on the materials. The amount of waste that ends up in the oceans has increased so much to the point where it already takes miles of m^3 [190]. In 2019, around 60–90% of marine litter was plastic [198], the majority of which comes from plastic bottles, shopping bags, bottle caps, food containers, cigarette butts, and among others. A certain amount of studies state how plastic is present even in our food and the fatal consequences on health [179, 190, 207]. In 2007, microplastics were just recognized as an emerging pollutant in the United States [185], and these microplastic were later studied to know their effects in marine fauna and, therefore, in humans through the trophic chain [74, 162, 179, 207]. Research has concluded that it is a fact that humans eat microplastics [190], plus the sources and amount are diverse and are often found on marine animals, tap water, honey, sugar, beer, and among others [179, 184], we even breathe plastic, especially in urban areas [106]. About 90% of the plastic we

Table 1 Identification of resins and their uses

Code	Material	Initials	Main uses	Recycling
01 PET	Polyethylene terephthalate	PET	Water bottles, containers, dispensers, food jars, fibers for clothing and rugs, certain shampoo bottles	Easy
02 HDPE	High-density polyethylene	HDPE	Shampoo bottles, detergent and milk bottles, freezer bags, ice cream containers, food containers, storage boxes, toys, garden furniture	Easy
03 PVC	Polyvinyl chloride	PVC	Bank cards, window and door frames, pipes, cable sheathing, faux leather	Really hard
04 LDPE	Low-density polyethylene	LDPE	Bags, trays, containers, food packaging film, bubble wrap flexible bottles, cable insulation	Feasible
05 PP	Polypropylene	PP	Microwave dishes, bottle caps, potato chip bags, straws, lunch boxes, coolers, fabric and carpet fibers, tarps, diapers	Feasible
06 PS	Polystyrene	PS	Cutlery, plates, cups, food trays, packing filler, yogurt containers, clothespins, insulation	Hard
07	Others	Others	Water bottles, containers, dispensers, food jars, laundry and carpet fibers, certain shampoo bottles, etc	Really hard

Source Meneses [147]

ingest is eliminated through the excretory system [190], and the effects depend on the type of polymer, associated chemicals, and dosage [74, 162, 190, 207].

The effects of the microplastics in the ecosystems are different from the humans but still just as worrying. The plastic that ends up in the sea begins to degrade over time, waves, and UV rays to tiny fragments [204], making it easy for marine animals to consume them by accident. Depending on their kind of feeding, animals can be direct consumers, ingesting microplastics through sediments or suspended particles, or they can be indirect consumers by ingesting their prey, like in the case of predators or scavengers. The size of the microplastic is essential to determine the damage, size more than 50 μm are incapable of crossing the intestinal barrier in the marine animal but size less than 50 μm are more than capable of crossing, and because there are not any enzymatic pathways capable of digesting plastic, it passes without been digested or absorbed. Although, it has been observed that the animals that eat nano plastics can pass it to their offspring, which implies serious consequences for the entire marine ecosystem like physical obstruction, damage to feeding appendages, leaching of chemical components in organisms, and others.

Cleaning the beaches to reduce the microplastic that affects the animals was an effort far from enough to reduce the damage. At the rate at which the waste of plastic in the sea is increasing, by 2050, there is be more plastic than fish [120]. It is essential to mention that there are two kinds of marine litter, the one that sinks and the one that floats. This last type is the one that can float for a significant amount of time before sinking and is commonly transported from one place to another far away because of currents and wind [139].

3 Evolution of the Plastic Problem During the COVID-19 Pandemic

Despite all of the current measures trying to reduce single-use plastic to reduce environmental pollution, the COVID-19 pandemic brought a new risk of contamination by using PPE (personal protection equipment). According to de Sousa [97], the COVID-19 pandemic has generated an enormous amount of plastic waste from hospitals, PPE use, and vaccination sites. The repercussions of increased plastic use during the pandemic are causing a lot of environmental issues making plastics coming from masks and other protective gear the "villains" instead of being the "heroes" that they are for protecting the population from getting infected with the disease [97]. Many countries worldwide promoted PPE and other protective products to ensure that the virus spread in their territories was not as damaging and did not elevate the number of cases so that the hospitals and clinics would not collapse due to lack of capacity. According to Shams et al. [188], there has been a significant increase in masks, gloves, hand sanitizer, packages of saline solution, protective medical suits, and more products during the COVID-19 pandemic. The improper disposal of these elements could increase plastic waste in many countries and cities, which could cause several effects on their ecosystems. In the case of South Korea, there were approximately 295 tons of medical waste, coming mainly from hospitals and isolated life treatment centers, related to the international context of the pandemic from February 2020 to the beginning of March of the same year. In April, the Ministry of Environment of the country reported that 20 tons of waste related to COVID-19 were generated every day [178]. In the case of Thailand, China, the Philippines, Vietnam, and Indonesia, there has been a combined waste production of 58,000 tons of plastic waste; these countries combined, it is essential to note that the countries mentioned product approximately 50% of plastic waste that goes into the world's oceans [125]. Benson et al. [69] detailed the plastic waste generated by the COVID-19 pandemic.

The estimated daily use of face masks and plastic waste generated by the COVID-19 pandemic can be appreciated. This data comes from countries before they started managing all of the plastic waste they were producing. The top 5 include China, India, the United States, Brazil, and Indonesia.

According to Benson et al. [69], some products being wasted, such as masks, gloves, face shields, and gloves, are made from non-woven materials such as polypropylene (a type of plastic), vinyl, and neoprene. If these products are wasted without proper care, they could potentially end up in marine and terrestrial ecosystems and contaminate those environments. The increasing use of plastics raises concerns about the leakage of this type of waste that goes to marine environments, this factor is especially relevant in coastal sites with a high population with high plastic waste emissions but has low recycling rates [91]. According to Selvaranjan et al. [217], when masks are not disposed of correctly, they can end up causing harm to marine animals since they can swallow masks or get them tied up in their legs, and also they present damage to birds because the strings can get tied in their extremities or beaks presenting a threat to their lives. Figure 3 shows the main plastic products from medical sites during the COVID-19 pandemic wasted in the environment.

If this issue is not solved soon, its repercussions are likely to cause negative impacts on marine life, as mentioned before, and humans. According to Shams et al. [188], the mismanagement of plastic waste could end up being potentially hazardous to human health both in short and in the long term. Also, with time, plastics degrade into microplastics which end up causing much damage in the future to different environments such as maritime and terrestrial ecosystems [188]. It is an urgent matter to find ways to manage plastic waste that is responsible for the environment and does

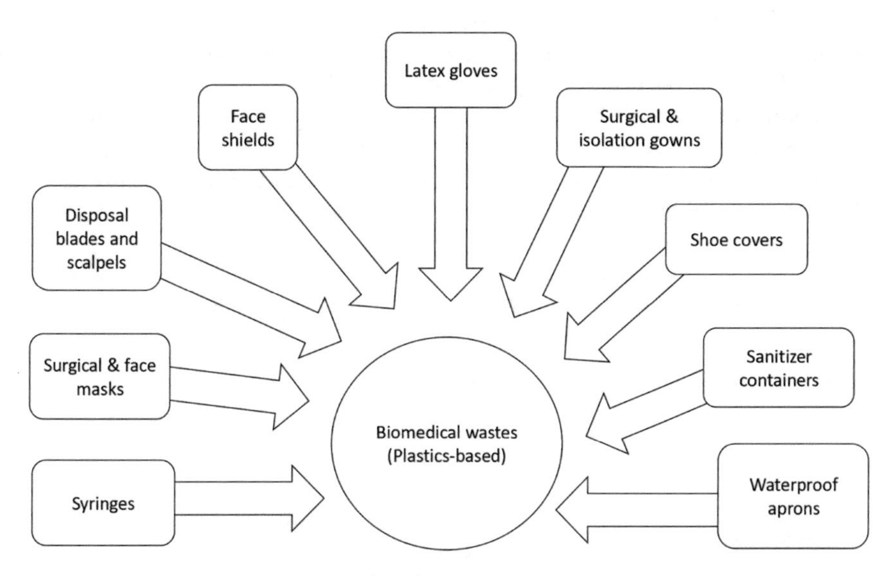

Fig. 3 Types of plastics based biomedical wastes originated during the COVID-19 pandemic. *Source* Cordova et al. [91]

generate negative impacts in these areas in the future so that the ecosystems are not in danger of increasing pollution [115].

4 Intermediate Solutions to Reduce Overproduction of Plastics

4.1 Circular Economy

The classic linear model that we have today of extraction-production-use-discharge of material and energy flow of the economic system turns out to be unsustainable [117]. The materials cycle initiative has been from the beginning of industrialization and has been put into practice linked to the idea that it reduces the negative impact on the environment and promotes novel trade opportunities throughout the origin of industrialization [103, 104] (Fig. 4).

CE is born primarily in the literature through 3 main "actions", which are called the Principles of the 3Rs: Reduction, Reuse, and Recycling establishes his definition of CE based on the vision of the Ellen Macarthur Foundation [108] mentions that a circular economy is a system made to restore and regenerate. Likewise, it is defined as an economic system that replaces the idea of "end of life" with reduction, the option of reuse, recycling, and recovery of materials in the production, distribution, or consumption processes [133]. According to a current report from the United Nations Environment Program in 2018, titled Re-defining Value—The Manufacturing Revolution, the CE could minimize industrial waste between 80 and 99% in some sectors. In the same way, it would reduce its emissions between 79 and 99%, with optimal results that drive its application (Fig. 5).

The flow designed by the circular economy means an innovative version of the life cycle, as it adopts its principles of—appropriate material choices when designing products and establishing appropriate recovery systems. Moreover, it also means improving competitiveness and the efficiency of available resources. In the case

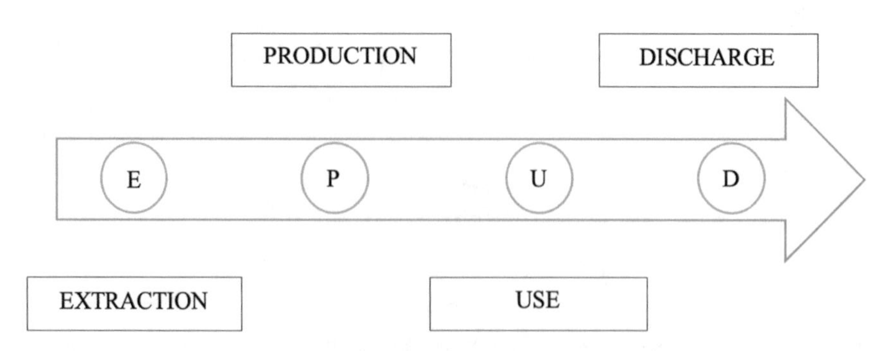

Fig. 4 Linear economy. *Source* REPSOL [177]

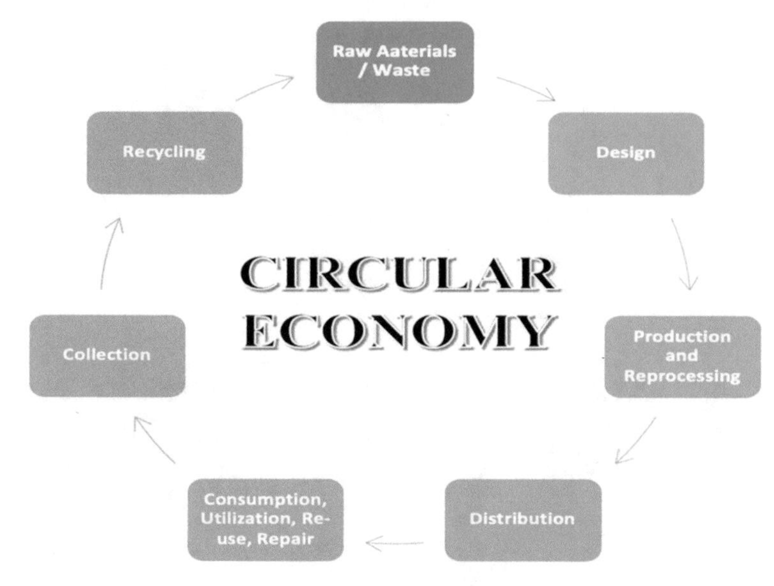

Fig. 5 Circular economy. *Source* Stahel [193]

of plastics, due to their characteristics—light, versatile and durable—plastics help to save essential resources such as energy and water in strategic sectors such as packaging, building and construction, transportation, and renewable energy, among others. In addition, plastic applications in packaging contribute to reducing food waste.

4.2 GloLitter Project

The GloLitter Project, led by the United Nations Food and Agriculture Organization (FAO) and the International Maritime Organization, seeks to reduce the use of plastics in these industries and identify opportunities to recycle them, and better protect the fragile marine environment well as human lives and livelihoods. The initiative helps the industry implement best practices to prevent and reduce plastic marine litter, including lost or discarded fishing gear, to safeguard coastal and global marine resources.

Also, it examines the availability and status of port facilities and seeks to raise awareness among the shipping and fishing sectors, including seafarers and fishers, and encourage the marking of fishing gear so that it can be traced back to its owner if discarded or lost at sea. It is expected from the GloLitter Project that experts from its organizers FAO and IMO work with the 30 partner countries to provide technical assistance and training while facilitating communication and equipping them with tools such as guidance documents, training materials, and strategies to help

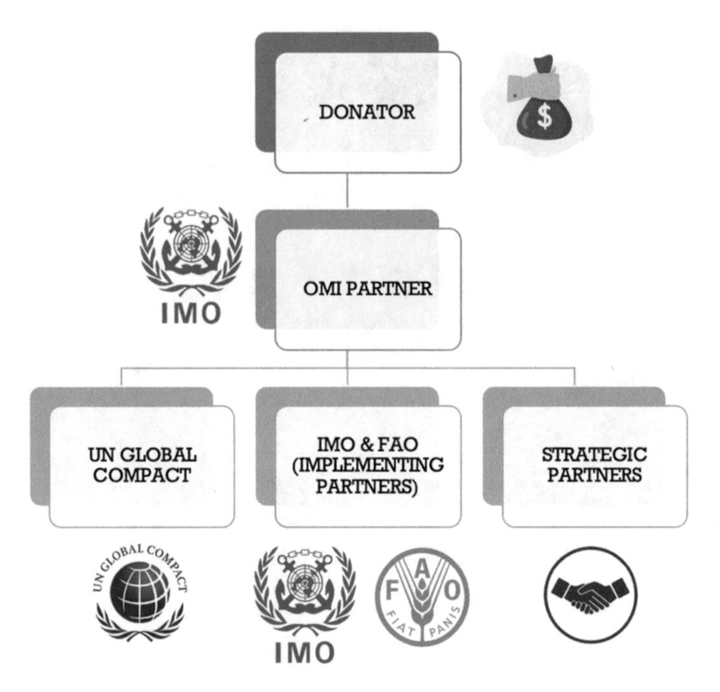

Fig. 6 GloLitter structure. *Source* FAO [113]

enforce existing regulations. Compliance with relevant FAO instruments, including the Voluntary Guidelines for the Marking of Fishing Gear and the Global Ghost Net Initiative, also be promoted (Fig. 6).

Ten countries have been confirmed as lead partners in this global effort and 20 nations as GloLitter project partners. As agreed, the lead partners exercise leadership roles in their respective regions to champion national actions that support the IMO Plan of Action to Address Plastic Marine Litter from Ships and the FAO Voluntary Guidelines for the Marking of Fishing Gear. Both to the project.

4.3 Peru Case: Restricting the Use of Tecnopor (Expanded Polystyrene)

In 2017, it was approved Peru the Law on Integrated Solid Waste Management, with the primary objective of minimizing the generation of solid waste at source and promoting its recovery and valorization through processes such as the recycling of plastics, metals, glass, and others, and the conversion of organic waste into compost, which boost a modern recycling industry, including small recyclers in this value chain. From the entry into force of this law would prohibit:

1. The acquisition, use, entry, or commercialization, as appropriate, of polymer-based bags; polymer-based straws such as drinking straws and containers or containers of expanded polystyrene for beverages and food for human consumption, in Natural Protected Areas, as well as in the entities of the state administration.
2. The use of polymeric-based bags or wrappers in printed advertising, newspapers, magazines, or other written press formats; receipts for public or private services; and any information directed to consumers, users, or citizens in general.

The general director of Environmental Quality of MINAM mentioned that the expectations concerning the New Law were met. Approximately 30% of single-use plastic was reduced, as reported by the manufacturers of the same; however, the measures proposed by the Peruvian government in recent years to reduce plastic were slowed down by the arrival of COVID-19, thus remaining inefficient measures since its entry into force [151]. According to the Ministry of Environment, there was a 50% setback. It is essential to bear in mind that although the reduction of this type of material is beneficial to the environment, it continues to have inevitable repercussions on business and forces people to rethink their standard practices; this is why despite being a necessary measure to apply, there are various conflicts of interest.

5 Green Small Businesses and Their Initiatives on Sustainability

Entrepreneurship is a discipline that must be followed accordingly to see good results, but most importantly, it is a new way of creating jobs and opportunities that eventually lead to more significant economic growth [105].

In recent years, entrepreneurship has grown more as time has progressed, and there are new findings regarding this topic. According to Sergi et al. [187], entrepreneurship is one of the most dynamic phenomenons in the economic sector, but it is mainly driven by globalization and crises that the economic system could suffer. Also, modern economic theory states that entrepreneurship is significant for economic growth and impacts this issue [187]. Entrepreneurship is driven by economic factors since, for it to happen, there have to be some resources, coming from the government, for example, to support the newly independent business owners and make sure that they are contributing to the economic growth of the country and bringing benefits in the long term. Some authors state that there is more than one type of entrepreneurship. Aulet and Murray [59] say that entrepreneurship is divided into two categories: one is innovation-driven entrepreneurship and small business entrepreneurship.

New business trends and ways of working have emerged due to growing awareness about climate change and environmental issues. One of them is green entrepreneurship. According to Allen and Malin [27], green entrepreneurship or eco entrepreneurship is a new kind of entrepreneur mixing a great sense of doing business with the consciousness of sustainability in the light of the recent environmental awareness movements. Anderson [51] says that both entrepreneurship and environmentalism are based on a perception of value. The attitudes which inform environmental concern create areas of value that can be exploited entrepreneurially. "Environmental Entrepreneurs" not only recognize opportunity but construct entire organizations to capture and fix change in society.

Green entrepreneurship is still a relatively new concept that is gaining more visibility in the face of corporate awareness regarding the environmental impact that business activities can have [98]. According to Hussain et al. [127], green entrepreneurship brings many advantages such as reducing deforestation, improving and maintaining ecological systems, and many more that have caught the attention of governments to make policies that include some of these points to ensure economic and social development.

Nowadays, businesses implement a high level of innovation to creatively create outstanding and unique products while looking for the environment. From paper to plastic, sustainable entrepreneurship can recycle waste, especially plastics, and transform them into products customers are willing to purchase. The growth of these environmentally friendly tendencies can also be seen in start-ups and small companies worldwide. "*Bolsos ecológicos del Perú*" is a small Peruvian business located in Lima—Peru, created in 2011 by Miguel Medina. This small company recycles plastic bottles to create products such as ecological merchandising such as bags and accessories distributed to other companies that share the same concerns about the environmental situation.

In 2015, the United Nations adopted the Sustainable Development Goals, which are a set of objectives to generate a universal call to protect the planet, end poverty, and ensure that by the year 2030, all people enjoy prosperity and peace [199]. Since SDGs were introduced six years ago, many companies such as "*Bolsos ecológicos del Perú*" have started to develop new business models and integrate the SDGs as an opportunity to redesign their corporate organization and generate new and innovative solutions based on what the SDGs say [78]. This company, for example, is mainly focused on contributing to the achievement of two specific SDGs: goal 12: responsible consumption and production and goal 13: climate action [107].

According to their website Bolsos Ecológicos del Peru [73], their business model consists of recollecting, reprocessing the plastic bottles until the creation of 100% polyester fiber. Then, they use the fabrics to create products such as bags, eco-pots, pencils, and more. Later, these products would be offered the goods with personalized models and sold to corporate clients who would distribute them as merchandise to the final customer (Fig. 7).

In the same way as the previous company, many others worldwide use similar processes to create new sustainable products based on 100% recycled fabrics, such as "Batoko" from the United Kingdom and "Montsenu" from South Korea. The case of "Batoko", is a small and independent swimwear company created in 2017, inspired by the enormous amounts of litter found in the beaches on the northwest of England. This brand produces swimsuits made 100% out of recycled plastics

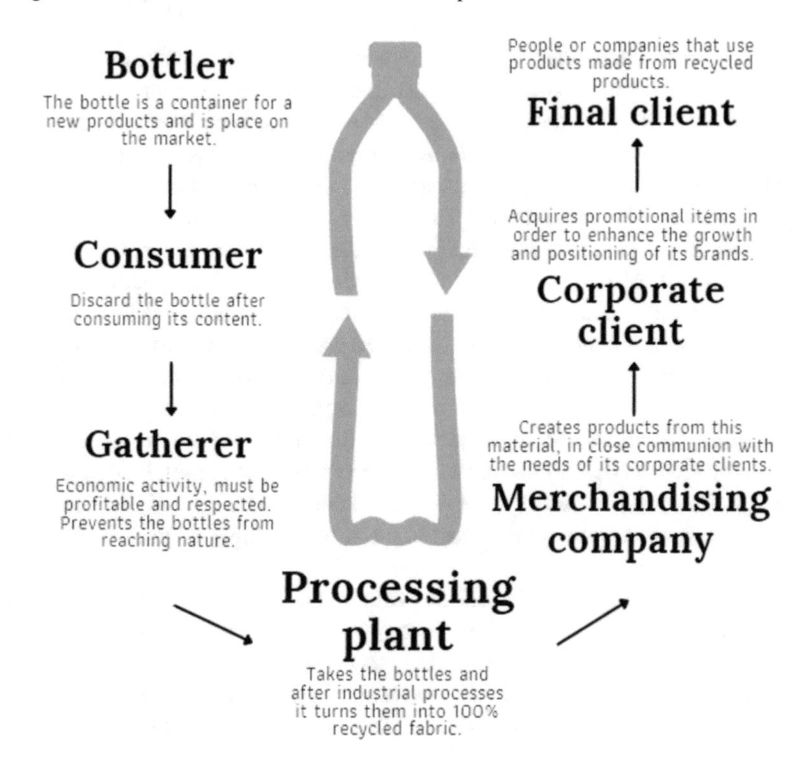

Fig. 7 "Bolsos ecológicos del Perú's" manufacturing process. *Source* Bolsos Ecológicos del Peru [73]

and fewer designs than a standard company in the same industry to keep it sustainable, avoiding overproduction. Moreover, this entrepreneurship assures its positive impact on the environment by donating annually to projects to take care of the marine ecosystem and even with their supply chain composed of two suppliers, both located in China, the biggest recycler in the world and holding several certificates corresponding to an adequate creation of fabrics made of recycled materials [67]. Finally, their meticulous measurements avoid waste or waste of unnecessary energy, thus following a complete production with a sustainable and eco-friendly orientation.

Another example is "Montsenu" from South Korea, whose mission is to spread a responsible and sustainable consumption of clothes by using recycled and natural materials to create innovative designs of clothes using technology in favor of a better sense of fashion that includes the message of the brand of human life, environment, and social issues. The core of this company is sustainability and Looking good. This green entrepreneurship aims to become 100% carbon neutral by 2025 [153]. This case shows how small companies make sustainability the foundation of their business and keep planning to get better at it in the future.

Furthermore, "MicGalaw" is a South-African small business created by Phumudzo Muthanyi and Mbali Mokgosi, that stands out as a street-style accessories shop that differentiates from the rest because of their exclusive handmade products, locally sourced, sustainable, and based 80% on recycled materials [149]. For "MicGalaw", sustainability is not necessarily its core but a competitive advantage for their business. The products such as handbags and jewelry have drawn the attention of many people of the region, not just because of their designs but because their production happens to be beneficial for the environment and the young community by creating new jobs and a collective environmental awareness [149].

The two founders of the brand stated that their process evolved from experimenting for at least 3 years to attend to the dumping sites of the city where they searched for different kinds of plastic that were perfect for reshaping and easily combined with other fabrics. Nowadays, the actual process is similar to the previous one. First, the plastic is collected by dumping sites or donated by other companies, then said plastics are moved to the manufacturing area where the patterns are created, the plastics are melted and turned into the fabric until they can be combined with the other pieces of new fabrics [173].

Finally, the last example of small sustainable businesses is "Precious plastics", a project created in 2012 by Brad Scott in Queensland. The former corporate boss creates endless products, from housewares to guitars, using the plastic bottle caps considered waste by other companies. In a small company, the resources are provided by other companies to dispose of the rubbish; however, some of them ask for other products in return for the materials [206]. All five small companies mentioned so far have one special characteristic in common, they all contribute to reducing plastic waste in their regions and the impulse of sustainability in commerce. Those exemplify that companies can be successful even by adopting eco-friendly practices and sustainable solutions through the years.

The term of sustainability has become more prominent in the last few years. This concept comes from forestry, where it means never harvesting more than what the forest can offer you, and also it is used in the economy, where many experts use this concept to study the distribution of scarce resources [136]. There is a debate about the definition of this term since many authors have their perspectives [155]. Sustainability's inspiration comes from the Brundtland report that the UN did in 1987 [136]. It is linked with the concept of sustainable development, and this concept is used to describe how a society or country can use its resources in a non-damaging way so that future generations can have an optimal life condition and not suffer from this lack of essential resources and have their basic needs fulfilled [148]. The concept of sustainability has been embraced by the five companies previously presented as a way to counteract the damages caused by the unmeasured production of goods, especially those containing plastics. In different ways, either recycling bottles to turn them into fabrics or melting plastic bottle caps to turn them into tables, all these businesses are looking to reduce what was once considered waste and give them other uses.

The policy-makers need to develop different kind of strategies for sustainability in each city and country which include the citizens [9, 12, 14, 16, 22, 36, 45–47,

55, 63, 85, 102, 118, 131, 144, 157, 180, 194, 203, 209, 213], organizations [5, 23, 25, 29, 41, 68, 71, 75, 76, 137, 141, 168, 183, 189, 196, 211, 212] and companies [6, 15, 19, 35, 61, 66, 92, 99, 143, 159, 169, 172, 192, 202] and activities as sports [89, 122, 129].

Also, it is included in sustainabilty the activities of tourism [10, 13, 20, 79–81, 114, 158, 167, 175], education [4, 16, 23, 34, 38, 42, 44, 84, 126, 138, 174], circular economy approach [2, 11, 26, 43, 65, 132, 140, 142, 201], prices [53, 54, 57, 60, 123, 124, 135, 138, 152, 186], hospitality [3, 21, 111, 119, 128, 145, 165, 195, 208].

Finally, due to innovation process some efforts are valued as intellectual property [7, 28, 33, 52, 62, 64, 110, 150], health [1, 8, 17, 18, 24, 30–32, 37, 39, 48–50, 77, 86, 93, 101, 109, 146, 156, 171, 181, 210, 214–216], and research [87, 134, 160, 163, 164, 166, 170, 176, 182].

Sustainability has three main pillars that explain what it is trying to achieve globally and its main objective more detailedly. The three pillars are Environmental Sustainability, Economic Sustainability, and Social Sustainability. The first pillar, environmental sustainability, refers to future generations being able to cover their needs by taking care of the environment and ecological spaces so that they can be enjoyed in the future without suffering damages from pollution or other dangers [155]. In the case of the company "Batoko", the company is focused on its revenues and preserving the environment by recycling and reducing the amounts of garbage, especially plastics found in the ocean.

On the other hand, the second pillar, economic sustainability, refers to a production system that makes goods for consumption that does not put resource availability and takes care of scarce resources that might be hard to make up for once they are used ultimately [148]. An excellent example of this pillar is "Precious plastics", this small business uses other companies' trash to create their products; repurposing something that was destined to be thrown away is now turned into innovative products.

Finally, the third pillar, social sustainability, is a term that often refers to how communities, societies, and individuals live together and make objectives for them to achieve, but taking into account the boundaries of planet earth and the environment as a whole as well as the physical boundaries of their places [88]. "MicGalaw" is a perfect example of how a business can adopt this pillar to enhance the community. This small business has been a big part of how people from the region can acquire positive perspectives on taking care of the environment, especially while teaching young people how to recycle and that the products originating from these practices can be suitable for their necessities eco-friendly (Fig. 8).

Recent suggested lectures about green approach

- Waste reduction and carbon footprint [121]
- Material selection for circularity and footprints [154]
- 3D Print, circularity, and footprints [95]
- Virtual tourism [95]
- Leadership for sustainability in crisis time [40]
- Virtual education and circularity [90]
- Circular Economy for packaging [83]

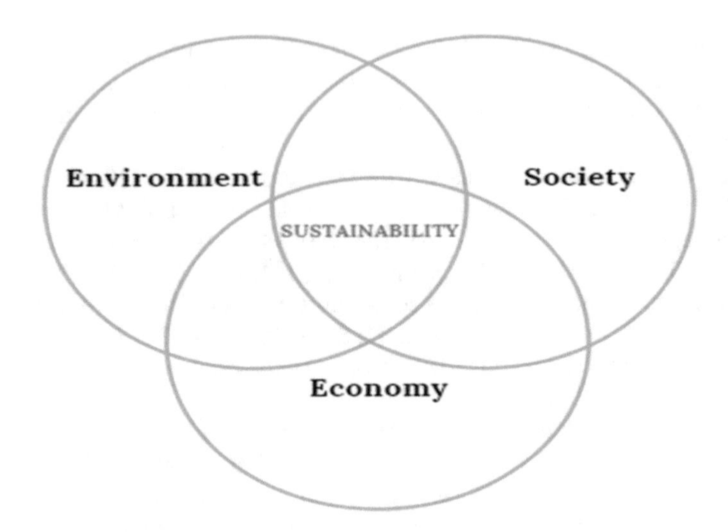

Fig. 8 Three pillars of sustainability. *Source* Soapboxie [191]

- Students oriented to circular learning [96]
- Food and circular economy [94]
- Water footprint and food supply chain management [130]
- Waste footprint [56]
- Measuring circular economy [100]
- Carbon footprint [112].

Closing Remarks

Something essential to consider is that even though there is sizeable plastic contamination due to the pandemic situation, pollution because of plastic is an issue that has been around for many years. This problem spreads since plastics tend to wear down to microplastics or translate into products of single-use plastics that are not biodegradable and end up polluting diverse ecosystems such as the marine and terrestrial environments or finishing their days in a landfill. The main concern surrounding plastic waste is correctly disposing of this product to avoid contamination and prevent environmental damage.

It can be concluded that before the COVID-19 pandemic started, there was already a significant concern surrounding the pollution generated by distinct types of plastic around the world. There is still a present problem that must be addressed to prevent the effects that it could have on the well-being of society since there is already evidence that microscopic plastics are in some foods, and these materials are causing irreversible damage to the distinct ecosystems of the planet. On the bright side, some businesses have been using compostable materials to reduce their pollution in the environment, and if more companies decided to use these tactics, there could be a positive impact on the problem.

The COVID-19 pandemic has brought with it another problem besides the high contagious rate that is the increase of plastic pollution because of PPE use. Globally, a large amount of plastic waste generates damage to the ecosystem. If these problems are not solved quickly, the consequences of this eventuality are likely to be felt in the long term and cause problems to humans.

Some solutions have been planned to solve the plastic pollution problem worsened by the COVID-19 pandemic. One of them is a circular economy that focuses on reducing the impacts on the environment. If these solutions are implemented correctly to combat the ongoing problem, it could help better the situation that is currently happening around the world and make the possible damaging effects in the long term reduce their negative consequences on the environment. It is also essential to keep the concept of sustainability and its three pillars in mind when applying or designing a possible solution to fight this problem.

References

1. Aaron L (2020) The oft forgotten part of nutrition: lessons from an integrated approach to understand consumer food safety behaviors. Curr Dev Nutr 4(Supplement_2):1322–1322. http://doi.org/10.1093/cdn/nzaa059_039
2. Abad-Segura E, Batlles-delaFuente A, González-Zamar M-D, Belmonte-Ureña LJ (2021) Implications for sustainability of the joint application of bioeconomy and circular economy: a worldwide trend study. Sustainability 13(13):7182. https://doi.org/10.3390/su13137182
3. Abad-Segura E, Cortés-García FJ, Belmonte-Ureña LJ (2019) The sustainable approach to corporate social responsibility: a global analysis and future trends. Sustainability 11(19):5382. https://doi.org/10.3390/su11195382
4. Abad-Segura E, González-Zamar M-D (2021) Sustainable economic development in higher education institutions: a global analysis within the SDGs framework. J Clean Prod 294:126133. http://doi.org/10.1016/j.jclepro.2021.126133
5. Abad-Segura E, González-Zamar M-D, Infante-Moro JC, Ruipérez García G (2020) Sustainable management of digital transformation in higher education: global research trends. Sustainability 12(5):2107. https://doi.org/10.3390/su12052107
6. Abad-Segura E, Morales ME, Cortés-García FJ, Belmonte-Ureña LJ (2020) Industrial processes management for a sustainable society: global research analysis. Processes 8(5). http://doi.org/10.3390/pr8050631
7. Abdaljaleel M, Singer EJ, Yong WH (2019) Sustainability in biobanking. In: Yong WH (ed) Biobanking: methods and protocols. Springer, New York, pp 1–6. http://doi.org/10.1007/978-1-4939-8935-5_1
8. Aberilla JM, Gallego-Schmid A, Stamford L, Azapagic A (2020) Environmental sustainability of cooking fuels in remote communities: life cycle and local impacts. Sci Total Environ 713:136445. https://doi.org/10.1016/j.scitotenv.2019.136445
9. Acevedo-Duque Á, Gonzalez-Diaz R, Vargas EC, Paz-Marcano A, Muller-Pérez S, Salazar-Sepúlveda G et al (2021) Resilience, leadership and female entrepreneurship within the context of SMEs: evidence from Latin America. Sustainability 13(15):8129. http://doi.org/10.3390/su13158129
10. Acevedo-Duque Á, Vega-Muñoz A, Salazar-Sepúlveda G (2020) Analysis of hospitality, leisure, and tourism studies in Chile. Sustainability 12(18):7238. https://doi.org/10.3390/su12187238

11. Ada E, Sagnak M, Mangla SK, Kazancoglu Y (2021) A circular business cluster model for sustainable operations management. Int J Logistics Res Appl 1–19. http://doi.org/10.1080/13675567.2021.2008335
12. Adam A (2015) Challenges of public finance sustainability in the European Union. Procedia Econ Financ 23:298–302. https://doi.org/10.1016/S2212-5671(15)00507-9
13. Adam I, Agyeiwaah E, Dayour F (2021) Understanding the social identity, motivations, and sustainable behaviour among backpackers: a clustering approach. J Travel Tour Mark 38(2):139–154. https://doi.org/10.1080/10548408.2021.1887053
14. Adam JN, Adams T, Gerber J-D, Haller T (2021) Decentralization for increased sustainability in natural resource management? Two cautionary cases from Ghana. Sustainability 13(12). http://doi.org/10.3390/su13126885
15. Adam M (2018) The role of human resource management (HRM) for the implementation of sustainable product-service systems (PSS)—an analysis of fashion retailers. Sustainability 10(7). http://doi.org/10.3390/su10072518
16. Adams R, Martin S, Boom K (2018) University culture and sustainability: designing and implementing an enabling framework. J Clean Prod 171:434–445. https://doi.org/10.1016/j.jclepro.2017.10.032
17. Adesogan AT, Havelaar AH, McKune SL, Eilittä M, Dahl GE (2020) Animal source foods: sustainability problem or malnutrition and sustainability solution? Perspective matters. Glob Food Secur 25:100325. https://doi.org/10.1016/j.gfs.2019.100325
18. Adomah-Afari A, Chandler JA (2018) The role of government and community in the scaling up and sustainability of mutual health organisations: an exploratory study in Ghana. Soc Sci Med 207:25–37. https://doi.org/10.1016/j.socscimed.2018.04.044
19. Adomako S, Amankwah-Amoah J, Danso A, Konadu R, Owusu-Agyei S (2019) Environmental sustainability orientation and performance of family and nonfamily firms. Bus Strateg Environ 28(6):1250–1259. https://doi.org/10.1002/bse.2314
20. Adongo CA, Taale F, Adam I (2018) Tourists' values and empathic attitude toward sustainable development in tourism. Ecol Econ 150:251–263. https://doi.org/10.1016/j.ecolecon.2018.04.013
21. Agyeiwaah E (2019) Exploring the relevance of sustainability to micro tourism and hospitality accommodation enterprises (MTHAEs): evidence from home-stay owners. J Clean Prod 226:159–171. https://doi.org/10.1016/j.jclepro.2019.04.089
22. Al-Naqbi AK, Alshannag Q (2018) The status of education for sustainable development and sustainability knowledge, attitudes, and behaviors of UAE university students. Int J Sustain High Educ 19(3):566–588. https://doi.org/10.1108/IJSHE-06-2017-0091
23. Aleixo AM, Leal S, Azeiteiro UM (2018) Conceptualization of sustainable higher education institutions, roles, barriers, and challenges for sustainability: an exploratory study in Portugal. J Clean Prod 172:1664–1673. https://doi.org/10.1016/j.jclepro.2016.11.010
24. Ali M, de Azevedo ARG, Marvila MT, Khan MI, Memon AM, Masood F et al (2021) The influence of COVID-19-induced daily activities on health parameters—a case study in Malaysia. Sustainability 13(13):7465. http://doi.org/10.3390/su13137465
25. Aliabadi V, Ataei P, Gholamrezai S, Aazami M (2019) Components of sustainability of entrepreneurial ecosystems in knowledge-intensive enterprises: the application of fuzzy analytic hierarchy process. Small Enterp Res 26(3):288–306. https://doi.org/10.1080/13215906.2019.1671215
26. Alkhuzaim L, Zhu Q, Sarkis J (2021) Evaluating emergy analysis at the nexus of circular economy and sustainable supply chain management. Sustain Prod Consumption 25:413–424. https://doi.org/10.1016/j.spc.2020.11.022
27. Allen JC, Malin S (2008) Green entrepreneurship: a method for managing natural resources? Soc Nat Resour 21(9):828–844. https://doi.org/10.1080/08941920701612917
28. Alonso-Fradejas A (2021) The resource property question in climate stewardship and sustainability transitions. Land Use Policy 108:105529. https://doi.org/10.1016/j.landusepol.2021.105529

29. Alshuwaikhat HM, Adenle YA, Saghir B (2016) Sustainability assessment of higher education institutions in Saudi Arabia. Sustainability 8(8). http://doi.org/10.3390/su8080750
30. Álvarez-Risco A, Arellano EZ, Valerio EM, Acosta NM, Tarazona ZS (2013) Pharmaceutical care campaign as a strategy for implementation of pharmaceutical services: experience Peru. Pharm Care Esp 15(1):35–37
31. Alvarez-Risco A, Dawson J, Johnson W, Conteh-Barrat M, Aslani P, Del-Aguila-Arcentales S, Diaz-Risco S (2020) Ebola virus disease outbreak: a global approach for health systems. Revista Cubana de Farmacia 53(4):1–13, Article e491
32. Alvarez-Risco A, Del-Aguila-Arcentales S (2015) Prescription errors as a barrier to pharmaceutical care in public health facilities: experience Peru. Pharm Care Esp 17(6):725–731
33. Alvarez-Risco A, Del-Aguila-Arcentales S (2021a) A note on changing regulation in international business: the world intellectual property organization (wipo) and artificial intelligence. In: Progress in international business research, vol 15, pp 363–371. http://doi.org/10.1108/S1745-886220210000015020
34. Alvarez-Risco A, Del-Aguila-Arcentales S (2021b) Public policies and private efforts to increase women entrepreneurship based on STEM background. In: Contributions to management science, pp 75–87. http://doi.org/10.1007/978-3-030-83792-1_5
35. Alvarez-Risco A, Del-Aguila-Arcentales S (2022) Sustainable initiatives in international markets. In: Contributions to management science, pp 181–191. http://doi.org/10.1007/978-3-030-85950-3_10
36. Alvarez-Risco A, Del-Aguila-Arcentales S, Delgado-Zegarra J, Yáñez JA, Diaz-Risco S (2019) Doping in sports: findings of the analytical test and its interpretation by the public. Sport Sci Health 15(1):255–257. https://doi.org/10.1007/s11332-018-0484-8
37. Alvarez-Risco A, Del-Aguila-Arcentales S, Diaz-Risco S (2018) Pharmacovigilance as a tool for sustainable development of healthcare in Peru. PharmacoVigilance Rev 10(2):4–6
38. Alvarez-Risco A, Del-Aguila-Arcentales S, Rosen MA, García-Ibarra V, Maycotte-Felkel S, Martínez-Toro GM (2021) Expectations and interests of university students in covid-19 times about sustainable development goals: evidence from Colombia, Ecuador, Mexico, and Peru. Sustainability (Switzerland) 13(6), Article 3306. http://doi.org/10.3390/su13063306
39. Alvarez-Risco A, Del-Aguila-Arcentales S, Stevenson JG (2015) Pharmacists and mass communication for implementing pharmaceutical care. Am J Pharm Benefits 7(3):e125–e126
40. Alvarez-Risco A, Del-Aguila-Arcentales S, Villalobos-Alvarez D, Diaz-Risco S (2022) Leadership for sustainability in crisis time. In: Alvarez-Risco A, Muthu SS, Del-Aguila-Arcentales S (eds) Circular economy: impact on carbon and water footprint. Springer, Singapore, pp 41–64. http://doi.org/10.1007/978-981-19-0549-0_3
41. Alvarez-Risco A, Del-Aguila-Arcentales S, Yáñez JA, Alvarez-Risco A (2021) Telemedicine in Peru as a result of the COVID-19 pandemic: perspective from a country with limited internet access. Am J Trop Med Hyg 105(1):6–11. https://doi.org/10.4269/ajtmh.21-0255
42. Alvarez-Risco A, Del-Aguila-Arcentales S, Yáñez JA, Rosen MA, Mejia CR (2021) Influence of technostress on academic performance of university medicine students in Peru during the covid-19 pandemic. Sustainability (Switzerland) 13(16), Article 8949. http://doi.org/10.3390/su13168949
43. Alvarez-Risco A, Delgado-Zegarra J, Yáñez JA, Diaz-Risco S, Del-Aguila-Arcentales S (2018) Predation risk by gastronomic boom—case Peru. J Landscape Ecol (Czech Repub) 11(1):100–103. http://doi.org/10.2478/jlecol-2018-0003
44. Alvarez-Risco A, Estrada-Merino A, Anderson-Seminario MM, Mlodzianowska S, García-Ibarra V, Villagomez-Buele C, Carvache-Franco M (2021) Multitasking behavior in online classrooms and academic performance: case of university students in Ecuador during COVID-19 outbreak. Interact Technol Smart Educ 18(3):422–434. https://doi.org/10.1108/ITSE-08-2020-0160
45. Alvarez-Risco A, Mejia CR, Delgado-Zegarra J, Del-Aguila-Arcentales S, Arce-Esquivel AA, Valladares-Garrido MJ et al (2020) The Peru approach against the COVID-19 infodemic: insights and strategies. Am J Trop Med Hyg 103(2):583–586. http://doi.org/10.4269/ajtmh.20-0536

46. Alvarez-Risco A, Mlodzianowska S, García-Ibarra V, Rosen MA, Del-Aguila-Arcentales S (2021) Factors affecting green entrepreneurship intentions in business university students in covid-19 pandemic times: case of ecuador. Sustainability (Switzerland) 13(11), Article 6447. http://doi.org/10.3390/su13116447
47. Alvarez-Risco A, Mlodzianowska S, Zamora-Ramos U, Del-Aguila-Arcentales S (2021) Green entrepreneurship intention in university students: the case of Peru. Entrepreneurial Bus Econ Rev 9(4):85–100. http://doi.org/10.15678/EBER.2021.090406
48. Alvarez-Risco A, Quiroz-Delgado D, Del-Aguila-Arcentales S (2016) Pharmaceutical care in hypertension patients in a Peruvian hospital [Article]. Indian J Public Health Res Dev 7(3):183–188. https://doi.org/10.5958/0976-5506.2016.00153.4
49. Alvarez-Risco A, Turpo-Cama A, Ortiz-Palomino L, Gongora-Amaut N, Del-Aguila-Arcentales S (2016) Barriers to the implementation of pharmaceutical care in pharmacies in Cusco, Peru. Pharm Care Esp 18(5):194–205
50. Alvarez-Risco A, Van Mil JWF (2007) Pharmaceutical care in community pharmacies: practice and research in Peru. Ann Pharmacother 41(12):2032–2037. https://doi.org/10.1345/aph.1K117
51. Anderson AR (1998) Cultivating the Garden of Eden: environmental entrepreneuring. J Organ Chang Manag 11(2):135–144. https://doi.org/10.1108/09534819810212124
52. Ang KL, Saw ET, He W, Dong X, Ramakrishna S (2021) Sustainability framework for pharmaceutical manufacturing (PM): a review of research landscape and implementation barriers for circular economy transition. J Clean Prod 280:124264. https://doi.org/10.1016/j.jclepro.2020.124264
53. Angulo-Mosquera LS, Alvarado-Alvarado AA, Rivas-Arrieta MJ, Cattaneo CR, Rene ER, García-Depraect O (2021) Production of solid biofuels from organic waste in developing countries: a review from sustainability and economic feasibility perspectives. Sci Total Environ 795:148816. http://doi.org/10.1016/j.scitotenv.2021.148816
54. Apcho-Ccencho LV, Cuya-Velásquez BB, Rodríguez DA, Anderson-Seminario MLM, Alvarez-Risco A, Estrada-Merino A, Mlodzianowska S (2021) The impact of international price on the technological industry in the United States and China during times of crisis: commercial war and covid-19. In: Advances in business and management forecasting, vol 14, pp 149–160
55. Aragon-Correa JA, Marcus AA, Rivera JE, Kenworthy AL (2017) Sustainability management teaching resources and the challenge of balancing planet, people, and profits. Acad Manag Learn Educ 16(3):469–483. https://doi.org/10.5465/amle.2017.0180
56. Arias-Meza M, Alvarez-Risco A, Cuya-Velásquez BB, de las Mercedes Anderson-Seminario M, Del-Aguila-Arcentales S (2022) Fashion and textile circularity and waste footprint. In: Alvarez-Risco A, Muthu SS, Del-Aguila-Arcentales S (eds) Circular economy: impact on carbon and water footprint. Springer, Singapore, pp 181–204. http://doi.org/10.1007/978-981-19-0549-0_9
57. Aschemann-Witzel J, Giménez A, Ares G (2018) Convenience or price orientation? Consumer characteristics influencing food waste behaviour in the context of an emerging country and the impact on future sustainability of the global food sector. Glob Environ Chang 49:85–94. https://doi.org/10.1016/j.gloenvcha.2018.02.002
58. ASTM International Standards Worldwide (2010) Identificación de resinas [Identification of resins]. https://www.astm.org/SNEWS/SPANISH/SPND10/d2095_spnd10.html
59. Aulet W, Murray F (2013) A tale of two entrepreneurs: understanding differences in the types of entrepreneurship in the economy. Available at SSRN 2259740
60. Aydın B, Alvarez MD (2020) Understanding the tourists' perspective of sustainability in cultural tourist destinations. Sustainability 12(21):8846. https://doi.org/10.3390/su12218846
61. Bamgbade JA, Kamaruddeen AM, Nawi MNM, Adeleke AQ, Salimon MG, Ajibike WA (2019) Analysis of some factors driving ecological sustainability in construction firms. J Clean Prod 208:1537–1545. https://doi.org/10.1016/j.jclepro.2018.10.229
62. Bannerman S (2020) The world intellectual property organization and the sustainable development agenda. Futures 122:102586. https://doi.org/10.1016/j.futures.2020.102586

63. Barr S, Gilg A, Shaw G (2011) 'Helping people make better choices': exploring the behaviour change agenda for environmental sustainability. Appl Geogr 31(2):712–720. https://doi.org/10.1016/j.apgeog.2010.12.003
64. Barragán-Ocaña A, Silva-Borjas P, Olmos-Peña S, Polanco-Olguín M (2020) Biotechnology and bioprocesses: their contribution to sustainability. Processes 8(4). http://doi.org/10.3390/pr8040436
65. Barros MV, Salvador R, do Prado GF, de Francisco AC, Piekarski CM (2021) Circular economy as a driver to sustainable businesses. Clean Environ Syst 2:100006. https://doi.org/10.1016/j.cesys.2020.100006
66. Batista AA, Francisco AC (2018) Organizational sustainability practices: a study of the firms listed by the corporate sustainability index. Sustainability 10(1). http://doi.org/10.3390/su10010226
67. Batoko (2021) We are rubbish. Literally. Retrieved 12/02/2021 from https://www.batoko.com
68. Ben Youssef A, Boubaker S, Omri A (2018) Entrepreneurship and sustainability: the need for innovative and institutional solutions. Technol Forecast Soc Chang 129:232–241. https://doi.org/10.1016/j.techfore.2017.11.003
69. Benson NU, Bassey DE, Palanisami T (2021) COVID pollution: impact of COVID-19 pandemic on global plastic waste footprint. Heliyon 7(2):e06343. https://doi.org/10.1016/j.heliyon.2021.e06343
70. Binelli A, Della Torre C, Nigro L, Riccardi N, Magni S (2022) A realistic approach for the assessment of plastic contamination and its ecotoxicological consequences: a case study in the metropolitan city of Milan (N. Italy). Sci Total Environ 806:150574. http://doi.org/10.1016/j.scitotenv.2021.150574
71. Bokpin GA (2017) Foreign direct investment and environmental sustainability in Africa: the role of institutions and governance. Res Int Bus Financ 39:239–247. http://doi.org/10.1016/j.ribaf.2016.07.038
72. Bollaín Pastor C, Vicente Agulló D (2020) Presence of microplastics in water and their potential impact on public health [Presencia de microplásticos en aguas y su potencial impacto en la salud pública]. Rev Esp Salud Publica 93:e201908064
73. Bolsos Ecológicos del Peru (2021) Libretas, carpetas y lapices con semillas. Retrieved 12/02/2021 from https://www.bolsoseco.com
74. Bouwmeester H, Hollman PC, Peters RJ (2015) Potential health impact of environmentally released micro-and nanoplastics in the human food production chain: experiences from nanotoxicology. Environ Sci Technol 49(15):8932–8947
75. Brito RM, Rodríguez C, Aparicio JL (2018) Sustainability in teaching: an evaluation of university teachers and students. Sustainability 10(2):439. https://doi.org/10.3390/su10020439
76. Brown HS, de Jong M, Levy DL (2009) Building institutions based on information disclosure: lessons from GRI's sustainability reporting. J Clean Prod 17(6):571–580. https://doi.org/10.1016/j.jclepro.2008.12.009
77. Brown KA, Harris F, Potter C, Knai C (2020) The future of environmental sustainability labelling on food products. Lancet Planet Health 4(4):e137–e138. https://doi.org/10.1016/S2542-5196(20)30074-7
78. Calabrese A, Costa R, Gastaldi M, Levialdi Ghiron N, Villazon Montalvan RA (2021) Implications for sustainable development goals: a framework to assess company disclosure in sustainability reporting. J Clean Prod 319:128624. https://doi.org/10.1016/j.jclepro.2021.128624
79. Carvache-Franco M, Alvarez-Risco A, Carvache-Franco O, Carvache-Franco W, Estrada-Merino A, Villalobos-Alvarez D (2021) Perceived value and its influence on satisfaction and loyalty in a coastal city: a study from Lima, Peru. J Policy Res Tourism Leisure Events. http://doi.org/10.1080/19407963.2021.1883634
80. Carvache-Franco M, Alvarez-Risco A, Carvache-Franco W, Carvache-Franco O, Estrada-Merino A, Rosen MA (2021) Coastal cities seen from loyalty and their tourist motivations: a study in Lima, Peru. Sustainability (Switzerland) 13(21), Article 11575. http://doi.org/10.3390/su132111575

81. Carvache-Franco M, Carvache-Franco O, Carvache-Franco W, Alvarez-Risco A, Estrada-Merino A (2021) Motivations and segmentation of the demand for coastal cities: a study in Lima, Peru. Int J Tourism Res 23(4):517–531. https://doi.org/10.1002/jtr.2423
82. Castañeta G, Gutiérrez AF, Nacaratte F, Manzano CA (2020) Microplastics: a growing pollutant in all environmental spheres, its characteristics and potential public health risks from exposure [Microplásticos: un contaminante que crece en todas las esferas ambientales, sus características y posibles riesgos para la salud pública por exposición]. Rev Boliv Quím 37(3):142–157
83. Castillo-Benancio S, Alvarez-Risco A, Esquerre-Botton S, Leclercq-Machado L, Calle-Nole M, Morales-Ríos F et al (2022) Circular economy for packaging and carbon footprint. In: Alvarez-Risco A, Muthu SS, Del-Aguila-Arcentales S (eds) Circular economy: impact on carbon and water footprint. Springer, Singapore, pp 115–138. http://doi.org/10.1007/978-981-19-0549-0_6
84. Chafloque-Cespedes R, Alvarez-Risco A, Robayo-Acuña PV, Gamarra-Chavez CA, Martinez-Toro GM, Vicente-Ramos W (2021) Effect of sociodemographic factors in entrepreneurial orientation and entrepreneurial intention in university students of Latin American business schools. In: Contemporary issues in entrepreneurship research, vol 11, pp 151–165. http://doi.org/10.1108/S2040-724620210000011010
85. Chafloque-Céspedes R, Vara-Horna A, Asencios-Gonzales Z, López-Odar D, Alvarez-Risco A, Quipuzco-Chicata L et al (2020) Academic presenteeism and violence against women in schools of business and engineering in Peruvian universities. Lecturas de Economia 93:127–153. http://doi.org/10.17533/udea.le.n93a340726
86. Chen X, Zhang SX, Jahanshahi AA, Alvarez-Risco A, Dai H, Li J, Ibarra VG (2020) Belief in a COVID-19 conspiracy theory as a predictor of mental health and well-being of health care workers in Ecuador: cross-sectional survey study. JMIR Public Health Surveill 6(3), Article e20737. http://doi.org/10.2196/20737
87. Chung SA, Olivera S, Román BR, Alanoca E, Moscoso S, Terceros BL et al (2021) Themes of scientific production of the Cuban journal of pharmacy indexed in scopus (1967–2020). Rev Cubana Farmacia 54(1), Article e511
88. Colantonio A (2009) Social sustainability: a review and critique of traditional versus emerging themes and assessment methods. Retrieved 01/01/2022 from http://eprints.lse.ac.uk/3586
89. Constantin P-N, Stanescu R, Stanescu M (2020) Social entrepreneurship and sport in Romania: how can former athletes contribute to sustainable social change? Sustainability 12(11):4688. https://doi.org/10.3390/su12114688
90. Contreras-Taica A, Alvarez-Risco A, Arias-Meza M, Campos-Dávalos N, Calle-Nole M, Almanza-Cruz C et al (2022) Virtual education: carbon footprint and circularity. In: Alvarez-Risco A, Muthu SS, Del-Aguila-Arcentales S (eds) Circular economy: impact on carbon and water footprint. Springer, Singapore, pp 265–285. http://doi.org/10.1007/978-981-19-0549-0_13
91. Cordova MR, Nurhati IS, Riani E, Nurhasanah, Iswari MY (2021) Unprecedented plastic-made personal protective equipment (PPE) debris in river outlets into Jakarta Bay during COVID-19 pandemic. Chemosphere 268:129360. https://doi.org/10.1016/j.chemosphere.2020.129360
92. Cruz-Torres W, Alvarez-Risco A, Del-Aguila-Arcentales S (2021) Impact of enterprise resource planning (ERP) implementation on performance of an education enterprise: a structural equation modeling (SEM) [Article]. Stud Bus Econ 16(2):37–52. https://doi.org/10.2478/sbe-2021-0023
93. Cruz JP, Alshammari F, Felicilda-Reynaldo RFD (2018) Predictors of Saudi nursing students' attitudes towards environment and sustainability in health care. Int Nurs Rev 65(3):408–416. https://doi.org/10.1111/inr.12432
94. Cuya-Velásquez BB, Alvarez-Risco A, Gomez-Prado R, Juarez-Rojas L, Contreras-Taica A, Ortiz-Guerra A et al (2022) Circular economy for food loss reduction and water footprint. In: Alvarez-Risco A, Muthu SS, Del-Aguila-Arcentales S (eds) Circular economy: impact on

carbon and water footprint. Springer, Singapore, pp 65–91. http://doi.org/10.1007/978-981-19-0549-0_4
95. De-la-Cruz-Diaz M, Alvarez-Risco A, Jaramillo-Arévalo M, de las Mercedes Anderson-Seminario M, Del-Aguila-Arcentales S (2022) 3D print, circularity, and footprints. In: Alvarez-Risco A, Muthu SS, Del-Aguila-Arcentales S (eds) Circular economy: impact on carbon and water footprint. Springer, Singapore, pp 93–112. http://doi.org/10.1007/978-981-19-0549-0_5
96. de las Mercedes Anderson-Seminario M, Alvarez-Risco A (2022) Better students, better companies, better life: circular learning. In: Alvarez-Risco A, Muthu SS, Del-Aguila-Arcentales S (eds) Circular economy: impact on carbon and water footprint. Springer, Singapore, pp 19–40. http://doi.org/10.1007/978-981-19-0549-0_2
97. de Sousa FDB (2021) Plastic and its consequences during the COVID-19 pandemic. Environ Sci Pollut Res 28(33):46067–46078. https://doi.org/10.1007/s11356-021-15425-w
98. Dean TJ, McMullen JS (2007) Toward a theory of sustainable entrepreneurship: reducing environmental degradation through entrepreneurial action. J Bus Ventur 22(1):50–76. https://doi.org/10.1016/j.jbusvent.2005.09.003
99. Del-Aguila-Arcentales S, Alvarez-Risco A (2013) Human error or burnout as explanation for mistakes in pharmaceutical laboratories. Accred Qual Assur 18(5):447–448. https://doi.org/10.1007/s00769-013-1000-0
100. Del-Aguila-Arcentales S, Alvarez-Risco A, Muthu SS (2022) Measuring circular economy. In: Alvarez-Risco A, Muthu SS, Del-Aguila-Arcentales S (eds) Circular economy: impact on carbon and water footprint. Springer, Singapore, pp 3–17. http://doi.org/10.1007/978-981-19-0549-0_1
101. Delgado-Zegarra J, Alvarez-Risco A, Yáñez JA (2018) Indiscriminate use of pesticides and lack of sanitary control in the domestic market in Peru. Rev Panam Salud Publica/Pan Am J Public Health 42, Article e3. http://doi.org/10.26633/RPSP.2018.3
102. Deslatte A, Swann WL (2019) Elucidating the linkages between entrepreneurial orientation and local government sustainability performance. Am Rev Public Adm 50(1):92–109. https://doi.org/10.1177/0275074019869376
103. Desrochers P (2002) Regional development and inter-industry recycling linkages: some historical perspectives. Entrep Reg Dev 14(1):49–65. https://doi.org/10.1080/08985620110096627
104. Desrochers P (2004) Industrial symbiosis: the case for market coordination. J Clean Prod 12(8):1099–1110. https://doi.org/10.1016/j.jclepro.2004.02.008
105. Diandra D, Azmy A (2020) Understanding definition of entrepreneurship. Int J Manag Acc Econ 7(5):235–241
106. Dris R, Gasperi J, Saad M, Mirande C, Tassin B (2016) Synthetic fibers in atmospheric fallout: a source of microplastics in the environment? Mar Pollut Bull 104(1):290–293. https://doi.org/10.1016/j.marpolbul.2016.01.006
107. Economía Verde (2021) Ecological bags from Peru [Bolsos ecológicos del Peru]. Retrieved 12/02/2021 from https://economiaverde.pe/pymes/bolsos-ecologicos
108. Ellen Macarthur Foundation (2021) Circular economy diagram. Retrieved 12/03/2021 from https://ellenmacarthurfoundation.org/circular-economy-diagram
109. Enciso-Zarate A, Guzmán-Oviedo J, Sánchez-Cardona F, Martínez-Rohenes D, Rodríguez-Palomino JC, Alvarez-Risco A et al (2016) Evaluation of contamination by cytotoxic agents in colombian hospitals. Pharm Care Esp 18(6):241–250
110. Eppinger E, Jain A, Vimalnath P, Gurtoo A, Tietze F, Hernandez Chea R (2021) Sustainability transitions in manufacturing: the role of intellectual property. Curr Opin Environ Sustain 49:118–126. https://doi.org/10.1016/j.cosust.2021.03.018
111. Ertuna B, Karatas-Ozkan M, Yamak S (2019) Diffusion of sustainability and CSR discourse in hospitality industry. Int J Contemp Hosp Manag 31(6):2564–2581. https://doi.org/10.1108/IJCHM-06-2018-0464
112. Esquerre-Botton S, Alvarez-Risco A, Leclercq-Machado L, de las Mercedes Anderson-Seminario M, Del-Aguila-Arcentales S (2022) Food loss reduction and carbon footprint

practices worldwide: a benchmarking approach of circular economy. In: Alvarez-Risco A, Muthu SS, Del-Aguila-Arcentales S (eds) Circular economy: impact on carbon and water footprint. Springer, Singapore, pp 161–179. http://doi.org/10.1007/978-981-19-0549-0_8
113. FAO (2020) Programa de Asociaciones GloLitter. Retrieved 12/03/2021 from https://www.fao.org/responsible-fishing/marking-of-fishing-gear/glolitter-partnerships-programme/es
114. Figueroa-Domecq C, Kimbu A, de Jong A, Williams AM (2020) Sustainability through the tourism entrepreneurship journey: a gender perspective. J Sustain Tourism 1–24. http://doi.org/10.1080/09669582.2020.1831001
115. Flores P (2020) La problemática del consumo de plásticos durante la pandemia de la COVID-19. South Sustain 1(2):e016–e016
116. Frias JPGL, Nash R (2019) Microplastics: finding a consensus on the definition. Mar Pollut Bull 138:145–147. https://doi.org/10.1016/j.marpolbul.2018.11.022
117. Frosch RA, Gallopoulos NE (1989) Strategies for manufacturing. Sci Am 261(3):144–153. http://www.jstor.org/stable/24987406
118. Galleli B, Teles NEB, dos Santos JAR, Freitas-Martins MS, Hourneaux Junior F (2021) Sustainability university rankings: a comparative analysis of UI green metric and the times higher education world university rankings. Int J Sustain High Educ (ahead-of-print). http://doi.org/10.1108/IJSHE-12-2020-0475
119. Gerdt S-O, Wagner E, Schewe G (2019) The relationship between sustainability and customer satisfaction in hospitality: an explorative investigation using eWOM as a data source. Tour Manage 74:155–172. https://doi.org/10.1016/j.tourman.2019.02.010
120. Geyer R, Jambeck Jenna R, Law Kara L (2017) Production, use, and fate of all plastics ever made. Sci Adv 3(7):e1700782. https://doi.org/10.1126/sciadv.1700782
121. Gómez-Prado R, Alvarez-Risco A, Sánchez-Palomino J, de las Mercedes Anderson-Seminario M, Del-Aguila-Arcentales S (2022) Circular economy for waste reduction and carbon footprint. In: Alvarez-Risco A, Muthu SS, Del-Aguila-Arcentales S (eds) Circular economy: impact on carbon and water footprint. Springer, Singapore, pp 139–159. http://doi.org/10.1007/978-981-19-0549-0_7
122. González-Serrano MH, Añó Sanz V, González-García RJ (2020) Sustainable sport entrepreneurship and innovation: a bibliometric analysis of this emerging field of research. Sustainability 12(12):5209. https://doi.org/10.3390/su12125209
123. Grewal J, Hauptmann C, Serafeim G (2021) Material sustainability information and stock price informativeness. J Bus Ethics 171(3):513–544. https://doi.org/10.1007/s10551-020-04451-2
124. Hall MR (2019) The sustainability price: expanding environmental life cycle costing to include the costs of poverty and climate change. Int J Life Cycle Assess 24(2):223–236. https://doi.org/10.1007/s11367-018-1520-2
125. Haque MS, Sharif S, Masnoon A, Rashid E (2021) SARS-CoV-2 pandemic-induced PPE and single-use plastic waste generation scenario. Waste Manag Res 39(1_suppl):3–17. http://doi.org/10.1177/0734242X20980828
126. Hermann RR, Bossle MB (2020) Bringing an entrepreneurial focus to sustainability education: a teaching framework based on content analysis. J Clean Prod 246:119038. http://doi.org/10.1016/j.jclepro.2019.119038
127. Hussain I, Nazir M, Hashmi SB, Di Vaio A, Shaheen I, Waseem MA, Arshad A (2021) Green and sustainable entrepreneurial intentions: a mediation-moderation perspective. Sustainability 13(15). http://doi.org/10.3390/su13158627
128. Jones P, Comfort D (2020) The COVID-19 crisis and sustainability in the hospitality industry. Int J Contemp Hosp Manag 32(10):3037–3050. https://doi.org/10.1108/IJCHM-04-2020-0357
129. Jones P, Ratten V, Hayduk T (2020) Sport, fitness, and lifestyle entrepreneurship. Int Entrepreneurship Manag J 16(3):783–793. https://doi.org/10.1007/s11365-020-00666-x
130. Juarez-Rojas L, Alvarez-Risco A, Campos-Dávalos N, de las Mercedes Anderson-Seminario M, Del-Aguila-Arcentales S (2022) Water footprint in the textile and food supply chain management: trends to become circular and sustainable. In: Alvarez-Risco A, Muthu SS, Del-Aguila-Arcentales S (eds) Circular economy: impact on carbon and water footprint. Springer, Singapore, pp 225–243. http://doi.org/10.1007/978-981-19-0549-0_11

131. Khan S, Henderson C (2020) How Western Michigan University is approaching its commitment to sustainability through sustainability-focused courses. J Clean Prod 253:119741. https://doi.org/10.1016/j.jclepro.2019.119741
132. Khan SAR, Razzaq A, Yu Z, Miller S (2021) Industry 4.0 and circular economy practices: a new era business strategies for environmental sustainability. Bus Strategy Environ 30(8):4001–4014. http://doi.org/10.1002/bse.2853
133. Kirchherr J, Reike D, Hekkert M (2017) Conceptualizing the circular economy: an analysis of 114 definitions. Resour Conserv Recycl 127:221–232. https://doi.org/10.1016/j.resconrec.2017.09.005
134. Kong L, Liu Z, Wu J (2020) A systematic review of big data-based urban sustainability research: state-of-the-science and future directions. J Clean Prod 273:123142. https://doi.org/10.1016/j.jclepro.2020.123142
135. Kozlowski A, Searcy C, Bardecki M (2018) The reDesign canvas: fashion design as a tool for sustainability. J Clean Prod 183:194–207. http://doi.org/10.1016/j.jclepro.2018.02.014
136. Kuhlman T, Farrington J (2010) What is sustainability? Sustainability 2(11). http://doi.org/10.3390/su2113436
137. Larrán Jorge M, Andrades Peña FJ, Herrera Madueño J (2019) An analysis of university sustainability reports from the GRI database: an examination of influential variables. J Environ Planning Manage 62(6):1019–1044. https://doi.org/10.1080/09640568.2018.1457952
138. Leiva-Martinez MA, Anderson-Seminario MLM, Alvarez-Risco A, Estrada-Merino A, Mlodzianowska S (2021) Price variation in lower goods as of previous economic crisis and the contrast of the current price situation in the context of covid-19 in Peru. In: Advances in business and management forecasting, vol 14, pp 161–166
139. Löhr A, Savelli H, Beunen R, Kalz M, Ragas A, Van Belleghem F (2017) Solutions for global marine litter pollution. Curr Opin Environ Sustain 28:90–99. https://doi.org/10.1016/j.cosust.2017.08.009
140. Lopez-Odar D, Alvarez-Risco A, Vara-Horna A, Chafloque-Cespedes R, Sekar MC (2020) Validity and reliability of the questionnaire that evaluates factors associated with perceived environmental behavior and perceived ecological purchasing behavior in Peruvian consumers [Article]. Soc Responsib J 16(3):403–417. https://doi.org/10.1108/SRJ-08-2018-0201
141. Lozano R, Barreiro-Gen M, Lozano FJ, Sammalisto K (2019) Teaching sustainability in European higher education institutions: assessing the connections between competences and pedagogical approaches. Sustainability 11(6):1602. https://doi.org/10.3390/su11061602
142. Manavalan E, Jayakrishna K (2019) An analysis on sustainable supply chain for circular economy. Procedia Manuf 33:477–484. https://doi.org/10.1016/j.promfg.2019.04.059
143. Mani V, Gunasekaran A, Delgado C (2018) Supply chain social sustainability: standard adoption practices in Portuguese manufacturing firms. Int J Prod Econ 198:149–164. https://doi.org/10.1016/j.ijpe.2018.01.032
144. Manzoor SR, Ho JSY, Al Mahmud A (2021) Revisiting the 'university image model' for higher education institutions' sustainability. J Mark High Educ 31(2):220–239. https://doi.org/10.1080/08841241.2020.1781736
145. Martinez-Martinez A, Cegarra-Navarro J-G, Garcia-Perez A, Wensley A (2019) Knowledge agents as drivers of environmental sustainability and business performance in the hospitality sector. Tour Manage 70:381–389. https://doi.org/10.1016/j.tourman.2018.08.030
146. Mejía-Acosta N, Alvarez-Risco A, Solís-Tarazona Z, Matos-Valerio E, Zegarra-Arellano E, Del-Aguila-Arcentales S (2016) Adverse drug reactions reported as a result of the implementation of pharmaceutical care in the Institutional Pharmacy DIGEMID—Ministry of Health. Pharm Care Esp 18(2):67–74
147. Meneses LI (2020) Impacts and consequences of the prohibition of single-use plastic. Universidad Militar Nueva Granada, Bogotá, Colombia. https://core.ac.uk/reader/344700660
148. Mensah J (2019) Sustainable development: meaning, history, principles, pillars, and implications for human action: literature review. Cogent Soc Sci 5(1):1653531. https://doi.org/10.1080/23311886.2019.1653531
149. MicGalaw (2021) Mic Galaw. Retrieved 12/03/2021 from http://www.micgalaw.business.site

150. Michelino F, Cammarano A, Celone A, Caputo M (2019) The linkage between sustainability and innovation performance in IT hardware sector. Sustainability 11(16):4275. https://doi.org/10.3390/su11164275
151. MINAM (2019) MINAM: single-use plastic consumption decreased by 30% in the last year [MINAM: Consumo de plástico de un solo uso se redujo en 30% en el último año]. Retrieved 12/03/2021 from https://www.gob.pe/institucion/minam/noticias/76261-minam-consumo-de-plastico-de-un-solo-uso-se-redujo-en-30-en-el-ultimo-ano
152. Mohamued EA, Ahmed M, Pypłacz P, Liczmańska-Kopcewicz K, Khan MA (2021) Global oil price and innovation for sustainability: the impact of R&D spending, oil price and oil price volatility on GHG emissions. Energies 14(6). http://doi.org/10.3390/en14061757
153. Montsenu (2021) Montsenu. Retrieved 12/02/2021 from https://montsenu.com
154. Morales-Ríos F, Alvarez-Risco A, Castillo-Benancio S, de las Mercedes Anderson-Seminario M, Del-Aguila-Arcentales S (2022) Material selection for circularity and footprints. In: Alvarez-Risco A, Muthu SS, Del-Aguila-Arcentales S (eds) Circular economy: impact on carbon and water footprint. Springer, Singapore, pp 205–221. http://doi.org/10.1007/978-981-19-0549-0_10
155. Morelli J (2011) Environmental sustainability: a definition for environmental professionals. J Environ Sustain 1(1):2
156. Mousa SK, Othman M (2020) The impact of green human resource management practices on sustainable performance in healthcare organisations: a conceptual framework. J Clean Prod 243:118595. https://doi.org/10.1016/j.jclepro.2019.118595
157. Muñoz P, Cohen B (2018) Sustainable entrepreneurship research: taking stock and looking ahead. Bus Strategy Environ 27(3):300–322. http://doi.org/10.1002/bse.2000
158. Mzembe AN, Lindgreen A, Idemudia U, Melissen F (2020) A club perspective of sustainability certification schemes in the tourism and hospitality industry. J Sustain Tour 28(9):1332–1350. https://doi.org/10.1080/09669582.2020.1737092
159. Niemann CC, Dickel P, Eckardt G (2020) The interplay of corporate entrepreneurship, environmental orientation, and performance in clean-tech firms—a double-edged sword. Bus Strategy Environ 29(1):180–196. http://doi.org/10.1002/bse.2357
160. Norström AV, Cvitanovic C, Löf MF, West S, Wyborn C, Balvanera P, Österblom H (2020) Principles for knowledge co-production in sustainability research. Nat Sustain 3(3):182–190. http://doi.org/10.1038/s41893-019-0448-2
161. Obbard RW, Sadri S, Wong YQ, Khitun AA, Baker I, Thompson RC (2014) Global warming releases microplastic legacy frozen in Arctic Sea ice. Earth's Future 2(6):315–320. https://doi.org/10.1002/2014EF000240
162. Ogonowski M, Gerdes Z, Gorokhova E (2018) What we know and what we think we know about microplastic effects—a critical perspective. Curr Opin Environ Sci Health 1:41–46. https://doi.org/10.1016/j.coesh.2017.09.001
163. Olawumi TO, Chan DWM (2018) A scientometric review of global research on sustainability and sustainable development. J Clean Prod 183:231–250. http://doi.org/10.1016/j.jclepro.2018.02.162
164. Omoloso O, Mortimer K, Wise WR, Jraisat L (2021) Sustainability research in the leather industry: a critical review of progress and opportunities for future research. J Clean Prod 285:125441. http://doi.org/10.1016/j.jclepro.2020.125441
165. Ozturkoglu Y, Sari FO, Saygili E (2021) A new holistic conceptual framework for sustainability oriented hospitality innovation with triple bottom line perspective. J Hosp Tour Technol 12(1):39–57. https://doi.org/10.1108/JHTT-02-2019-0022
166. Peçanha Enqvist J, West S, Masterson VA, Haider LJ, Svedin U, Tengö M (2018) Stewardship as a boundary object for sustainability research: linking care, knowledge and agency. Landsc Urban Plan 179:17–37. https://doi.org/10.1016/j.landurbplan.2018.07.005
167. Peeters P (2018) Why space tourism will not be part of sustainable tourism. Tour Recreat Res 43(4):540–543. https://doi.org/10.1080/02508281.2018.1511942
168. Peña Miguel N, Corral Lage J, Mata Galindez A (2020) Assessment of the development of professional skills in university students: sustainability and serious games. Sustainability 12(3). http://doi.org/10.3390/su12031014

169. Pham H, Kim S-Y (2019) The effects of sustainable practices and managers' leadership competences on sustainability performance of construction firms. Sustain Prod Consumption 20:1–14. https://doi.org/10.1016/j.spc.2019.05.003
170. Plewnia F, Guenther E (2018) Mapping the sharing economy for sustainability research. Manag Decis 56(3):570–583. https://doi.org/10.1108/MD-11-2016-0766
171. Quispe-Cañari JF, Fidel-Rosales E, Manrique D, Mascaró-Zan J, Huamán-Castillón KM, Chamorro–Espinoza SE et al (2021) Self-medication practices during the COVID-19 pandemic among the adult population in Peru: a cross-sectional survey. Saudi Pharm J 29(1):1–11. http://doi.org/10.1016/j.jsps.2020.12.001
172. Rajesh R (2020) Exploring the sustainability performances of firms using environmental, social, and governance scores. J Clean Prod 247:119600. https://doi.org/10.1016/j.jclepro.2019.119600
173. Rakhetsi (202) Meet the South African duo turning plastic trash into 'leather' high fashion. Retrieved 12/02/2021 from https://www.globalcitizen.org/en/content/south-african-women-turn-plastic-leather/?template=next
174. Ramísio PJ, Pinto LMC, Gouveia N, Costa H, Arezes D (2019) Sustainability strategy in higher education institutions: lessons learned from a nine-year case study. J Clean Prod 222:300–309. https://doi.org/10.1016/j.jclepro.2019.02.257
175. Ransfield AK, Reichenberger I (2021) Māori indigenous values and tourism business sustainability. Altern Int J Indigenous Peoples 17(1):49–60. http://doi.org/10.1177/1177180121994680
176. Rau H, Goggins G, Fahy F (2018) From invisibility to impact: recognising the scientific and societal relevance of interdisciplinary sustainability research. Res Policy 47(1):266–276. https://doi.org/10.1016/j.respol.2017.11.005
177. REPSOL (2021) Circular economy [Economia circular]. Retrieved 12/04/2021 from https://www.repsol.com/es/sostenibilidad/economia-circular/index.cshtml
178. Rhee S-W (2020) Management of used personal protective equipment and wastes related to COVID-19 in South Korea. Waste Manage Res 38(8):820–824. https://doi.org/10.1177/0734242X20933343
179. Rist S, Hartmann NB (2018) Aquatic ecotoxicity of microplastics and nanoplastics: lessons learned from engineered nanomaterials. In: Wagner M, Lambert S (eds) Freshwater microplastics: emerging environmental contaminants? Springer International Publishing, pp 25–49. http://doi.org/10.1007/978-3-319-61615-5_2
180. Rojas-Osorio M, Alvarez-Risco A (2019) Intention to use smartphones among Peruvian university students. Int J Interact Mobile Technol 13(3):40–52. https://doi.org/10.3991/ijim.v13i03.9356
181. Román BR, Moscoso S, Chung SA, Terceros BL, Álvarez-Risco A, Yáñez JA (2020) Treatment of COVID-19 in Peru and Bolivia, and self-medication risks. Rev Cubana Farmacia 53(2):1–20, Article e435
182. Sakao T, Brambila-Macias SA (2018) Do we share an understanding of transdisciplinarity in environmental sustainability research? J Clean Prod 170:1399–1403. http://doi.org/10.1016/j.jclepro.2017.09.226
183. Salmerón-Manzano E, Manzano-Agugliaro F (2018) The higher education sustainability through virtual laboratories: the Spanish University as case of study. Sustainability 10(11). http://doi.org/10.3390/su10114040
184. Schymanski D, Goldbeck C, Humpf H-U, Fürst P (2018) Analysis of microplastics in water by micro-Raman spectroscopy: release of plastic particles from different packaging into mineral water. Water Res 129:154–162. https://doi.org/10.1016/j.watres.2017.11.011
185. Sedlak D (2017) Three lessons for the microplastics voyage. ACS Publications
186. Serafeim G (2020) Public sentiment and the price of corporate sustainability. Financ Anal J 76(2):26–46. https://doi.org/10.1080/0015198X.2020.1723390
187. Sergi BS, Popkova EG, Bogoviz AV, Ragulina JV (2019) Entrepreneurship and economic growth: the experience of developed and developing countries. In: Sergi BS, Scanlon CC (eds) Entrepreneurship and development in the 21st century. Emerald Publishing Limited, pp 3–32. http://doi.org/10.1108/978-1-78973-233-720191002

188. Shams M, Alam I, Mahbub MS (2021) Plastic pollution during COVID-19: plastic waste directives and its long-term impact on the environment. Environ Adv 5:100119. https://doi.org/10.1016/j.envadv.2021.100119
189. Shuqin C, Minyan L, Hongwei T, Xiaoyu L, Jian G (2019) Assessing sustainability on Chinese university campuses: development of a campus sustainability evaluation system and its application with a case study. J Build Eng 24:100747. https://doi.org/10.1016/j.jobe.2019.100747
190. Smith M, Love DC, Rochman CM, Neff RA (2018) Microplastics in seafood and the implications for human health. Curr Environ Health Rep 5(3):375–386. https://doi.org/10.1007/s40572-018-0206-z
191. Soapboxie (2020) The environmental, economic, and social components of sustainability. Retrieved 12/03/2021 from https://soapboxie.com/social-issues/The-Environmental-Economic-and-Social-Components-of-Sustainability
192. Sroufe R, Gopalakrishna-Remani V (2018) Management, social sustainability, reputation, and financial performance relationships: an empirical examination of U.S. firms. Organ Environ 32(3):331–362. http://doi.org/10.1177/1086026618756611
193. Stahel WR (2016) Circular economy: a new relationship whit our goods and material swould save resource sandenergyandcreate local jobs. Nature Publishing Group
194. Sung E, Kim H, Lee D (2018) Why do people consume and provide sharing economy accommodation? A sustainability perspective. Sustainability 10(6):2072. https://doi.org/10.3390/su10062072
195. Thirumalesh Madanaguli A, Kaur P, Bresciani S, Dhir A (2021) Entrepreneurship in rural hospitality and tourism. A systematic literature review of past achievements and future promises. Int J Contemp Hospitality Manag 33(8):2521–2558. http://doi.org/10.1108/IJCHM-09-2020-1121
196. Trosper RL (2002) Northwest coast indigenous institutions that supported resilience and sustainability. Ecol Econ 41(2):329–344. https://doi.org/10.1016/S0921-8009(02)00041-1
197. UN Environment Program (2021) From pollution to solution. https://www.unep.org/interactive/pollution-to-solution
198. United Nations (2019) Microplastics, microbeads and single-use plastics poisoning sea life and affecting humans. Retrieved 12/0372021 from https://news.un.org/en/story/2019/11/1050511
199. United Nations (2021) The SDGS in action. Retrieved 12/03/2021 from https://www.undp.org/sustainable-development-goals
200. Van Rensburg ML, Nkomo SL, Dube T (2020) The ‘plastic waste era’; social perceptions towards single-use plastic consumption and impacts on the marine environment in Durban, South Africa. Appl Geogr 114:102132. https://doi.org/10.1016/j.apgeog.2019.102132
201. Veleva V, Bodkin G (2018) Corporate-entrepreneur collaborations to advance a circular economy. J Clean Prod 188:20–37. http://doi.org/10.1016/j.jclepro.2018.03.196
202. Vismara S (2019) Sustainability in equity crowdfunding. Technol Forecast Soc Change 141:98–106. http://doi.org/10.1016/j.techfore.2018.07.014
203. Vizcardo D, Salvador LF, Nole-Vara A, Dávila KP, Alvarez-Risco A, Yáñez JA, Mejia CR (2022) Sociodemographic predictors associated with the willingness to get vaccinated against COVID-19 in Peru: a cross-sectional survey. Vaccines 10(1), Article 48. http://doi.org/10.3390/vaccines10010048
204. Von Moos N, Burkhardt-Holm P, Köhler A (2012) Uptake and effects of microplastics on cells and tissue of the blue mussel Mytilus edulis L. after an experimental exposure. Environ Sci Technol 46(20):11327–11335
205. Walker TR, McGuinty E, Charlebois S, Music J (2021) Single-use plastic packaging in the Canadian food industry: consumer behavior and perceptions. Humanit Soc Sci Commun 8(1):80. https://doi.org/10.1057/s41599-021-00747-4
206. Whetham B (2020) Brad Scott's precious plastic creations are turning heads, as he turns trash into homewares. Retrieved 12/03/2021 from https://www.abc.net.au/news/2020-06-26/nescafe-comes-calling-brad-scott-plastic-recycled-homewares/12380828

207. Wright SL, Kelly FJ (2017) Plastic and human health: a micro issue? Environ Sci Technol 51(12):6634–6647. https://doi.org/10.1021/acs.est.7b00423
208. Yan J, Kim S, Zhang SX, Foo MD, Alvarez-Risco A, Del-Aguila-Arcentales S, Yáñez JA (2021) Hospitality workers' COVID-19 risk perception and depression: a contingent model based on transactional theory of stress model. Int J Hospitality Manag 95, Article 102935. http://doi.org/10.1016/j.ijhm.2021.102935
209. Yáñez JA, Alvarez-Risco A, Delgado-Zegarra J (2020) Covid-19 in Peru: from supervised walks for children to the first case of Kawasaki-like syndrome. The BMJ 369, Article m2418. http://doi.org/10.1136/bmj.m2418
210. Yáñez JA, Jahanshahi AA, Alvarez-Risco A, Li J, Zhang SX (2020) Anxiety, distress, and turnover intention of healthcare workers in Peru by their distance to the epicenter during the COVID-19 crisis. Am J Trop Med Hyg 103(4):1614–1620. https://doi.org/10.4269/ajtmh.20-0800
211. Yáñez S, Uruburu Á, Moreno A, Lumbreras J (2019) The sustainability report as an essential tool for the holistic and strategic vision of higher education institutions. J Clean Prod 207:57–66. https://doi.org/10.1016/j.jclepro.2018.09.171
212. Yarime M, Tanaka Y (2012) The issues and methodologies in sustainability assessment tools for higher education institutions: a review of recent trends and future challenges. J Educ Sustain Dev 6(1):63–77. https://doi.org/10.1177/097340821100600113
213. Zamora-Polo F, Sánchez-Martín J (2019) Teaching for a better world. Sustainability and sustainable development goals in the construction of a change-maker university. Sustainability 11(15). http://doi.org/10.3390/su11154224
214. Zhang SX, Chen J, Afshar Jahanshahi A, Alvarez-Risco A, Dai H, Li J, Patty-Tito RM (2021) Succumbing to the COVID-19 pandemic—healthcare workers not satisfied and intend to leave their jobs. Int J Mental Health Addict. https://doi.org/10.1007/s11469-020-00418-6
215. Zhang SX, Chen J, Jahanshahi AA, Alvarez-Risco A, Dai H, Li J, Patty-Tito RM (2021) Correction to: succumbing to the COVID-19 pandemic—healthcare workers not satisfied and intend to leave their jobs. Int J Mental Health Addict. https://doi.org/10.1007/s11469-021-00502-5
216. Zhang SX, Sun S, Afshar Jahanshahi A, Alvarez-Risco A, Ibarra VG, Li J, Patty-Tito RM (2020) Developing and testing a measure of COVID-19 organizational support of healthcare workers—results from Peru, Ecuador, and Bolivia. Psychiatry Res 291. http://doi.org/10.1016/j.psychres.2020.113174
217. Selvaranjan K, Navaratnam S, Rajeev P, Ravintherakumaran N (2021) Environmental challenges induced by extensive use of face masks during COVID-19: A review and potential solutions. Environmental Challenges, 3, 100039. https://doi.org/10.1016/j.envc.2021.100039

Redefining Entrepreneurship: The Incorporation of CSR and Positive Corporate Image as Business Strategies in Green Entrepreneurialism

Flavio Morales-Rios, Aldo Alvarez-Risco, Sharon Esquerre-Botton, Sarahit Castillo-Benancio, María de las Mercedes Anderson-Seminario, Shyla Del-Aguila-Arcentales, and Francis Julca-Zamalloa

Abstract Green entrepreneurship proposes practices focused on mitigating environmental issues through innovation to harmonize economic success and ecological conservation. The proposed topic aims to determine the role of corporate social responsibility (CSR) and corporate image (CI) as plausible critical strategies in developing green entrepreneurialism to improve corporate performance and maintain a long-term competitive edge. A collection of academic evidence was carried out on CSR and CI in the positioning process of sustainable enterprises in various countries.

Keywords Green entrepreneurship · Sustainability · Corporate social responsibility · Green corporate image · Greenwashing · Green marketing

1 Introduction

The increase in environmental emissions has had consequences that have been reflected in the environment, both in flora, fauna, and society itself, thus generating a change in consumers' priorities and, therefore, manufacturers and sellers [68]. Consequently, over the last decades, monetary growth is not the only objective of companies [109]; given the increased attention to social and environmental aspects, both large and medium along with small-scale enterprises, including entrepreneurship, seek to increase their potential through the development of Corporate Social Responsibility (CSR) practices in an attempt to transform their business into a sustainable and responsible project [9, 54]. In addition, those businesses that prefer to remain traditional, profit-focused eventually take the route to sustainability since the current solutions reflect lower costs and risks [51]. Moreover, in the eyes of consumers, a brand with a positive corporate image (CI) is more attractive. Businesses must know

F. Morales-Rios · A. Alvarez-Risco · S. Esquerre-Botton · S. Castillo-Benancio · M. de las Mercedes Anderson-Seminario · F. Julca-Zamalloa
Universidad de Lima, Lima, Peru

S. Del-Aguila-Arcentales (✉)
Escuela de Posgrado, Universidad San Ignacio de Loyola, Lima, Peru
e-mail: sdelaguila@usil.edu.pe

A. Alvarez-Risco et al. (eds.), *Footprint and Entrepreneurship*, Environmental Footprints and Eco-design of Products and Processes,
https://doi.org/10.1007/978-981-19-8895-0_6

how to develop and communicate it correctly to achieve preference, generating reputation and competitiveness [51, 63]. This chapter has been developed to recognize the importance of incorporating CSR practices and positive corporate development in the various enterprises that, to integrate these aspects, decide to take a green path. Thus, the structure is a sum of past information related to traditional and green entrepreneurship, the implementation of CSR over the last years and in current times, the effect of a green corporate image in business, and the exposition of cases that exemplify the application of all the above mentioned.

2 Traditional Entrepreneurship to Green Entrepreneurship

A key factor relevant to a country's entrepreneurial health is having many active entrepreneurs and ventures [57]. In many cases, it depends on the competitiveness of enterprises with the constantly changing markets and changing needs of contemporary society [61, 79]. In this way, new opportunities are explored by entrepreneurs, and in the process, they may encounter many difficulties [101], although their main objective is to persevere in the business, so they must anticipate future changes and transform themselves [83, 84].

Traditional ventures are classified according to their entrepreneurs as novice entrepreneurs, regular entrepreneurs who create a new product or service to sell their business and acquire new ones subsequently, and finally, there are portfolio entrepreneurs who create clusters [3]. Entrepreneurs must identify opportunities to generate new business based on their background and experience [19, 101]. Also, for a company to be competitive, it needs attention, good planning, and well-analyzed decision-making based on good practices [11, 61, 108]. That is why long-term improvements become advantages that benefit not only the company but also the environment [61].

A new way of thinking breaks paradigms and links two concepts such as innovation and entrepreneurship. There are different types of characteristics of traditional entrepreneurship, but two of the most relevant ones are related to entrepreneurial teams [39]. On the one hand, there are the financial interests, and all must be shareholders [95], on the other hand, the management of the company and there is a higher probability of having an increased growth and subsistence in the market [65]. In this way, sets of people are more active in participating in the evolutionary development of the business, as it has an acquisitive interest [39]. The world is changing to focus on greater sustainability and ecological concern [52, 83, 84]. Therefore, green entrepreneurship solves social, economic, and environmental problems [68].

There are several categories of entrepreneurship, one of them is social entrepreneurship, which is a broad, popular, and recent trend in different sectors and types of entrepreneurships [73, 90, 100]. It is also characterized by identifying the various problems of society to promote a social impact through awareness-raising [47, 66, 67, 71, 114]. In addition, there are two subcategories of social entrepreneurship [90]. The first is limited, as it focuses on non-profit-making organizations such

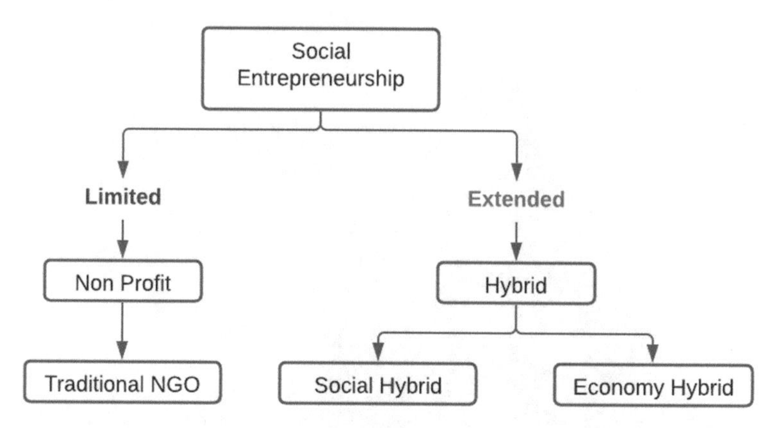

Fig. 1 Social entrepreneurship model. *Source* Rahim and Mohtar [98]

as NGOs [47, 90]. And the second is a little bit more extensive, because it covers the social and economic sphere, and is concentrated in companies that have lucrative purposes to different degrees [12, 47, 53, 98]. In addition, there can always be companies that are more focused on the social aspect and others on the economic aspect, considering them mixed [47, 98] (Fig. 1).

2.1 *Theoretical Consideration of CSR*

Corporate social responsibilities (CSR) are linked to social standards and environmental protection [10], which organizations should prioritize [115], becoming more efficient by achieving relevant financial performance [58]. To analyze the concept of CSR in depth, we will draw on Archie Carroll's definition of sustainable business, shown below [111]. In which the base is the economic responsibility that must be sustainable and solid [117], since it is fundamental in business, giving support to the whole infrastructure of the pyramid [31, 42]. In the next link is the legal sphere, due to the fact that society demands that laws are complied with by working ethically, thus providing legitimacy in law for business growth [33, 42]. This goes hand in hand with ethical responsibility, in which there is a degree of expectation above the minimum levels demanded by the rules [40, 111]. Finally, there is philanthropic responsibility in which there is always an expectation to do the right thing [32, 117]. It is important to stress that all four aspects of CSR are ethically motivated [33, 42] (Fig. 2).

Fig. 2 Carroll's pyramid. *Source* Carroll [33]

2.2 Differences Between Social Entrepreneurship and CSR

Social entrepreneurship is still at an early stage, so the topic is very attractive to many researchers [29, 50]. Also, companies are more focused on social changes and use digital platforms to legitimize themselves, thus overcoming the underfunding of communication with customers that existed in the traditional model [35]. Furthermore, various definitions have been offered [35, 47], but all are based on four important factors, which are operational management, characteristics of the entrepreneurs, resources used as raw material, and the mission of the organization [43]. The purpose of social entrepreneurship is value creation [89], which has a social impact and is its major differentiator [35]. Although the mission is established, it does not exclude that part of its objective is also economic by coupling a sustainable strategy [43] balancing both aspects, this gives rise to several logical institutions competing with each other [21]. To do this, entrepreneurs recognize an inequality that may still affect a certain number of people, to subsequently identify an alternative that improves the situation for those involved in the social system [85], which is why measurable objectives and projected results must be defined from the outset [7, 28].

It is important to carry out an analysis in which CSR and social entrepreneurship can be compared, which arises from the increase in activities related to both concepts [13]. As we have seen in the previous paragraphs, the social entrepreneur focuses his or her enterprise on positive effects and can be given the role of an agent of change [47, 94], without forgetting the benefits for the current and future society [47], this being a basic difference [111], furthermore, trade-offs can be found in the rates of return applied to market share and customer satisfaction [13, 102].

2.3 *Theoretical Consideration of CI*

Organizations conduct strategic communication among stakeholders to promote positive perceptions and discourage negative ones [72, 87]. Organizational image management theory stems from an individual perspective of image management and self-presentation notions, alluding to the steps required to establish, preserve, or even, re-establish a desirable image in the eyes of a firm's interest groups [86, 87]. To enhance their likelihood of succeeding in the market, corporations are compelled to maintain a positive image among their clients [60, 96, 97]. The dilemma for corporations relies on being able to comprehend the volatility of the competitive landscape, as well as to adapt their image according to market flux, in a responsive manner [24, 60, 86]. Organizational image management is a dialectic process wherein the corporate image surges as a construct developed by the repeated interaction of a firm's both internal and external stakeholders [86, 87, 96]. Therefore, enterprises are faced with the need of proactively interacting with stakeholders following a dialogic schema to cultivate certain images and discourage others [86, 87, 118]. Considering a company's executives might advocate and promote a certain perception of their CI, however, that perception shall first attain the conviction and validation of the public to prevail [24, 103]. For instance, an entity's corporate image can be ultimately understood as a social conceptualization originated from the deliberate inter-exchange of information between business executives and the audience [24, 86, 87, 97, 118].

After defining the notion of the corporate image via the theory of organizational image management, it seems appropriate to elaborate on the theoretical conceptualization of a closely related term known as corporate identity [24]. Corporate identity represents the pure essence of an organization, highlighting the personality and characteristics that the conglomerate provides to the public; as well as the offer, mission and vision that represents the company [15]. On the other hand, corporate image is the public's perception of the company based on the ideas, principles, and values of each person [5, 16, 18, 27, 64, 75]. These terms can also be influenced by the business culture, the company's reputation, and the quality of the work environment [76]. However, both terms are linked and depend on business performance and the cultural environment of the moment [88].

3 Implementing CSR in Green Entrepreneurialism

Recent studies have determined "the green wave" in a bifurcation of two elements which are entrepreneurship and innovation [49]. Green entrepreneurship is a new proposal, which has emerged due to various associations' environmental concerns and because society demands greater transparency in its reporting [14, 110]. Moreover, sustainability and corporate social responsibility (CSR) are nowadays two primary characteristics for value creation in all areas [22, 107], and their success lies in the low pollution of their environment [105]. Social media has boosted the concept of green entrepreneurship among young people with environmental campaigns [68], so CSR-based information about a company has a transactional view [99]. Also, the pandemic has made CSR play a key role in companies [30, 106], as they receive positive feedback from their target audience and both internal and external stakeholders [80].

There are two types of CSR-related activities. Firstly, there are the external ones, and these are the social tasks in an environmental framework that build the company's originality and reputation concerning external stakeholders [25, 31]. Secondly, we find the internal ones focused on its employees' psychological and physiological well-being [25, 104, 113]. In this last point, it is essential to highlight that the opportunities are equal, and the health and safety are relevant of each member of the organization, as well as positively improving the reciprocal commitment of the employees with the company [81].

4 Business Cases that Incorporate CSR as a Strategy in the World

Given the growing concern for implementing sustainability, integrating action not only by policies but also by individual parties such as companies [81], the strategies implemented by businesses are becoming more conscious when talking about their impact in social and environmental terms [109] thus significantly evolving the concept of CSR at the corporate level, involving organizations from all sectors and extending to a global level [92] (Table 1).

The application of CSR not only seeks to encourage the implementation of activities with a positive impact on society and the environment, considering animal, human, and plant health, but also integrates the search for innovative solutions

Table 1 Incorporation of CSR as business strategies

Country/region	Business strategies	Sources
China	Convergence of mass entrepreneurship and innovation as a new economic driver	Low and Bu [81]
Indonesia	Integration of responsible and sustainable training and management of stakeholders	Indarti [70]
Latin America	Integration of social and economic growth activities that are inclusive, sustainable, and eco-friendly	Acevedo-Duque et al. [1]
Poland	CSR communication with stakeholders, CSR-knowledge management and CSR-strategy	Stawicka [109]
China	Integration of CSR, transactional and transformational leadership in search of sustainable performance improvement	Dai et al. [44]
Chile	CSR practices in different media and their impact on the economic performance of companies	Fernandez et al. [59]

that allow the growth of the economy, the creation and development of sustainable processes, and the modeling of competitive advantages [1, 81], these results will be reflected in aspects such as an increase in the strategic positioning of the company that applies it, in addition to also show improvements in business efficiency and improve in turn the adaptation to the dynamics of the environment in which it operates [92]. According to Samaibekova et al. [102], CSR integrates brand value and engagement with job creation for sustainable domestic results, additionally, to captivate the entire workforce, companies must also include digitalization as a strategy in their CSR practices [81]. However, when talking about small businesses, they may face greater technological challenges compared to large companies that have a large presence in the same industry, this aspect is due to the possibility of developing a poor knowledge management in the subject (difficulties to have access to the Internet, fear of failure, generate little presence or activity on the web, or simply not knowing how to develop specific strategies for the use of technology) [109] in a complementary manner, mention may be made of what has been expressed by Lozano [82], who mentions that, in a complementary way to CSR strategies, such business systems are combined with technological techniques and solutions. In brief, the cases observed show that companies, regardless of their sectors, that opt for a sustainable path and therefore contribute to the overall growth of CSR, are committed to actions that reflect a positive impact on social transformation, integrating the community, industries and the environment [59, 92].

5 Green Corporate Image (GCI)

The concept of corporate image (CI) can be defined as the intended overall impression an organization has regarding its various stakeholders [24, 96, 103]. For instance, a company's corporate image is a very distinguishing trait in the eyes of both internal and external participants and their relationship with the entity in question [24, 60, 103]. Organizations devote significant resources, such as capital, time, and personnel, to establish a positive CI [60, 96, 97]. Previous research reveals that organizations with a positive CI tend to achieve a higher position in the market and achieve a competitive advantage over their competitors, as they maintain a beneficial perception in the eyes of the public [24, 60, 96, 97].

Several studies have looked at the green corporate image (GCI) in various businesses [20, 96]. Green corporate image, in turn, refers to how stakeholders can perceive a firm's environmental principles and practices [6, 20, 36]. GCI has been shown to have a variety of effects on a range of different organizational aspects, including employee work-life, managerial and business performance, motivation, purchasing intentions, etc. [6, 24, 36, 96].

Field studies suggest a correlation between green corporate image and consumer satisfaction, along with brand loyalty and organizational performance [6, 36, 60, 116]. An increasing number of businesses are attempting to promote themselves as sustainable and environmentally aware brands [6, 23, 36]. Thus, to remain competitive in an ever-shifting market, businesses are pressured to adapt their marketing plan and advertising strategies to appear more sustainable, as modern trends dictate [23, 24, 96]. However, this has also influenced the adoption of harmful corporate practices such as greenwashing [23]. Greenwashing is known as the deliberate dissemination of positive information from a company while concealing harmful practices to project a deceitfully favorable corporate image [17, 23]. The methodology of greenwashing is intended to make a firm, product, or brand appear more environmentally sustainable than it is through various marketing tools conveying misinformation, in order to create a positive corporate image among the public [17, 23, 36].

The notion of the corporate image, however, is inextricably tied to the firm's environmental legitimacy [24, 60, 116]. Since environmental legitimacy is highly dependent on the impression of the firm's environmental performance, business executives may increase the firm's integrity by acting on such perceptions through the enhancement of the company's green image [6, 24, 36, 60, 116]. As a result, for businesses that operate in ecologically sustainable industries featuring significant social and environmental implications, cultivating a solid green corporate image in the perception of their stakeholders is an inherent requirement for development and growth [6, 23, 24, 60, 116].

6 Business Cases that Use GCI as a Strategy in the World

Progressively, companies started stressing the recyclability, sustainability, and environmentally innocuous nature of their goods, encouraging ecologically responsible consumer habits, and investigating green market prospects to foster a green corporate image and business culture [6, 23, 36]. Green advertisement for maintaining the long-term viability of marketing campaigns soon gained popularity, becoming commonplace [6, 24, 36, 60, 116] (Table 2).

Environmental awareness has risen quickly in the last decade, as businesses have progressively begun to include environmental protection principles into their research and development agenda, product manufacturing, and advertisement [2, 20, 37, 78]. In turn, consumers also pay close attention to whether products are environmentally friendly and whether brands adopt eco-friendly practices [6, 112]. The introduction to the concept of green marketing, this term refers to how businesses approach the environment, integrating it as one of their core values, working towards a GCI, to establish a brand that is more valuable to the public [37, 93, 116].

In recent years, green marketing has flourished [56, 112]. Due to customers' growing knowledge and expectations of environmental preservation and increased market rivalry, several businesses have begun to use green marketing as a strategy for growth [56, 93, 112, 116]. As a result, various firms have turned to green marketing as a critical component for surviving in intensely competitive industries, encouraging the audience to acquire green goods and services, fostering brand loyalty through

Table 2 Examples of GCI as a business strategy

Country	Business strategies	Sources
Starbucks	Starbucks has made a solid commitment to environmental conservation and green marketing. Their efforts to increase the recyclability of their coffee cups have been linked with an in-store water and energy conservation initiative to create a green corporate image	Tsai et al. [112]
Ziploc	The new series of eco-friendly bags include less plastic, less paper usage in the packaging, and are labeled as being manufactured using sustainable wind energy	Brannan et al. [26]
Unilever	Unilever is known for its social and environmental initiatives, which is a marketing strategy and a component of the company's commitment to implementing environmental and social responsibility	Widyastuti et al. [116]
IBM	IBM's public initiative of 21 goals for environmental sustainability	IBM Corp [69]
Tesla	Tesla wants to be a more environmentally friendly corporation, thus conducting and disclosing a "well-to-wheel" investigation, which concluded that the Roadster still has twice the energy efficiency of popular hybrid cars and emits a third of the CO_2	Adén and Barray [2]

a GCI, raising market share and cash flow, and ensuring economic sustainability [6, 37, 96, 112, 116].

Recent suggested lectures about green approach

- Waste reduction and carbon footprint [62]
- Material selection for circularity and footprints [91]
- 3D Print, circularity, and footprints [45]
- Virtual tourism [45]
- Leadership for sustainability in crisis time [4]
- Virtual education and circularity [38]
- Circular Economy for packaging [34]
- Students oriented to circular learning [46]
- Food and circular economy [41]
- Water footprint and food supply chain management [74]
- Waste footprint [8]
- Measuring circular economy [48]
- Carbon footprint [55].

7 Closing Remarks

Since the emergence of the concept of green and sustainable, the products and services offered have faced a change, including all the dimensions that go through the same to be conducted from sender to consumer (e.g., selection of raw material, clean production, sustainable logistics, transportation regulations, and disposal) [77, 82, 109]. Among the reasons for responsible and sustainable consumption and activities to continue to increase, it is given to the effects of environmental and social levels that consumption entails [82]; in this way, the consumer is aware of what they are looking for and prefer, altering their behavior and purchase intention [63], which the brands evaluate to generate preference and encourage reputation and loyalty [6]. Thus, to lessen the negative impact on our surroundings, different corporate activities are modified with a more sustainable vision, creating new business model opportunities, including ventures which also offer innovative changes in the business mentality and the way to serve the changing market to ensure consumer satisfaction, and maintain the balance between environmental, social, and economical care [68, 77, 82–84]. In this sense, companies have been implementing CSR strategies for some decades now, and they also seek to work on their corporate image so that consumers are kept informed of what the companies offer [6, 63].

References

1. Acevedo-Duque Á, Gonzalez-Diaz R, Vega-Muñoz A, Fernández Mantilla MM, Ovalles-Toledo LV, Cachicatari-Vargas E (2021) The role of b companies in tourism towards recovery from the crisis covid-19 inculcating social values and responsible entrepreneurship in Latin America. Sustainability 13(14):7763
2. Adén E, Barray A (2008) Go green in the automotive industry: open and networked innovation applied by Tesla Motors and Renault. In: Handelshögskolan BBS
3. Alarcón A, Brunet I (2004) Teorías sobre la figura del emprendedor. Recuperado de. https://ddd.uab.cat/pub/papers/02102862n73/02102862n73p81.pdf
4. Alvarez-Risco A, Del-Aguila-Arcentales S, Villalobos-Alvarez D, Diaz-Risco S (2022) Leadership for sustainability in crisis time. In: Alvarez-Risco A, Muthu SS, Del-Aguila-Arcentales S (eds) Circular economy: impact on carbon and water footprint. Springer, Singapore, pp 41–64. http://doi.org/10.1007/978-981-19-0549-0_3
5. Alvesson M (1990) Organization: from substance to image? Organ Stud 11(3):373–394
6. Amores-Salvadó J, Martín-de Castro G, Navas-López JE (2014) Green corporate image: moderating the connection between environmental product innovation and firm performance. J Clean Prod 83:356–365
7. André K, Pache A-C (2016) From caring entrepreneur to caring enterprise: addressing the ethical challenges of scaling up social enterprises. J Bus Ethics 133(4):659–675
8. Arias-Meza M, Alvarez-Risco A, Cuya-Velásquez BB, de las Mercedes Anderson-Seminario M, Del-Aguila-Arcentales S (2022) Fashion and textile circularity and waste footprint. In: Alvarez-Risco A, Muthu SS, Del-Aguila-Arcentales S (eds) Circular economy: impact on carbon and water footprint. Springer, Singapore, pp 181–204. http://doi.org/10.1007/978-981-19-0549-0_9
9. Ashrafi M, Magnan GM, Adams M, Walker TR (2020) Understanding the conceptual evolutionary path and theoretical underpinnings of corporate social responsibility and corporate sustainability. Sustainability 12(3). http://doi.org/10.3390/su12030760
10. Asmeri R, Alvionita T, Gunardi A (2017) CSR disclosures in the mining industry: empirical evidence from listed mining firms in Indonesia. Indonesian J Sustain Acc Manag 1(1):16–22
11. Audretsch DB, Keilbach MC, Lehmann EE (2006) Entrepreneurship and economic growth. Oxford University Press

12. Austin J (2006) Corporate social entrepreneurship: a new vision of CSR. Retrieved 01/01/2022 from https://www.hbs.edu/faculty/Pages/item.aspx?num=20663
13. Austin J, Stevenson H, Wei-Skillern J (2006) Social and commercial entrepreneurship: same, different, or both? Entrep Theory Pract 30(1):1–22
14. Aversano N, Nicolò G, Sannino G, Tartaglia Polcini P (2020) The integrated plan in Italian public universities: new patterns in intellectual capital disclosure. Meditari Accountancy Res 28(4):655–679. https://doi.org/10.1108/MEDAR-07-2019-0519
15. Balmer JM (1995) Corporate branding and connoisseurship. J Gen Manag 21(1):24–46
16. Balmer JM (1998) Corporate identity and the advent of corporate marketing. J Mark Manag 14(8):963–996
17. Baran T, Kiziloglu M (2018) Effect of greenwashing advertisements on organizational image. In: Organizational culture and behavioral shifts in the green economy. IGI Global, pp 59–77
18. Barich H, Kotler P (1991) A framework for marketing image management. MIT Sloan Manag Rev 32(2):94
19. Baron RA, Franklin RJ, Hmieleski KM (2013) Why entrepreneurs often experience low, not high, levels of stress: the joint effects of selection and psychological capital. J Manag 42(3):742–768. https://doi.org/10.1177/0149206313495411
20. Bathmanathan V, Hironaka C (2016) Sustainability and business: what is green corporate image? In: IOP conference series: earth and environmental science
21. Battilana J, Dorado S (2010) Building sustainable hybrid organizations: the case of commercial microfinance organizations. Acad Manag J 53(6):1419–1440
22. Bezhani I (2010) Intellectual capital reporting at UK universities. J Intellect Cap 11(2):179–207. https://doi.org/10.1108/14691931011039679
23. Bowen F, Aragon-Correa JA (2014) Greenwashing in corporate environmentalism research and practice: the importance of what we say and do, vol 27. Sage Publications, Los Angeles, CA, pp 107–112
24. Bozkurt M (2018) Corporate image, brand and reputation concepts and their importance for tourism establishments. Int J Contemp Tourism Res 2(2):60–66
25. Brammer S, Millington A, Rayton B (2007) The contribution of corporate social responsibility to organizational commitment. Int J Human Resour Manag 18(10):1701–1719
26. Brannan DB, Heeter J, Bird L (2012) Made with renewable energy: how and why companies are labeling consumer products. Contract 303:275–3000
27. Bromley DB (1993) Reputation, image and impression management. Wiley
28. Byus K, Deis DR (2015) Disadvantaged consumers on the southern border: the social entrepreneurs' challenge. J Ethics Entrepreneurship 5(2):55
29. Calic G, Mosakowski E (2016) Kicking off social entrepreneurship: how a sustainability orientation influences crowdfunding success. J Manage Stud 53(5):738–767
30. Campbell JL (2018) 2017 Decade award invited article reflections on the 2017 decade award: corporate social responsibility and the financial crisis. Acad Manag Rev 43(4):546–556
31. Carroll AB (1979) A three-dimensional conceptual model of corporate performance. Acad Manag Rev 4(4):497–505
32. Carroll AB (1999) Corporate social responsibility: evolution of a definitional construct. Bus Soc 38(3):268–295
33. Carroll AB (2016) Carroll's pyramid of CSR: taking another look. Int J Corp Soc Responsib 1(1):1–8
34. Castillo-Benancio S, Alvarez-Risco A, Esquerre-Botton S, Leclercq-Machado L, Calle-Nole M, Morales-Ríos F et al (2022) Circular economy for packaging and carbon footprint. In: Alvarez-Risco A, Muthu SS, Del-Aguila-Arcentales S (eds) Circular economy: impact on carbon and water footprint. Springer, Singapore, pp 115–138. http://doi.org/10.1007/978-981-19-0549-0_6
35. Chandna V (2022) Social entrepreneurship and digital platforms: Crowdfunding in the sharing-economy era. Bus Horiz 65(1):21–31. https://doi.org/10.1016/j.bushor.2021.09.005
36. Chang N-J, Fong C-M (2010) Green product quality, green corporate image, green customer satisfaction, and green customer loyalty. Afr J Bus Manage 4(13):2836–2844

37. Choudhary A, Gokarn S (2013) Green marketing: a means for sustainable development. J Arts Sci Commer 4(3):3
38. Contreras-Taica A, Alvarez-Risco A, Arias-Meza M, Campos-Dávalos N, Calle-Nole M, Almanza-Cruz C et al (2022) Virtual education: carbon footprint and circularity. In: Alvarez-Risco A, Muthu SS, Del-Aguila-Arcentales S (eds) Circular economy: impact on carbon and water footprint. Springer, Singapore, pp 265–285. http://doi.org/10.1007/978-981-19-0549-0_13
39. Cooney TM (2005) What is an entrepreneurial team? vol 23. Sage Publications, pp 226–235
40. Crane A, McWilliams A, Matten D, Moon J, Siegel DS (2008) The Oxford handbook of corporate social responsibility. OUP, Oxford
41. Cuya-Velásquez BB, Alvarez-Risco A, Gomez-Prado R, Juarez-Rojas L, Contreras-Taica A, Ortiz-Guerra A et al (2022) Circular economy for food loss reduction and water footprint. In: Alvarez-Risco A, Muthu SS, Del-Aguila-Arcentales S (eds) Circular economy: impact on carbon and water footprint. Springer, Singapore, pp 65–91. http://doi.org/10.1007/978-981-19-0549-0_4
42. D'Avanzo E, Franch M, Borgonovi E (2021) Ethics and sustainable management. An empirical modelling of Carroll's pyramid for the Italian landscape. Sustainability 13(21):12057
43. Dacin MT, Dacin PA, Tracey P (2011) Social entrepreneurship: a critique and future directions. Organ Sci 22(5):1203–1213
44. Dai Y, Abdul-Samad Z, Chupradit S, Nassani AA, Haffar M, Michel M (2021) Influence of CSR and leadership style on sustainable performance: moderating impact of sustainable entrepreneurship and mediating role of organizational commitment. Econ Res Ekonomska Istraživanja 1–23
45. De-la-Cruz-Diaz M, Alvarez-Risco A, Jaramillo-Arévalo M, de las Mercedes Anderson-Seminario M, Del-Aguila-Arcentales S (2022) 3D print, circularity, and footprints. In: Alvarez-Risco A, Muthu SS, Del-Aguila-Arcentales S (eds) Circular economy: impact on carbon and water footprint. Springer, Singapore, pp 93–112. http://doi.org/10.1007/978-981-19-0549-0_5
46. de las Mercedes Anderson-Seminario M, Alvarez-Risco A (2022) Better students, better companies, better life: circular learning. In: Alvarez-Risco A, Muthu SS, Del-Aguila-Arcentales S (eds) Circular economy: impact on carbon and water footprint. Springer, Singapore, pp 19–40. http://doi.org/10.1007/978-981-19-0549-0_2
47. Dees JG (2009) Social ventures as learning laboratories in innovations: technology, governance, and globalization. MIT Press Journal
48. Del-Aguila-Arcentales S, Alvarez-Risco A, Muthu SS (2022) Measuring circular economy. In: Alvarez-Risco A, Muthu SS, Del-Aguila-Arcentales S (eds) Circular economy: impact on carbon and water footprint. Springer, Singapore, pp 3–17. http://doi.org/10.1007/978-981-19-0549-0_1
49. Demirel P, Li QC, Rentocchini F, Tamvada JP (2019) Born to be green: new insights into the economics and management of green entrepreneurship. Small Bus Econ 52(4):759–771
50. Dwivedi A, Weerawardena J (2018) Conceptualizing and operationalizing the social entrepreneurship construct. J Bus Res 86:32–40
51. Dyllick T, Muff K (2015) Clarifying the meaning of sustainable business: introducing a typology from business-as-usual to true business sustainability. Organ Environ 29(2):156–174. https://doi.org/10.1177/1086026615575176
52. Ebrahimi P, Mirbargkar SM (2017) Green entrepreneurship and green innovation for SME development in market turbulence. Eurasian Bus Rev 7(2):203–228
53. Emerson J, Twersky F (1996) New social entrepreneurs: the success, challenge and lessons of non-profit enterprise creation. Am J Ind Bus Manag 6(26):2016
54. Esposito B, Sessa MR, Sica D, Malandrino O (2021) Corporate social responsibility engagement through social media. Evidence from the University of Salerno. Adm Sci 11(4). http://doi.org/10.3390/admsci11040147
55. Esquerre-Botton S, Alvarez-Risco A, Leclercq-Machado L, de las Mercedes Anderson-Seminario M, Del-Aguila-Arcentales S (2022) Food loss reduction and carbon footprint

practices worldwide: a benchmarking approach of circular economy. In: Alvarez-Risco A, Muthu SS, Del-Aguila-Arcentales S (eds) Circular economy: impact on carbon and water footprint. Springer, Singapore, pp 161–179. http://doi.org/10.1007/978-981-19-0549-0_8
56. Fatmawati I, Alikhwan MA (2021) How does green marketing claim affect brand image, perceived value, and purchase decision? In: E3S web of conferences
57. Fazio G (2010) Competition and entrepreneurship as engines of growth. UCL (University College London). https://discovery.ucl.ac.uk/id/eprint/624499/
58. Fernández-Gago R, Cabeza-García L, Nieto M (2018) Independent directors' background and CSR disclosure. Corp Soc Responsib Environ Manag 25(5):991–1001
59. Fernandez LV, Jara-Bertin M, Pineaur FV (2015) Social responsibility practices, corporate reputation and financial performance/Prácticas de responsabilidad social, reputación corporativa y desempeño financiero/Praticas de responsabilidade social, reputacao corporativa e desempenho financeiro. RAE 55(3):329–345
60. Flavián C, Guinaliu M, Torres E (2005) The influence of corporate image on consumer trust: a comparative analysis in traditional versus internet banking. Internet Res 15(4):447–470. https://doi.org/10.1108/10662240510615191
61. García AC, Cuervo Á, Ribeiro D, Roig S (2007) Entrepreneurship: concepts, theory and perspective. Springer
62. Gómez-Prado R, Alvarez-Risco A, Sánchez-Palomino J, de las Mercedes Anderson-Seminario M, Del-Aguila-Arcentales S (2022) Circular economy for waste reduction and carbon footprint. In: Alvarez-Risco A, Muthu SS, Del-Aguila-Arcentales S (eds) Circular economy: impact on carbon and water footprint. Springer, Singapore, pp 139–159. http://doi.org/10.1007/978-981-19-0549-0_7
63. Gómez-Rico M, Molina-Collado A, Santos-Vijande ML, Molina-Collado MV, Imhoff B (2022) The role of novel instruments of brand communication and brand image in building consumers' brand preference and intention to visit wineries. Curr Psychol. https://doi.org/10.1007/s12144-021-02656-w
64. Gray ER, Balmer JM (1998) Managing corporate image and corporate reputation. Long Range Plan 31(5):695–702
65. Harper DA (2008) Towards a theory of entrepreneurial teams. J Bus Ventur 23(6):613–626
66. Helm S (2007) Social entrepreneurship: defining the nonprofit behavior and creating an instrument for measurement. University of Missouri-Kansas City
67. Henton D, Melville J, Walesh K (1997) The age of the civic entrepreneur: restoring civil society and building economic community. Natl Civ Rev 86(2):149–157
68. Hussain I, Nazir M, Hashmi SB, Di Vaio A, Shaheen I, Waseem MA, Arshad A (2021) Green and sustainable entrepreneurial intentions: a mediation-moderation perspective. Sustainability 13(15). http://doi.org/10.3390/su13158627
69. IBM Corp (2020) IBM 31st annual environmental report. Retrieved 01/01/2022 from https://www.ibm.com/ibm/environment/annual/IBMEnvReport_2020.pdf
70. Indarti S (2020) The effects of education and training, management supervision on development of entrepreneurship attitude and growth of small and micro enterprise. Int J Organ Anal 29(1):16–34. https://doi.org/10.1108/IJOA-09-2019-1890
71. Jeffs L (2006) Social entrepreneurs and social enterprises—do they have a future in New Zealand? In: 51st ICSB world conference, Melbourne. http://www.communityresearch.org.nz/wp-content/uploads/formidable/jeffs3.pdf
72. Jo Hatch M, Schultz M (1997) Relations between organizational culture, identity and image. Eur J Mark 31(5/6):356–365. https://doi.org/10.1108/eb060636
73. Johnson S (2003) Social entrepreneurship literature review. New Acad Rev 2:42–56
74. Juarez-Rojas L, Alvarez-Risco A, Campos-Dávalos N, de las Mercedes Anderson-Seminario M, Del-Aguila-Arcentales S (2022) Water footprint in the textile and food supply chain management: trends to become circular and sustainable. In: Alvarez-Risco A, Muthu SS, Del-Aguila-Arcentales S (eds) Circular economy: impact on carbon and water footprint. Springer, Singapore, pp 225–243. http://doi.org/10.1007/978-981-19-0549-0_11

75. Kennedy SH (1977) Nurturing corporate images. Eur J Mark 11(3):119–164. https://doi.org/10.1108/EUM0000000005007
76. Khan MG, Khan O (2016) Corporate identity, corporate branding and brand image. http://urn.kb.se/resolve?urn=urn:nbn:se:lnu:diva-57516
77. Khan MI, Khalid S, Zaman U, José AE, Ferreira P (2021) Green paradox in emerging tourism supply chains: achieving green consumption behavior through strategic green marketing orientation, brand social responsibility, and green image. Int J Environ Res Public Health 18(18):9626
78. Ko E, Hwang YK, Kim EY (2013) Green marketing' functions in building corporate image in the retail setting. J Bus Res 66(10):1709–1715
79. KritiKoS AS (2014) Entrepreneurs and their impact on jobs and economic growth. IZA World of Labor. http://doi.org/10.15185/izawol.8
80. Lins KV, Servaes H, Tamayo A (2017) Social capital, trust, and firm performance: the value of corporate social responsibility during the financial crisis. J Financ 72(4):1785–1824
81. Low MP, Bu M (2022) Examining the impetus for internal CSR practices with digitalization strategy in the service industry during COVID-19 pandemic. Bus Ethics Environ Responsib 31(1):209–223. http://doi.org/10.1111/beer.12408
82. Lozano R (2015) A holistic perspective on corporate sustainability drivers. Corp Soc Responsib Environ Manag 22(1):32–44
83. Márquez FOS, Ortiz GER (2021a) Disruptive thinking and entrepreneurship: fictions and disenchantments. Glob Bus Rev 0972150920988647
84. Márquez FOS, Ortiz GER (2021b) Entrepreneurship and Bédard's theoretical rhombus: an extended proposal for management field. Glob Bus Rev 0972150920988649
85. Martin RL, Osberg S (2007) Social entrepreneurship: the case for definition. https://www.ngobiz.org/picture/File/Social%20Enterpeuneur-The%20Case%20of%20Definition.pdf
86. Massey J (2011) Organizational image management. The practice of organizational communication. McGraw-Hill, New York
87. Massey JE (2016) A theory of organizational image management. Int J Manag Appl Sci 2(1):1–6
88. Melewar TC, Jenkins E (2002) Defining the corporate identity construct. Corp Reput Rev 5(1):76–90
89. Miller TL, Grimes MG, McMullen JS, Vogus TJ (2012) Venturing for others with heart and head: How compassion encourages social entrepreneurship. Acad Manag Rev 37(4):616–640
90. Mohtar S, Rahim H (2014). Social entrepreneurship, entrepreneurial leadership and organizational performance: a mediation conceptual framework. Aust J Basic Appl Sci 8(23):184, 190
91. Morales-Ríos F, Alvarez-Risco A, Castillo-Benancio S, de las Mercedes Anderson-Seminario M, Del-Aguila-Arcentales S (2022) Material selection for circularity and footprints. In: Alvarez-Risco A, Muthu SS, Del-Aguila-Arcentales S (eds) Circular economy: impact on carbon and water footprint. Springer, Singapore, pp 205–221. http://doi.org/10.1007/978-981-19-0549-0_10
92. Nevárez VL, Féliz BDZ (2019) Social responsibility in the dimensions of corporate citizenship. A case study in agricultural manufacturing [La responsabilidad social en las dimensiones de la ciudadanía corporativa. Un estudio de caso en la manufactura agrícola]. CIRIEC-España, revista de economía pública, social y cooperativa (97):179–211
93. Öztürk R (2020) Green marketing. In: Current and historical debates in social sciences: field studies and analysis, p 195
94. Partzsch L, Ziegler R (2011) Social entrepreneurs as change agents: a case study on power and authority in the water sector. Int Environ Agreements Polit Law Econ 11(1):63–83
95. Pinzón N, Montero J, González-Pernía JL (2021) The influence of individual characteristics on getting involved in an entrepreneurial team: the contingent role of individualism. Int Entrepreneurship Manag J 1–38
96. Polinkevych O, Kamiński R (2018) Corporate image in behavioral marketing of business entities. Innov Mark 14(1):33–40

97. Poon Teng Fatt J, Wei M, Yuen S, Suan W (2000) Enhancing corporate image in organisations. Manag Res News 23(5/6):28–54. https://doi.org/10.1108/01409170010782037
98. Rahim HL, Mohtar S (2015) Social entrepreneurship: a different perspective. Int Acad Res J Bus Technol 1(1):9–15
99. Reilly AH, Hynan KA (2014) Corporate communication, sustainability, and social media: it's not easy (really) being green. Bus Horiz 57(6):747–758
100. Roberts D, Woods C (2005) Changing the world on a shoestring: the concept of social entrepreneurship. Univ Auckland Bus Rev 7(1):45–51
101. Saiz-Álvarez JM, Cuervo-Arango C, Coduras A (2013) Entrepreneurial strategy, innovation, and cognitive capabilities: what role for intuitive SMEs? J Small Bus Strateg 23(2):29–40
102. Samaibekova Z, Choyubekova G, Isabaeva K, Samaibekova A (2021) Corporate sustainability and social responsibility. In: E3S web of conferences
103. Schmitt BH, Simonson A, Marcus J (1995) Managing corporate image and identity. Long Range Plan 28(5):82–92
104. Shen J, Jiuhua Zhu C (2011) Effects of socially responsible human resource management on employee organizational commitment. Int J Human Resour Manag 22(15):3020–3035
105. Sheppard J, Mahdad M (2021) Unpacking hybrid organizing in a born green entrepreneurial company. Sustainability 13(20):11353
106. Siano A, Vollero A, Conte F, Amabile S (2017) "More than words": expanding the taxonomy of greenwashing after the Volkswagen scandal. J Bus Res 71:27–37
107. Siboni B, del Sordo C, Pazzi S (2013) Sustainability reporting in state universities: an investigation of Italian pioneering practices. Int J Soc Ecol Sustain Dev (IJSESD) 4(2):1–15
108. Socorro Márquez FO, Reyes Ortiz GE (2021) The disruptive triad and entrepreneurship: a theoretical model. J Innov Entrepreneurship 10(1):1–21
109. Stawicka E (2021) Sustainable development in the digital age of entrepreneurship. Sustainability 13(8). http://doi.org/10.3390/su13084429
110. Teles D, Schachtebeck C (2019) Entrepreneurial orientation in South African social enterprises. Entrepreneurial Bus Econ Rev 7(3):83–97. http://doi.org/10.15678/EBER.2019.070305
111. Thörnqvist C, Kilstam J (2021) Aligning corporate social responsibility with the United Nations' sustainability goals: trickier than it seems? Economics 9(1):161–177
112. Tsai P-H, Lin G-Y, Zheng Y-L, Chen Y-C, Chen P-Z, Su Z-C (2020) Exploring the effect of Starbucks' green marketing on consumers' purchase decisions from consumers' perspective. J Retail Consum Serv 56:102162
113. Turker D (2009) How corporate social responsibility influences organizational commitment. J Bus Ethics 89(2):189–204
114. Waddock SA, Post JE (1991) Social entrepreneurs and catalytic change. Public Adm Rev 393–401
115. Wei J, Xiong R, Hassan M, Shoukry AM, Aldeek FF, Khader JA (2021) Entrepreneurship, corporate social responsibilities, and innovation impact on banks' financial performance. Front Psychol 12. https://doi.org/10.3389/fpsyg.2021.680661
116. Widyastuti S, Said M, Siswono S, Firmansyah DA (2019) Customer trust through green corporate image, green marketing strategy, and social responsibility: a case study. Eur Res Stud J 22(2):83–99
117. Wong T-A, Reevany BM (2019) Understanding corporate social responsibility (CSR) among micro businesses using social capital theory. Int J Bus Soc 20(2):675–690
118. Yen NTH (2014) Corporate image in the context of organizational transformation: an integrative theoretical model. J Econ Dev 16(3):96–116

Creation of Sustainable Enterprises from the Female Directionality

Berdy Briggitte Cuya-Velásquez, Aldo Alvarez-Risco, María de las Mercedes Anderson-Seminario, and Shyla Del-Aguila-Arcentales

Abstract Women's sustainable entrepreneurship has increased significantly due to all parties' demands and the positive contributions to sustainable development (SDG). Entrepreneurs are motivated to carry out their sustainable enterprises by benefiting society, the planet, and the economy. However, women entrepreneurs face various barriers imposed by a patriarchal power system. The main barriers are gender discrepancy, little government support, lack of financing, and individual capacities. Additionally, the family, especially the husband, promotes or limits the creation and execution of women's businesses. For this reason, this research investigates female leadership, business legitimacy, motivations, and barriers when creating female enterprises. The information was compiled from a vast number of articles related to sustainable female entrepreneurship found in the Scopus database.

Keywords Entrepreneurs · Entrepreneurship · Woman · Women · Sustainable development goals · SDG · Family · Sustainability · Circularity

1 Introduction

In the business world, sustainability has increasingly become an essential aspect for the economic growth of the business and the country, as well as for the care of the planet and human beings [238, 251], which are motivated mainly by altruistic attitudes and extrinsic rewards [266]. To do this, sustainable entrepreneurs encompass the three dimensions of SDGs [83] which is because the theories of sustainability and entrepreneurship are linked [258], since the success of entrepreneurship depends on the capabilities of the company and the incorporation of sustainability [145, 281].

The increased interest in sustainable enterprises is due to the creation of specialized magazines on the subject and the increased demand for research by businesses

B. B. Cuya-Velásquez · A. Alvarez-Risco · M. de las Mercedes Anderson-Seminario
Universidad de Lima, Lima, Peru

S. Del-Aguila-Arcentales (✉)
Escuela de Posgrado, Universidad San Ignacio de Loyola, Lima, Peru
e-mail: sdelaguila@usil.edu.pe

A. Alvarez-Risco et al. (eds.), *Footprint and Entrepreneurship*, Environmental Footprints and Eco-design of Products and Processes,
https://doi.org/10.1007/978-981-19-8895-0_7

and national and international institutions [251]. However, [10] found a negative relationship between business cognitions and sustainability because entrepreneurs sometimes do not consider sustainability issues. Barrachina Fernández et al. [76] found that awareness of the female sex is essential to creating sustainable businesses, but they also pointed out that further study is necessary due to the lack of knowledge and research on the subject.

Women are more likely to worry about climate change than men [249]. Since men are more indifferent to climate change [182, 225, 260]. However, [239] pointed out that men with more experience carry out community actions. On the other hand, gender and entrepreneurship accentuate patriarchal business dominance but leave gender challenges at entrepreneurship.

Women entrepreneurs present many barriers or challenges that make it impossible to create or maintain their businesses. Some barriers include insufficient capital [210], sociocultural [148], and little government support [168]. In some countries such as Iran, women act according to norms of femininity dominated by a male leadership structure, while cultural factors in society limit Western women.

In this sense, there is gender discrimination because the businesses developed by women are smaller than those of men [76]. In Europe, 37% of women are managers, 28% work in the administrative area, while only a fifth are in executive positions. [134] determined that in the UK, 45% of companies are led by women. In the Arab region, there is a minority number of female business leaders [8], evidencing more excellent leadership perceived by male entrepreneurs [176]. Currently, there is little research on female entrepreneurship in emerging countries [110], social and cultural diversity [79].

2 Entrepreneurship and Sustainable Entrepreneurship Gender and Sustainable Development (SDG)

The sustainable strategies developed with entrepreneurs for sustainability need consider population [13, 17, 19, 21, 31, 47, 56–58, 67, 75, 99, 124, 142, 174, 195, 206, 233, 247, 265, 272, 276], organizations [5, 32, 36, 40, 52, 84, 88–90, 184, 191, 222, 236, 242, 256, 274, 275], and companies [6, 20, 24, 46, 73, 80, 111, 121, 193, 211, 223, 229, 243, 264] and activities as sports [108, 152, 171].

Also, it is included in sustainability actions of tourism [14, 18, 25, 93–95, 138, 207, 221, 231], education [4, 21, 32, 45, 53–55, 98, 162, 186, 230], circular economy approach [2, 15, 37, 78, 123, 175, 190, 192, 261], prices [65, 66, 70, 71, 153, 160, 180, 186, 202, 241], and hospitality [3, 28, 132, 147, 170, 197, 216, 252, 271].

Then, new entrepreneurship initiatives are related to intellectual property [9, 39, 44, 63, 74, 77, 131, 200], health [1, 11, 22, 23, 35, 41–43, 48–50, 50, 59–61, 91, 103, 112, 123, 130, 199, 204, 227, 234, 277–279], and research [107, 179, 212, 214, 215, 220, 224, 232, 235].

Sustainable entrepreneurship is essential since it helps solve problems related to the lack of sustainability in the three dimensions of SDG [127, 146, 246]. The concept of sustainable entrepreneurship is associated with the SDG, which encompasses problems of the planet, society, and the economy [144]. This type of entrepreneurship generates economic growth [143], produces employment [137, 240], innovation [38], and is a way out of poverty [140].

Women entrepreneurs contribute substantially to sustainable entrepreneurship because they are considered part of economic growth and prosperity [157]. For this reason, some policies help increase female entrepreneurship [85]. In some cases, social enterprises provide economic freedom and recognition to women [209].

Al-Lawati et al. [30] and Gustavsson [158] found outstanding concepts of female entrepreneurship. The entrepreneurial spirit of women refers to female empowerment contributing to economic growth and female emancipation [64, 262]. Innovation is part of the business skills of women entrepreneurs, as well as creativity, opportunity identification, leadership skills, marketing, and social awareness [27]. Therefore, female entrepreneurs must generate value for clients through a sustainable entrepreneurial spirit that involves innovations within the business [208] at the micro and macro levels [113].

However, there is also a gender discrepancy [16, 210] in business performance [168]. From a psychological perspective, gender stereotypes affect men and women [129]. The female role has changed, but the gap between male and female leadership persists [72, 218]. For example, in the Middle East and South Korea, women face problems of patriarchy regarding power and stereotypes about female skills or abilities [105]. The leadership and development of women in India and Korea are affected by the male dominance imparted in these countries [101]. Therefore, the inclusion of sustainable female entrepreneurship is seen as an alternative to reduce social inequality and damage to the planet [206].

It is essential to include women in sustainable enterprises because women entrepreneurs achieve the Sustainable Development Goals (SDGs). Goal five of the SDGs includes gender equality and female empowerment [26, 226, 259]. Therefore, companies must integrate diversity and inclusion strategies to drive sustainability [81]. The SDGs have three dimensions: social, economic, and environmental [126, 185, 205, 237, 255]. These three dimensions are the triple bottom line [139]. Goal number five is in the social dimension [134] and is an elementary goal for the SDGs [250].

Authors [167] described three African-based, women-led social enterprises promoting sustainability through social investments. The companies are Heartfelt Project, Bright Kids Uganda, and Chikumbusos, located in South Africa, Uganda, and Lusaka. The Heartfelt Project company is led by Martha Letsoalo, who previously worked as a domestic worker. Currently, it provides employment to fifteen women through its education and training workshop, in addition its craft products are designed by felt and beading and are sold locally and internationally.

Bright Kids Uganda is led by Victoria Nalongo, a fisherman's daughter. Her business is a care center and school that educates defenseless children and empowers physically and psychologically abused women. In addition, it provides microcredits to women looking to start a business. Finally, the Chikumbuso enterprise is led by Gertrude, who was abandoned; for this reason, she created a place of resilience to educate children and provide jobs for women.

3 Motivators and Barriers of Female Entrepreneurship

Motivation is essential when starting a business [187]. Women entrepreneurs' primary motivators are confidence in their entrepreneurial skills [213], stopping depending on traditional jobs [198]. For [168], business motivation and optimism positively influence the performance of female entrepreneurs, and achievement motivation significantly impacts entrepreneurial intention [280] and a pro-business culture.

According to Daulerio [116], the main intrinsic motivation factors involved in creating a female business have appeared for approximately 30 years. Fig. 1 shows these three factors numbered 1–3, while around them are other motivational factors designed and summarized by [257].

Other motivators include financial freedom, independence, and personal growth [269]. Just as there are motivators to start a business, there are also barriers that cause women entrepreneurs to give up when starting their business or continuing with their business project [188, 213]; in the same way, there are barriers to achieving leadership positions [7].

Among the barriers female entrepreneurs face is the lack of knowledge regarding the relevance of digital transformation in the execution of ventures [245]. Digitization is relevant because sustainability is linked to technological capabilities and innovation [163, 169]. Women entrepreneurs have difficulties taking advantage of their limited resources when choosing the wrong partners for R&D [82, 166, 177]. Present innovation problems when conducting sustainable businesses due to limited access to financing [208] and knowledge exchange imply that they obtain fewer laboratories or places with the latest technologies than men [100].

Regarding the lack of financing, in Arabia, women require their father's consent to take out a loan, but sometimes they tend to rely on limited personal savings [198]. In this context, [118] found that female entrepreneurs are less likely to request a loan at the beginning of their business because they face financial restrictions or high-interest rates [69]. On the other hand, [141] determined that entrepreneur's women do not receive gender discrimination when applying for a loan but to signs of gender bias in the positive stage of a country's economy.

In northeastern Italy, female entrepreneurs present sustainability-related capital difficulties when starting their businesses. Likewise, female entrepreneurs, compared to male entrepreneurs, are more sensitive to the difficulty of investment required and finding the appropriate number of trained employees who can perform their duties effectively and efficiently [115]. Women entrepreneurs from emerging countries present various challenges when starting their businesses [217]. Latina entrepreneurs present feelings of fear, gender discrimination, poor financial management, and lack

Fig. 1 Factors of female business motivation. *Source* Daulerio [116] and ul Haq et al. [257]

of skills to help choose the right place for business development [13, 104]. In Oman, women entrepreneurs present sociocultural barriers [148], little commercial support from the government [150], and organizational problems [149]. In Zanzibar, Chile, France, and the UK, there are barriers related to gender power disparities within the fisheries sector. Additionally, women entrepreneurs in this sector have limited participation due to procedural injustices [159].

Chreim [106] mentioned that Asian entrepreneurs present gender barriers and increased entrepreneurial immigrants. [281] found that female entrepreneurs face finding reliable employees, insufficient capital, little government support, and high competition in China and Vietnam. Iranian women lack opportunities [168], and Bahrain women entrepreneurs lack knowledge and little experience in entrepreneurship [34]. Table 1 shows the types of barriers to female entrepreneurship according to the individual, organizational, and national levels investigated by different authors.

In that same context, [228] identified 14 barriers faced by female entrepreneurs: reduced interest in commercial activities, limited capital, few strategic practices, slow growth, few profits, high failure rate, lack of entrepreneurial skills, social skills

Table 1 Business barriers for women entrepreneurs

Barrier levels	Features
Individual barriers	Personality traits, level of education. Also, low skill, training, and experience
Organizational barriers	Financial or capital restrictions
National barriers	Difficulties interacting with men when they work

Source Faisal et al. [136] and Modarresi et al. [201], and Gupta and Mirchandani [156]

problems, limited management knowledge business, limited technological knowledge, education problems, difficulty in taking on risks, and lack of institutional and family support. Additionally, [136] found a range of barriers that inhibit female entrepreneurship growth, as shown in Fig. 2.

Fig. 2 Growth barriers for female entrepreneurs. *Source* Faisal et al. [136]

4 Key Factors for the Success of Female Entrepreneurs: The Family and Business Legitimacy

Family support significantly affects the success of female entrepreneurship [150, 268]. Women's businesses benefit from family support, which reduces the tension between the business and the family [270]. The family can provide support in different ways to entrepreneurs, but mainly they can be emotional support [173]. In contrast, family conflicts produce emotional exhaustion for female entrepreneurs. Therefore, family ties drive or limit female entrepreneurs [194, 228].

Gender ideology influences women's decisions about professional and family aspects [189], [263]), which is related to the conflict between work and family [181, 183] because women are still in charge of housework [97, 219]. Sometimes women seek to hide their obligations at home or avoid disclosing their workplaces to counteract the tensions that originate from the interpretations of social norms based on working women [267].

Women negotiate with their husbands to access business networks, while negotiations with male colleagues achieve personal and business legitimacy and are considered entrepreneurs [244]. Legitimacy also helps construct female business identity to close the gap between business legitimacy and femininity [248], contributing to changing institutional policies that discriminate against women entrepreneurs [253]. Additionally, business identity changes women entrepreneurs' context [155]. The SDGs can also foster the legitimacy of women entrepreneurs [154].

Jaim [165] found that the geographical location must be considered to consider the family as a positive or negative factor since the importance in non-Western countries differs from that of developed countries. For example, in Bangladesh, women need the help of their husbands to apply for bank loans since the norms are patriarchal [165]. Table 2 shows the classification made by the author to publicize the influence of each family aspect on female entrepreneurship in developing countries.

Recent suggested lectures about green approach

- Waste reduction and carbon footprint [151]
- Material selection for circularity and footprints [203]
- 3D print, circularity, and footprints [117]
- Virtual tourism [117]
- Leadership for sustainability in crisis time [51]
- Virtual education and circularity [109]
- Circular economy for packaging [96]
- Students oriented to circular learning [119]
- Food and circular economy [114]
- Water footprint and food supply chain management [172]
- Waste footprint [68]
- Measuring circular economy (Del-Aguila-[122]
- Carbon footprint [133].

Table 2 Aspects linking the family role and female entrepreneurship in developing countries

Aspects	Problems associated with the family	Effects on entrepreneurship
Housework and child-rearing	There are high expectations about housework, taking care of multiple children, or maternity difficulties	Reduced availability to run the business, stress when working, and sometimes running the business from home lead to fewer sales and revenue
The function of husbands	The spouses exercise their function with high signs of patriarchy, complexity to solve the family, and the existence of polygamy	Husbands as assistants in entrepreneurship or as a negative influence to develop the female business
Other family members	Many family members, relatives, are involved in legal problems or engaging in extreme religious practices	Dependence on relatives and negative influences of relatives in the creation of female entrepreneurship

Source Jaim [165]

5 Closing Remarks

Female entrepreneurship is a concept that has grown due to the demand of female entrepreneurs, researchers, governments, and institutions [249, 251], which is because it generates significant benefits for society, the economy, and the environment [127]. Women entrepreneurs show great social and environmental concern, as well as the innovation of their products [102], which allows them to be a vital objective of the SDGs that promotes gender inclusion [26, 81, 226]. Women entrepreneurs have different motivational factors in developing sustainable enterprises, such as life ambition, challenge and achievement, family responsibility, and breaking barriers [116, 257].

Entrepreneurs in developing countries create their companies out of the need for well-being and to take advantage of opportunities [62]. Gender equality has provided the possibility for women to be educated and have opportunities to prosper economically [87]. For example, [8] pointed out that women's education in the Arab region contributes to the ability to withstand gender discrimination. At the same time, [128] found that educated women can resist the barriers related to a patriarchal world of work. On the other hand, education increases the ability of women to generate income [125].

Women entrepreneurs are exposed to many barriers that prevent or stop the development of their companies [213]. Among the main barriers is lack of financing, gender problems, the prevalence of patriarchy in business, family problems, lack of government support, lack of technological knowledge, investment risk, among others [136, 157, 201, 228]. Therefore, gender-related entrepreneurship should be investigated as an essential aspect within the institutional environment [92] and economic development [120]. Berguiga and Adair [86] pointed out that the government must provide basic social protection coverage as a financing product to women who have MSMEs.

The authorities must promote sustainable businesses and create policies to encourage female entrepreneurship [217]. The government must provide financing, social support, create sustainable policies and business empowerment to women entrepreneurs [12, 135, 173], as well as foster a stereotype of impartial entrepreneurship concerning gender [164], and carry out gender parity policies in the face of the existence of gender discrimination [218]. However, [29] questioned the importance of implementing current policies that encourage more women to start a business since there is gender discrimination and little social support, while [254] argued that women must accept the presence of a discriminatory bias and understand that sometimes their lack of professional progress is due to their limitations. However, [102] determined that a positive institutional context reduces the uncertainties or barriers presented by female entrepreneurs.

On the other hand, [178] found that women entrepreneurs require self-efficacy and resilience to face business barriers. Additionally, they pointed out that female entrepreneurs see themselves as successful businesswomen despite patriarchal norms. Additionally, [213] found that failures make entrepreneurs make better decisions in their subsequent businesses. The same entrepreneurs can carry out community activities through forums to support other women who seek to undertake [161], and create a network exclusively for women who seek to grow professionally and individually, such as the Girlboss and Her network, which have a similar vision and mission [33]. Then, women could carry out collective and not individualized social enterprises [196].

All efforts by women must be supported by authorities, and the universities need to promote more entrepreneurship leaded by female students.

References

1. Aaron L (2020) The oft forgotten part of nutrition: lessons from an integrated approach to understand consumer food safety behaviors. Curr Develop Nutr 4(Supplement_2):1322–1322. https://doi.org/10.1093/cdn/nzaa059_039
2. Abad-Segura E, Batlles-delaFuente A, González-Zamar M-D, Belmonte-Ureña LJ (2021) Implications for sustainability of the joint application of bioeconomy and circular economy: a worldwide trend study. Sustainability 13(13):7182. https://doi.org/10.3390/su13137182
3. Abad-Segura E, Cortés-García FJ, Belmonte-Ureña LJ (2019) The Sustainable approach to corporate social responsibility: a global analysis and future trends. Sustainability 11(19):5382. https://doi.org/10.3390/su11195382
4. Abad-Segura E, González-Zamar M-D (2021) Sustainable economic development in higher education institutions: a global analysis within the SDGs framework. J Cleaner Prod 294:126133. https://doi.org/10.1016/j.jclepro.2021.126133
5. Abad-Segura E, González-Zamar M-D, Infante-Moro JC, Ruipérez García G (2020) Sustainable management of digital transformation in higher education: global research trends. Sustainability 12(5):2107. https://doi.org/10.3390/su12052107

6. Abad-Segura E, Morales M E, Cortés-García FJ, Belmonte-Ureña LJ (2020) Industrial processes management for a sustainable society: global research analysis. Processes 8(5). https://doi.org/10.3390/pr8050631
7. Abalkhail JM (2017) Women and leadership: challenges and opportunities in Saudi higher education. Career Dev Int 22(2):165–183. https://doi.org/10.1108/CDI-03-2016-0029
8. Abalkhail JM (2019) Women's career development in an Arab Middle Eastern context. Hum Resour Dev Int 22(2):177–199. https://doi.org/10.1080/13678868.2018.1499377
9. Abdaljaleel M, Singer EJ, Yong WH (2019) Sustainability in Biobanking. In: Yong WH (ed) Biobanking: methods and protocols. Springer New York, pp 1–6. https://doi.org/10.1007/978-1-4939-8935-5_1
10. Abdelnaeim SM, El-Bassiouny N (2020) The relationship between entrepreneurial cognitions and sustainability orientation: the case of an emerging market. J Entrepreneurship Emerg Economies 13(5):1033–1056. https://doi.org/10.1108/JEEE-03-2020-0069
11. Aberilla JM, Gallego-Schmid A, Stamford L, Azapagic A (2020) Environmental sustainability of cooking fuels in remote communities: life cycle and local impacts. Sci Total Environ 713:136445. https://doi.org/10.1016/j.scitotenv.2019.136445
12. Abuhussein T, Koburtay T (2021) Opportunities and constraints of women entrepreneurs in Jordan: an update of the 5Ms framework. Int J Entrep Behav Res 27(6):1448–1475. https://doi.org/10.1108/IJEBR-06-2020-0428
13. Acevedo-Duque Á, Gonzalez-Diaz R, Vargas EC, Paz-Marcano A, Muller-Pérez S, Salazar-Sepúlveda G, … D'Adamo I (2021) Resilience, leadership and female entrepreneurship within the context of smes: evidence from latin america. Sustainability 13(15):8129. https://doi.org/10.3390/su13158129
14. Acevedo-Duque Á, Vega-Muñoz A, Salazar-Sepúlveda G (2020) Analysis of hospitality, leisure, and tourism studies in chile. Sustainability 12(18):7238. https://doi.org/10.3390/su12187238
15. Ada E, Sagnak M, Mangla SK, Kazancoglu Y (2021) A circular business cluster model for sustainable operations management. Int J Logistics Res Appl:1–19. https://doi.org/10.1080/13675567.2021.2008335
16. Adachi T, Hisada T (2017) Gender differences in entrepreneurship and intrapreneurship: an empirical analysis. Small Bus Econ 48(3):447–486. https://doi.org/10.1007/s11187-016-9793-y
17. Adam A (2015) Challenges of public finance sustainability in the European Union. Procedia Econ Finance 23:298–302. https://doi.org/10.1016/S2212-5671(15)00507-9
18. Adam I, Agyeiwaah E, Dayour F (2021) Understanding the social identity, motivations, and sustainable behaviour among backpackers: a clustering approach. J Travel Tour Mark 38(2):139–154. https://doi.org/10.1080/10548408.2021.1887053
19. Adam JN, Adams T, Gerber J-D, Haller T (2021) Decentralization for increased sustainability in natural resource management? Two cautionary cases from Ghana. Sustainability 13(12). https://doi.org/10.3390/su13126885
20. Adam M (2018) The role of human resource management (HRM) for the implementation of sustainable product-service systems (PSS)—an analysis of fashion retailers. Sustainability 10(7). https://doi.org/10.3390/su10072518
21. Adams R, Martin S, Boom K (2018) University culture and sustainability: designing and implementing an enabling framework. J Clean Prod 171:434–445. https://doi.org/10.1016/j.jclepro.2017.10.032
22. Adesogan AT, Havelaar AH, McKune SL, Eilittä M, Dahl GE (2020) Animal source foods: sustainability problem or malnutrition and sustainability solution? Perspective matters. Glob Food Secur 25:100325. https://doi.org/10.1016/j.gfs.2019.100325
23. Adomah-Afari A, Chandler JA (2018) The role of government and community in the scaling up and sustainability of mutual health organisations: an exploratory study in Ghana. Soc Sci Med 207:25–37. https://doi.org/10.1016/j.socscimed.2018.04.044
24. Adomako S, Amankwah-Amoah J, Danso A, Konadu R, Owusu-Agyei S (2019) Environmental sustainability orientation and performance of family and nonfamily firms. Bus Strateg Environ 28(6):1250–1259. https://doi.org/10.1002/bse.2314

25. Adongo CA, Taale F, Adam I (2018) Tourists' values and empathic attitude toward sustainable development in tourism. Ecol Econ 150:251–263. https://doi.org/10.1016/j.ecolecon.2018.04.013
26. Agarwal B (2018) Gender equality, food security and the sustainable development goals. Curr Opin Environ Sustain 34:26–32. https://doi.org/10.1016/j.cosust.2018.07.002
27. Agarwal S, Lenka U, Singh K, Agrawal V, Agrawal AM (2020) A qualitative approach towards crucial factors for sustainable development of women social entrepreneurship: Indian cases. J Clean Prod 274:123135. https://doi.org/10.1016/j.jclepro.2020.123135
28. Agyeiwaah E (2019) Exploring the relevance of sustainability to micro tourism and hospitality accommodation enterprises (MTHAEs): evidence from home-stay owners. J Clean Prod 226:159–171. https://doi.org/10.1016/j.jclepro.2019.04.089
29. Ahl H, Marlow S (2021) Exploring the false promise of entrepreneurship through a postfeminist critique of the enterprise policy discourse in Sweden and the UK. Hum Relat 74(1):41–68. https://doi.org/10.1177/0018726719848480
30. Al-Lawati EH, Kohar UHA, Suleiman ESB (2021) Entrepreneurship studies in the sultanate of oman: a scoping review of 20 years publications. Stud Appl Econ 39(10). https://doi.org/10.25115/eea.v39i10.5974
31. Al-Naqbi AK, Alshannag Q (2018) The status of education for sustainable development and sustainability knowledge, attitudes, and behaviors of UAE University students. Int J Sustain High Educ 19(3):566–588. https://doi.org/10.1108/IJSHE-06-2017-0091
32. Aleixo AM, Leal S, Azeiteiro UM (2018) Conceptualization of sustainable higher education institutions, roles, barriers, and challenges for sustainability: an exploratory study in Portugal. J Clean Prod 172:1664–1673. https://doi.org/10.1016/j.jclepro.2016.11.010
33. Alexandersson A, Kalonaityte V (2021) Girl bosses, punk poodles, and pink smoothies: girlhood as enterprising femininity. Gend Work Organ 28(1):416–438. https://doi.org/10.1111/gwao.12582
34. Alexandre L, Kharabsheh R (2019) The evolution of female entrepreneurship in the Gulf Cooperation Council, the case of Bahrain. Int J Gend Entrep 11(4):390–407. https://doi.org/10.1108/IJGE-02-2019-0041
35. Ali M, de Azevedo ARG, Marvila MT, Khan MI, Memon AM, Masood F, … Haq IU (2021) The influence of COVID-19-induced daily activities on health parameters—a case study in Malaysia. Sustainability 13(13):7465. https://doi.org/10.3390/su13137465
36. Aliabadi V, Ataei P, Gholamrezai S, Aazami M (2019) Components of sustainability of entrepreneurial ecosystems in knowledge-intensive enterprises: the application of fuzzy analytic hierarchy process. Small Enterp Res 26(3):288–306. https://doi.org/10.1080/13215906.2019.1671215
37. Alkhuzaim L, Zhu Q, Sarkis J (2021) Evaluating emergy analysis at the nexus of circular economy and sustainable supply chain management. Sustain Prod Consumption 25:413–424. https://doi.org/10.1016/j.spc.2020.11.022
38. Almodóvar-González M, Fernández-Portillo A, Díaz-Casero JC (2020) Entrepreneurial activity and economic growth. a multi-country analysis. Eur Res Manage Bus Econ 26(1):9–17. https://doi.org/10.1016/j.iedeen.2019.12.004
39. Alonso-Fradejas A (2021) The resource property question in climate stewardship and sustainability transitions. Land Use Policy 108:105529. https://doi.org/10.1016/j.landusepol.2021.105529
40. Alshuwaikhat HM, Adenle YA, Saghir B (2016) Sustainability assessment of higher education institutions in Saudi Arabia. Sustainability 8(8). https://doi.org/10.3390/su8080750
41. Álvarez-Risco A, Arellano EZ, Valerio EM, Acosta NM, Tarazona ZS (2013) Pharmaceutical care campaign as a strategy for implementation of pharmaceutical services: experience Peru. Pharm Care Espana 15(1):35–37
42. Alvarez-Risco A, Dawson J, Johnson W, Conteh-Barrat M, Aslani P, Del-Aguila-Arcentales S, Diaz-Risco S (2020) Ebola virus disease outbreak: a global approach for health systems. Revista Cubana de Farmacia 53(4):1–13, Article e491

43. Alvarez-Risco A, Del-Aguila-Arcentales S (2015) Prescription errors as a barrier to pharmaceutical care in public health facilities: experience Peru. Pharm Care Espana 17(6):725–731
44. Alvarez-Risco A, Del-Aguila-Arcentales S (2021) A note on changing regulation in international business: the world intellectual property organization (wipo) and artificial intelligence. In: Progress in international business research, vol 15, pp 363–371. https://doi.org/10.1108/S1745-886220210000015020
45. Alvarez-Risco A, Del-Aguila-Arcentales S (2021b) Public policies and private efforts to increase women entrepreneurship based on STEM background. In: Contributions to management science, pp 75–87. https://doi.org/10.1007/978-3-030-83792-1_5
46. Alvarez-Risco A, Del-Aguila-Arcentales S (2022) Sustainable Initiatives in International Markets. In: Contributions to management science, pp. 181–191. https://doi.org/10.1007/978-3-030-85950-3_10
47. Alvarez-Risco A, Del-Aguila-Arcentales S, Delgado-Zegarra J, Yáñez JA, Diaz-Risco S (2019) Doping in sports: findings of the analytical test and its interpretation by the public. Sport Sci Health 15(1):255–257. https://doi.org/10.1007/s11332-018-0484-8
48. Alvarez-Risco A, Del-Aguila-Arcentales S, Diaz-Risco S (2018) Pharmacovigilance as a tool for sustainable development of healthcare in Peru. PharmacoVigilance Rev 10(2):4–6
49. Alvarez-Risco A, Del-Aguila-Arcentales S, Rosen MA, García-Ibarra V, Maycotte-Felkel S, Martínez-Toro GM (2021) Expectations and interests of university students in covid-19 times about sustainable development goals: evidence from colombia, ecuador, mexico, and peru. Sustainability (Switzerland) 13(6), Article 3306. https://doi.org/10.3390/su13063306
50. Alvarez-Risco A, Del-Aguila-Arcentales S, Stevenson JG (2015) Pharmacists and mass communication for implementing pharmaceutical care. Am J Pharm Benefits 7(3):e125–e126
51. Alvarez-Risco A, Del-Aguila-Arcentales S, Villalobos-Alvarez D, Diaz-Risco S (2022) Leadership for sustainability in crisis time. In: Alvarez-Risco A, Muthu SS, Del-Aguila-Arcentales S (eds) Circular economy: impact on carbon and water footprint. Springer Singapore, pp 41–64. https://doi.org/10.1007/978-981-19-0549-0_3
52. Alvarez-Risco A, Del-Aguila-Arcentales S, Yáñez JA, Alvarez-Risco A (2021) Telemedicine in Peru as a result of the COVID-19 pandemic: perspective from a country with limited internet access. Am J Trop Med Hyg 105(1):6–11. https://doi.org/10.4269/ajtmh.21-0255
53. Alvarez-Risco A, Del-Aguila-Arcentales S, Yáñez JA, Rosen MA, Mejia CR (2021) Influence of technostress on academic performance of university medicine students in peru during the covid-19 pandemic. Sustainability (Switzerland) 13(16), Article 8949. https://doi.org/10.3390/su13168949
54. Alvarez-Risco A, Delgado-Zegarra J, Yáñez JA, Diaz-Risco S, Del-Aguila-Arcentales S (2018) Predation risk by gastronomic boom—case Peru. J Landscape Ecol (Czech Republic), 11(1):100–103. https://doi.org/10.2478/jlecol-2018-0003
55. Alvarez-Risco A, Estrada-Merino A, Anderson-Seminario MM, Mlodzianowska S, García-Ibarra V, Villagomez-Buele C, Carvache-Franco M (2021) Multitasking behavior in online classrooms and academic performance: case of university students in Ecuador during COVID-19 outbreak. Interact Technol Smart Educ 18(3):422–434. https://doi.org/10.1108/ITSE-08-2020-0160
56. Alvarez-Risco A, Mejia CR, Delgado-Zegarra J, Del-Aguila-Arcentales S, Arce-Esquivel AA, Valladares-Garrido MJ, … Yáñez JA (2020) The Peru approach against the COVID-19 infodemic: Insights and strategies. Am J Trop Med Hyg 103(2):583–586. https://doi.org/10.4269/ajtmh.20-0536
57. Alvarez-Risco A, Mlodzianowska S, García-Ibarra V, Rosen MA, Del-Aguila-Arcentales S (2021) Factors affecting green entrepreneurship intentions in business university students in covid-19 pandemic times: case of ecuador. Sustainability (Switzerland) 13(11), Article 6447. https://doi.org/10.3390/su13116447
58. Alvarez-Risco A, Mlodzianowska S, Zamora-Ramos U, Del-Aguila-Arcentales S (2021) Green entrepreneurship intention in university students: the case of Peru. Entrepreneurship Bus Econ Rev 9(4):85–100. https://doi.org/10.15678/EBER.2021.090406

59. Alvarez-Risco A, Quiroz-Delgado D, Del-Aguila-Arcentales S (2016) Pharmaceutical care in hypertension patients in a peruvian hospital [Article]. Indian J Public Health Res Develop 7(3):183–188. https://doi.org/10.5958/0976-5506.2016.00153.4
60. Alvarez-Risco A, Turpo-Cama A, Ortiz-Palomino L, Gongora-Amaut N, Del-Aguila-Arcentales S (2016) Barriers to the implementation of pharmaceutical care in pharmacies in Cusco Peru. Pharm Care Espana 18(5):194–205
61. Alvarez-Risco A, Van Mil JWF (2007) Pharmaceutical care in community pharmacies: practice and research in Peru. Ann Pharmacother 41(12):2032–2037. https://doi.org/10.1345/aph.1K117
62. Amorós JE, Cristi O, Naudé W (2021) Entrepreneurship and subjective well-being: does the motivation to start-up a firm matter? J Bus Res 127:389–398. https://doi.org/10.1016/j.jbusres.2020.11.044
63. Ang KL, Saw ET, He W, Dong X, Ramakrishna S (2021) Sustainability framework for pharmaceutical manufacturing (PM): a review of research landscape and implementation barriers for circular economy transition. J Clean Prod 280:124264. https://doi.org/10.1016/j.jclepro.2020.124264
64. Anggadwita G, Luturlean BS, Ramadani V, Ratten V (2017) Socio-cultural environments and emerging economy entrepreneurship: women entrepreneurs in Indonesia. J Entrepreneurship Emerg Economies 9(1):85–96. https://doi.org/10.1108/JEEE-03-2016-0011
65. Angulo-Mosquera LS, Alvarado-Alvarado AA, Rivas-Arrieta MJ, Cattaneo CR, Rene ER, García-Depraect O (2021) Production of solid biofuels from organic waste in developing countries: a review from sustainability and economic feasibility perspectives. Sci Total Environ 795:148816. https://doi.org/10.1016/j.scitotenv.2021.148816
66. Apcho-Ccencho LV, Cuya-Velásquez BB, Rodríguez DA, Anderson-Seminario MLM, Alvarez-Risco A, Estrada-Merino A, Mlodzianowska S (2021) The impact of international price on the technological industry in the united states and china during times of crisis: commercial war and covid-19. In: Advances in business and management forecasting, vol 14, pp 149–160
67. Aragon-Correa JA, Marcus AA, Rivera JE, Kenworthy AL (2017) Sustainability management teaching resources and the challenge of balancing planet, people, and profits. Acad Manage Learn Educ 16(3):469–483. https://doi.org/10.5465/amle.2017.0180
68. Arias-Meza M, Alvarez-Risco A, Cuya-Velásquez BB, de las Mercedes Anderson-Seminario M, Del-Aguila-Arcentales S (2022) Fashion and textile circularity and waste footprint. In: Alvarez-Risco A, Muthu SS, Del-Aguila-Arcentales S (eds) Circular economy: impact on carbon and water footprint. Springer Singapore, pp 181–204. https://doi.org/10.1007/978-981-19-0549-0_9
69. Aristei D, Gallo M (2021) Are female-led firms disadvantaged in accessing bank credit? Evidence from transition economies. Int J Emerg Markets. https://doi.org/10.1108/IJOEM-03-2020-0286
70. Aschemann-Witzel J, Giménez A, Ares G (2018) Convenience or price orientation? Consumer characteristics influencing food waste behaviour in the context of an emerging country and the impact on future sustainability of the global food sector. Glob Environ Chang 49:85–94. https://doi.org/10.1016/j.gloenvcha.2018.02.002
71. Aydın B, Alvarez MD (2020) Understanding the tourists' perspective of sustainability in cultural tourist destinations. Sustainability 12(21):8846. https://doi.org/10.3390/su12218846
72. Badura KL, Grijalva E, Newman DA, Yan TT, Jeon G (2018) Gender and leadership emergence: a meta-analysis and explanatory model. Pers Psychol 71(3):335–367. https://doi.org/10.1111/peps.12266
73. Bamgbade JA, Kamaruddeen AM, Nawi MNM, Adeleke AQ, Salimon MG, Ajibike WA (2019) Analysis of some factors driving ecological sustainability in construction firms. J Clean Prod 208:1537–1545. https://doi.org/10.1016/j.jclepro.2018.10.229
74. Bannerman S (2020) The world intellectual property organization and the sustainable development agenda. Futures 122:102586. https://doi.org/10.1016/j.futures.2020.102586

75. Barr S, Gilg A, Shaw G (2011) 'Helping people make better choices': exploring the behaviour change agenda for environmental sustainability. Appl Geogr 31(2):712–720. https://doi.org/10.1016/j.apgeog.2010.12.003
76. Barrachina Fernández M, García-Centeno MdC, Calderón Patier C (2021) Women sustainable entrepreneurship: review and research agenda. Sustainability 13(21):12047. https://doi.org/10.3390/su132112047
77. Barragán-Ocaña A, Silva-Borjas P, Olmos-Peña S, Polanco-Olguín M (2020) Biotechnology and Bioprocesses: their contribution to sustainability. Processes 8(4). https://doi.org/10.3390/pr8040436
78. Barros MV, Salvador R, do Prado GF, de Francisco AC, Piekarski CM (2021) Circular economy as a driver to sustainable businesses. Cleaner Environ Syst 2:100006. https://doi.org/10.1016/j.cesys.2020.100006
79. Bastian BL, Sidani YM, El Amine Y (2018) Women entrepreneurship in the Middle East and North Africa: A review of knowledge areas and research gaps. Gend Manage Int J 33(1):14–29. https://doi.org/10.1108/GM-07-2016-0141
80. Batista AA, Francisco AC (2018) Organizational sustainability practices: a study of the firms listed by the corporate sustainability index. Sustainability 10(1). https://doi.org/10.3390/su10010226
81. Beba U, Church AH (2020) Changing the game for women leaders at PepsiCo: from local action to enterprise accountability. Consult Psychol J Pract Res 72(4):288. https://doi.org/10.1037/cpb0000169
82. Belderbos R, Gilsing V, Lokshin B, Carree M, Sastre JF (2018) The antecedents of new R&D collaborations with different partner types: on the dynamics of past R&D collaboration and innovative performance. Long Range Plan 51(2):285–302. https://doi.org/10.1016/j.lrp.2017.10.002
83. Belz FM, Binder JK (2017) Sustainable entrepreneurship: a convergent process model. Bus Strateg Environ 26(1):1–17. https://doi.org/10.1002/bse.1887
84. Ben Youssef A, Boubaker S, Omri A (2018) Entrepreneurship and sustainability: the need for innovative and institutional solutions. Technol Forecast Soc Chang 129:232–241. https://doi.org/10.1016/j.techfore.2017.11.003
85. Berglund K, Ahl H, Pettersson K, Tillmar M (2018) Women's entrepreneurship, neoliberalism and economic justice in the postfeminist era: a discourse analysis of policy change in Sweden. Gend Work Organ 25(5):531–556. https://doi.org/10.1111/gwao.12269
86. Berguiga I, Adair P (2021) Funding female entrepreneurs in North Africa: self-selection versus discrimination? MSMEs, the informal sector and the microfinance industry. Int J Gend Entrep 13(4):394–419. https://doi.org/10.1108/IJGE-10-2020-0171
87. Bloodhart B, Swim JK (2020) Sustainability and consumption: What's gender got to do with it. J Soc Issues 76(1):101–113. https://doi.org/10.1111/josi.12370
88. Bokpin GA (2017) Foreign direct investment and environmental sustainability in Africa: the role of institutions and governance. Res Int Bus Finance 39:239–247. https://doi.org/10.1016/j.ribaf.2016.07.038
89. Brito RM, Rodríguez C, Aparicio JL (2018) Sustainability in teaching: an evaluation of university teachers and students. Sustainability 10(2):439. https://doi.org/10.3390/su10020439
90. Brown HS, de Jong M, Levy DL (2009) Building institutions based on information disclosure: lessons from GRI's sustainability reporting. J Clean Prod 17(6):571–580. https://doi.org/10.1016/j.jclepro.2008.12.009
91. Brown KA, Harris F, Potter C, Knai C (2020) The future of environmental sustainability labelling on food products. The Lancet Planet Health 4(4):e137–e138. https://doi.org/10.1016/S2542-5196(20)30074-7
92. Brush CG, Greene PG, Welter F (2020) The Diana project: a legacy for research on gender in entrepreneurship. Int J Gend Entrep 12(1):7–25. https://doi.org/10.1108/IJGE-04-2019-0083
93. Carvache-Franco M, Alvarez-Risco A, Carvache-Franco O, Carvache-Franco W, Estrada-Merino A, Villalobos-Alvarez D (2021) Perceived value and its influence on satisfaction and

loyalty in a coastal city: a study from Lima, Peru. J Policy Res Tourism Leisure Events. https://doi.org/10.1080/19407963.2021.1883634

94. Carvache-Franco M, Alvarez-Risco A, Carvache-Franco W, Carvache-Franco O, Estrada-Merino A, Rosen MA (2021) Coastal cities seen from loyalty and their tourist motivations: a study in Lima, Peru. Sustainability (Switzerland) 13(21), Article 11575. https://doi.org/10.3390/su132111575
95. Carvache-Franco M, Carvache-Franco O, Carvache-Franco W, Alvarez-Risco A, Estrada-Merino A (2021) Motivations and segmentation of the demand for coastal cities: a study in Lima Peru. Int J Tourism Res 23(4):517–531. https://doi.org/10.1002/jtr.2423
96. Castillo-Benancio S, Alvarez-Risco A, Esquerre-Botton S, Leclercq-Machado L, Calle-Nole M, Morales-Ríos F, … Del-Aguila-Arcentales S (2022) Circular economy for packaging and carbon footprint. In: Alvarez-Risco A, Muthu SS, Del-Aguila-Arcentales S (eds) Circular economy: impact on carbon and water footprint. Springer Singapore, pp 115–138. https://doi.org/10.1007/978-981-19-0549-0_6
97. Cerrato J, Cifre E (2018) Gender inequality in household chores and work-family conflict. Front Psychol 9:1330. https://doi.org/10.3389/fpsyg.2018.01330
98. Chafloque-Cespedes R, Alvarez-Risco A, Robayo-Acuña PV, Gamarra-Chavez CA, Martinez-Toro GM, Vicente-Ramos W (2021) Effect of sociodemographic factors in entrepreneurial orientation and entrepreneurial intention in university students of latin american business schools. In: Contemporary issues in entrepreneurship research, vol 11, pp 151–165. https://doi.org/10.1108/S2040-724620210000011010
99. Chafloque-Céspedes R, Vara-Horna A, Asencios-Gonzales Z, López-Odar D, Alvarez-Risco A, Quipuzco-Chicata L, … Sánchez-Villagomez M (2020) Academic presenteeism and violence against women in schools of business and engineering in Peruvian universities. Lecturas de Economia 93:127–153. https://doi.org/10.17533/udea.le.n93a340726
100. Chatterjee C, Ramu S (2018) Gender and its rising role in modern Indian innovation and entrepreneurship. IIMB Manage Rev 30(1):62–72. https://doi.org/10.1016/j.iimb.2017.11.006
101. Chaudhuri S, Park S, Kim S (2019) The changing landscape of women's leadership in India and Korea from cultural and generational perspectives. Hum Resour Dev Rev 18(1):16–46. https://doi.org/10.1177/1534484318809753
102. Chávez Rivera ME, Fuentes Fuentes M, d. M., & Ruiz-Jiménez, J. M. (2021) Challenging the context: mumpreneurship, copreneurship and sustainable thinking in the entrepreneurial process of women–a case study in Ecuador. Academia Revista Latinoamericana de Administración 34(3):368–398. https://doi.org/10.1108/ARLA-07-2020-0172
103. Chen X, Zhang SX, Jahanshahi AA, Alvarez-Risco A, Dai H, Li J, Ibarra VG (2020) Belief in a COVID-19 conspiracy theory as a predictor of mental health and well-being of health care workers in Ecuador: Cross-sectional survey study. JMIR Public Health Surveill 6(3), Article e20737. https://doi.org/10.2196/20737
104. Cho E, Moon ZK, Bounkhong T (2019) A qualitative study on motivators and barriers affecting entrepreneurship among Latinas. Gend Manage Int J (34):4. https://doi.org/10.1108/GM-07-2018-0096
105. Cho Y, Kim S, You J, Han H, Kim M, Yoon S (2021) How South Korean women leaders respond to their token status: assimilation and resistance. Hum Resour Develop Int:377–400. https://doi.org/10.1080/13678868.2021.1885207
106. Chreim S, Spence M, Crick D, Liao X (2018) Review of female immigrant entrepreneurship research: past findings, gaps and ways forward. Eur Manage J 36(2):210–222. https://doi.org/10.1016/j.emj.2018.02.001
107. Chung SA, Olivera S, Román BR, Alanoca E, Moscoso S, Terceros BL, … Yáñez JA (2021) Themes of scientific production of the cuban journal of pharmacy indexed in scopus (1967–2020). Revista Cubana de Farmacia 54(1), Article e511
108. Constantin P-N, Stanescu R, Stanescu M (2020) Social entrepreneurship and sport in Romania: how can former athletes contribute to sustainable social change? Sustainability 12(11):4688. https://doi.org/10.3390/su12114688

109. Contreras-Taica A, Alvarez-Risco A, Arias-Meza M, Campos-Dávalos N, Calle-Nole M, Almanza-Cruz C, … Del-Aguila-Arcentales S (2022) Virtual education: carbon footprint and circularity. In: Alvarez-Risco A, Muthu SS, Del-Aguila-Arcentales S (eds) Circular economy: impact on carbon and water footprint. Springer Singapore, pp 265–285. https://doi.org/10.1007/978-981-19-0549-0_13
110. Corrêa VS, da Silva Brito FR, de Lima RM, Queiroz MM (2021) Female entrepreneurship in emerging and developing countries: a systematic literature review. Int J Gend Entrep. https://doi.org/10.1108/IJGE-08-2021-0142
111. Cruz-Torres W, Alvarez-Risco A, Del-Aguila-Arcentales S (2021) Impact of enterprise resource planning (ERP) implementation on performance of an education enterprise: a structural equation modeling (SEM) [Article]. Stud Bus Econ 16(2):37–52. https://doi.org/10.2478/sbe-2021-0023
112. Cruz JP, Alshammari F, Felicilda-Reynaldo RFD (2018) Predictors of Saudi nursing students' attitudes towards environment and sustainability in health care. Int Nurs Rev 65(3):408–416. https://doi.org/10.1111/inr.12432
113. Cukier W, Chavoushi ZH (2020) Facilitating women entrepreneurship in Canada: the case of WEKH. Gend Manage Int J 35(3):303–318. https://doi.org/10.1108/GM-11-2019-0204
114. Cuya-Velásquez BB, Alvarez-Risco A, Gomez-Prado R, Juarez-Rojas L, Contreras-Taica A, Ortiz-Guerra A, … Del-Aguila-Arcentales S (2022) Circular economy for food loss reduction and water footprint. In: Alvarez-Risco A, Muthu SS, Del-Aguila-Arcentales S (eds) Circular economy: impact on carbon and water footprint. Springer Singapore, pp 65–91. https://doi.org/10.1007/978-981-19-0549-0_4
115. Dal Mas F, Paoloni P (2019) A relational capital perspective on social sustainability; the case of female entrepreneurship in Italy. Meas Bus Excell 24(1):114–130. https://doi.org/10.1108/MBE-08-2019-0086
116. Daulerio PP Jr (2018) Intrinsic entrepreneurial motivation factors: gender differences. Int J Entrep Small Bus 34(3):362–380
117. De-la-Cruz-Diaz M, Alvarez-Risco A, Jaramillo-Arévalo M, de las Mercedes Anderson-Seminario M, Del-Aguila-Arcentales S (2022) 3D print, circularity, and footprints. In: Alvarez-Risco A, Muthu SS, Del-Aguila-Arcentales S (eds) Circular economy: impact on carbon and water footprint. Springer Singapore, pp 93–112. https://doi.org/10.1007/978-981-19-0549-0_5
118. De Andrés P, Gimeno R, de Cabo RM (2021) The gender gap in bank credit access. J Corp Finan 71:101782. https://doi.org/10.1016/j.jcorpfin.2020.101782
119. de las Mercedes Anderson-Seminario M, Alvarez-Risco A (2022) Better students, better companies, better life: circular learning. In: Alvarez-Risco A, Muthu SS, Del-Aguila-Arcentales S (eds) Circular economy: impact on carbon and water footprint. Springer Singapore, pp 19–40. https://doi.org/10.1007/978-981-19-0549-0_2
120. De Rosa M, McElwee G, Smith R (2019) Farm diversification strategies in response to rural policy: a case from rural Italy. Land Use Policy 81:291–301. https://doi.org/10.1016/j.landusepol.2018.11.006
121. Del-Aguila-Arcentales S, Alvarez-Risco A (2013) Human error or burnout as explanation for mistakes in pharmaceutical laboratories. Accred Qual Assur 18(5):447–448. https://doi.org/10.1007/s00769-013-1000-0
122. Del-Aguila-Arcentales S, Alvarez-Risco A, Muthu SS (2022) Measuring circular economy. In: Alvarez-Risco A, Muthu SS, Del-Aguila-Arcentales S (eds) Circular economy: impact on carbon and water footprint. Springer Singapore, pp 3–17. https://doi.org/10.1007/978-981-19-0549-0_1
123. Delgado-Zegarra J, Alvarez-Risco A, Yáñez JA (2018) Indiscriminate use of pesticides and lack of sanitary control in the domestic market in Peru. Revista Panamericana de Salud Publica/Pan Am J Public Health 42, Article e3. https://doi.org/10.26633/RPSP.2018.3
124. Deslatte A, Swann WL (2019) Elucidating the linkages between entrepreneurial orientation and local government sustainability performance. Am Rev Public Adm 50(1):92–109. https://doi.org/10.1177/0275074019869376

125. Dhanaraj S, Mahambare V (2019) Family structure, education and women's employment in rural India. World Dev 115:17–29. https://doi.org/10.1016/j.worlddev.2018.11.004
126. Diepolder CS, Weitzel H, Huwer J (2021) Competence frameworks of sustainable entrepreneurship: a systematic review. Sustainability 13(24):13734. https://doi.org/10.3390/su132413734
127. Douglas EJ, Shepherd DA, Venugopal V (2021) A multi-motivational general model of entrepreneurial intention. J Bus Ventur 36(4):106107. https://doi.org/10.1016/j.jbusvent.2021.106107
128. Dukhaykh S, Bilimoria D (2021) The factors influencing Saudi Arabian women's persistence in nontraditional work careers. Career Dev Int 26(5):720–746. https://doi.org/10.1108/CDI-04-2020-0089
129. Eagly AH (2018) The shaping of science by ideology: How feminism inspired, led, and constrained scientific understanding of sex and gender. J Soc Issues 74(4):871–888. https://doi.org/10.1111/josi.12291
130. Enciso-Zarate A, Guzmán-Oviedo J, Sánchez-Cardona F, Martínez-Rohenes D, Rodríguez-Palomino JC, Alvarez-Risco A, … Diaz-Risco S (2016) Evaluation of contamination by cytotoxic agents in colombian hospitals. Pharm Care Espana 18(6):241–250
131. Eppinger E, Jain A, Vimalnath P, Gurtoo A, Tietze F, Hernandez Chea R (2021) Sustainability transitions in manufacturing: the role of intellectual property. Curr Opin Environ Sustain 49:118–126. https://doi.org/10.1016/j.cosust.2021.03.018
132. Ertuna B, Karatas-Ozkan M, Yamak S (2019) Diffusion of sustainability and CSR discourse in hospitality industry. Int J Contemp Hosp Manage 31(6):2564–2581. https://doi.org/10.1108/IJCHM-06-2018-0464
133. Esquerre-Botton S, Alvarez-Risco A, Leclercq-Machado L, de las Mercedes Anderson-Seminario M, Del-Aguila-Arcentales S (2022) Food loss reduction and carbon footprint practices worldwide: a benchmarking approach of circular economy. In: Alvarez-Risco A, Muthu SS, Del-Aguila-Arcentales S (eds) Circular economy: impact on carbon and water footprint. Springer Singapore, pp 161–179. https://doi.org/10.1007/978-981-19-0549-0_8
134. Esteves AM, Genus A, Henfrey T, Penha-Lopes G, East M (2021) Sustainable entrepreneurship and the sustainable development goals: community-led initiatives, the social solidarity economy and commons ecologies. Bus Strateg Environ 30(3):1423–1435. https://doi.org/10.1002/bse.2706
135. Evans CA, Mayo LM, Quijada MA (2018) Women's empowerment and nonprofit sector development. Nonprofit Volunt Sect Q 47(4):856–871. https://doi.org/10.1177/0899764018764331
136. Faisal MN, Jabeen F, Katsioloudes MI (2017) Strategic interventions to improve women entrepreneurship in GCC countries: a relationship modeling approach. J Entrepreneurship Emerg Econ 9(2):161–180. https://doi.org/10.1108/JEEE-07-2016-0026
137. Ferreira J, Sousa BM, Gonçalves F (2018) Encouraging the subsistence artisan entrepreneurship in handicraft and creative contexts. J Enterprising Communities: people and places in the global economy 13(1/2):64–83. https://doi.org/10.1108/JEC-09-2018-0068
138. Figueroa-Domecq C, Kimbu A, de Jong A, Williams AM (2020) Sustainability through the tourism entrepreneurship journey: a gender perspective. J Sustain Tourism:1–24. https://doi.org/10.1080/09669582.2020.1831001
139. Filser M, Kraus S, Roig-Tierno N, Kailer N, Fischer U (2019) Entrepreneurship as catalyst for sustainable development: opening the black box. Sustainability 11(16):4503. https://doi.org/10.3390/su11164503
140. Fröcklin S, Jiddawi NS, de la Torre-Castro M (2018) Small-scale innovations in coastal communities. Ecol Soc 23(2). https://doi.org/10.5751/ES-10136-230234
141. Gali N, Niemand T, Shaw E, Hughes M, Kraus S, Brem A (2020) Social entrepreneurship orientation and company success: the mediating role of social performance. Technol Forecast Soc Chang 160:120230. https://doi.org/10.1016/j.techfore.2020.120230
142. Galleli B, Teles NEB, Santos JARd, Freitas-Martins MS, Hourneaux Junior F (2021) Sustainability university rankings: a comparative analysis of UI green metric and the times higher

education world university rankings. Int J Sustain High Educ, ahead-of-print(ahead-of-print). https://doi.org/10.1108/IJSHE-12-2020-0475

143. García-Rodríguez FJ, Gil-Soto E, Ruiz-Rosa I, Gutiérrez-Taño D (2017) Entrepreneurial potential in less innovative regions: the impact of social and cultural environment. Eur J Manage Bus Econ. https://doi.org/10.1108/EJMBE-07-2017-010
144. Gast J, Gundolf K, Cesinger B (2017) Doing business in a green way: a systematic review of the ecological sustainability entrepreneurship literature and future research directions. J Clean Prod 147:44–56. https://doi.org/10.1016/j.jclepro.2017.01.065
145. Gbadamosi A (2019) Women-entrepreneurship, religiosity, and value-co-creation with ethnic consumers: revisiting the paradox. J Strateg Mark 27(4):303–316. https://doi.org/10.1080/0965254X.2017.1344293
146. Genus A (2021) Sustainable entrepreneurship research in the 2020s: an introduction 30(3):1419–1422. https://doi.org/10.1002/bse.2705
147. Gerdt S-O, Wagner E, Schewe G (2019) The relationship between sustainability and customer satisfaction in hospitality: an explorative investigation using eWOM as a data source. Tour Manage 74:155–172. https://doi.org/10.1016/j.tourman.2019.02.010
148. Ghouse S, McElwee G, Meaton J, Durrah O (2017) Barriers to rural women entrepreneurs in Oman. Int J Entrepreneurship Behav Res 23(6):998–1016. https://doi.org/10.1108/IJEBR-02-2017-0070
149. Ghouse SM, Durrah O, McElwee G (2021) Rural women entrepreneurs in Oman: problems and opportunities. Int J Entrepreneurship Behav Res 27(7):1674–1695. https://doi.org/10.1108/IJEBR-03-2021-0209
150. Ghouse SM, McElwee G, Durrah O (2019) Entrepreneurial success of cottage-based women entrepreneurs in Oman. Int J Entrepreneurship Behav Res 25(3):480–498. https://doi.org/10.1108/IJEBR-10-2018-0691
151. Gómez-Prado R, Alvarez-Risco A, Sánchez-Palomino J, de las Mercedes Anderson-Seminario M, Del-Aguila-Arcentales S (2022) Circular economy for waste reduction and carbon footprint. In: Alvarez-Risco A, Muthu SS, Del-Aguila-Arcentales S (eds) Circular economy: impact on carbon and water footprint. Springer Singapore, pp 139–159. https://doi.org/10.1007/978-981-19-0549-0_7
152. González-Serrano MH, Añó Sanz V, González-García RJ (2020) Sustainable sport entrepreneurship and innovation: a bibliometric analysis of this emerging field of research. Sustainability 12(12):5209. https://doi.org/10.3390/su12125209
153. Grewal J, Hauptmann C, Serafeim G (2021) Material sustainability information and stock price informativeness. J Bus Ethics 171(3):513–544. https://doi.org/10.1007/s10551-020-04451-2
154. Günzel-Jensen F, Siebold N, Kroeger A, Korsgaard S (2020) Do the United Nations' sustainable development goals matter for social entrepreneurial ventures? A bottom-up perspective. J Bus Ventur Insights 13:e00162. https://doi.org/10.1016/j.jbvi.2020.e00162
155. Gupta N, Etzkowitz H (2021) Women founders in a high-tech incubator: negotiating entrepreneurial identity in the Indian socio-cultural context. Int J Gend Entrepreneurship 13(4). https://doi.org/10.1108/IJGE-11-2020-0181
156. Gupta N, Mirchandani A (2018) Investigating entrepreneurial success factors of women-owned SMEs in UAE. Manage Decis 56(1):219–232. https://doi.org/10.1108/MD-04-2017-0411
157. Gupta N, Mirchandani A (2018b) Investigating entrepreneurial success factors of women-owned SMEs in UAE. Manage Decis 56(1). https://doi.org/10.1108/MD-04-2017-0411
158. Gustavsson M (2021) The invisible (woman) entrepreneur? Shifting the discourse from fisheries diversification to entrepreneurship. Sociologia Ruralis. https://doi.org/10.1111/soru.12343
159. Gustavsson M, Frangoudes K, Lindström L, Ávarez MC, de la Torre Castro M (2021) Gender and blue justice in small-scale fisheries governance. Mar Policy 133:104743. https://doi.org/10.1016/j.marpol.2021.104743
160. Hall MR (2019) The sustainability price: expanding environmental life cycle costing to include the costs of poverty and climate change. Int J Life Cycle Assess 24(2):223–236. https://doi.org/10.1007/s11367-018-1520-2

161. Hallward M, Bekdash-Muellers H (2019) Success and agency: localizing women's leadership in Oman. Gend Manage: Int J 34(7):606–618. https://doi.org/10.1108/GM-11-2017-0162
162. Hermann RR, Bossle MB (2020) Bringing an entrepreneurial focus to sustainability education: a teaching framework based on content analysis. J Cleaner Prod 246:119038. https://doi.org/10.1016/j.jclepro.2019.119038
163. Hong SH (2021) Determinants of selection of R&D cooperation partners: insights from Korea. Sustainability 13(17):9637. https://doi.org/10.3390/su13179637
164. Hormiga E, Jaén I (2021) Why does she start up? The role of personal values in women's entrepreneurial intentions. Int J Entrep Small Bus 44(1):53–74
165. Jaim J (2021) Bank loans access for women business-owners in Bangladesh: obstacles and dependence on husbands. J Small Bus Manage 59(sup1):S16–S41. https://doi.org/10.1080/00472778.2020.1727233
166. Jee SJ, Sohn SY (2020) Patent-based framework for assisting entrepreneurial firms' R&D partner selection: leveraging their limited resources and managing the tension between learning and protection. J Eng Tech Manage 57:101575. https://doi.org/10.1016/j.jengtecman.2020.101575
167. Jeong BG, Compion S (2021) Characteristics of women's leadership in African social enterprises: the heartfelt project, bright kids Uganda and Chikumbuso. Emerald Emerg Markets Case Stud 11(2). https://doi.org/10.1108/EEMCS-11-2019-0305
168. Jha P, Makkad M, Mittal S (2018) Performance-oriented factors for women entrepreneurs–a scale development perspective. J Entrepreneurship Emerg Economies 10(2):329–360. https://doi.org/10.1108/JEEE-08-2017-0053
169. Jiang H, Gao S, Song Y, Sheng K, Amaratunga GA (2019) An empirical study on the impact of collaborative R&D networks on enterprise innovation performance based on the mediating effect of technology standard setting. Sustainability 11(24):7249. https://doi.org/10.3390/su11247249
170. Jones P, Comfort D (2020) The COVID-19 crisis and sustainability in the hospitality industry. Int J Contemp Hosp Manage 32(10):3037–3050. https://doi.org/10.1108/IJCHM-04-2020-0357
171. Jones P, Ratten V, Hayduk T (2020) Sport, fitness, and lifestyle entrepreneurship. Int Entrepreneurship Manage J 16(3):783–793. https://doi.org/10.1007/s11365-020-00666-x
172. Juarez-Rojas L, Alvarez-Risco A, Campos-Dávalos N, de las Mercedes Anderson-Seminario M, Del-Aguila-Arcentales S (2022) Water footprint in the textile and food supply chain management: trends to become circular and sustainable. In: Alvarez-Risco A, Muthu SS, Del-Aguila-Arcentales S (eds) Circular economy: impact on carbon and water footprint. Springer Singapore, pp 225–243. https://doi.org/10.1007/978-981-19-0549-0_11
173. Kaciak E, Welsh DH (2020) Women entrepreneurs and work–life interface: the impact of sustainable economies on success. J Bus Res 112:281–290. https://doi.org/10.1016/j.jbusres.2019.11.073
174. Khan S, Henderson C (2020) How Western Michigan University is approaching its commitment to sustainability through sustainability-focused courses. J Clean Prod 253:119741. https://doi.org/10.1016/j.jclepro.2019.119741
175. Khan SAR, Razzaq A, Yu Z, Miller S (2021) Industry 4.0 and circular economy practices: a new era business strategies for environmental sustainability. Bus Strategy Environ 30(8):4001–4014. https://doi.org/10.1002/bse.2853
176. Kim J-Y, Hsu N, Newman DA, Harms P, Wood D (2020) Leadership perceptions, gender, and dominant personality: the role of normality evaluations. J Res Pers 87:103984. https://doi.org/10.1016/j.jrp.2020.103984
177. Ko Y, Chung Y, Seo H (2020) Coopetition for sustainable competitiveness: R&D collaboration in perspective of productivity. Sustainability 12(19):7993. https://doi.org/10.3390/su12197993
178. Kogut CS, Mejri K (2021) Female entrepreneurship in emerging markets: challenges of running a business in turbulent contexts and times. Int J Gend Entrepreneurship. https://doi.org/10.1108/IJGE-03-2021-0052

179. Kong L, Liu Z, Wu J (2020) A systematic review of big data-based urban sustainability research: state-of-the-science and future directions. J Clean Prod 273:123142. https://doi.org/10.1016/j.jclepro.2020.123142
180. Kozlowski A, Searcy C, Bardecki M (2018) The reDesign canvas: fashion design as a tool for sustainability. J Cleaner Prod 183:194–207. https://doi.org/10.1016/j.jclepro.2018.02.014
181. Kuo PX, Volling BL, Gonzalez R (2018) Gender role beliefs, work–family conflict, and father involvement after the birth of a second child. Psychol Men Masculinity 19(2):243. https://doi.org/10.1037/men0000101
182. Lamb W, Mattioli G, Levi S, Roberts J, Capstick S, Creutzig F, … Steinberger J (2020) Discourses of climate delay. Glob Sustain 3(e17):1-5. https://doi.org/10.1017/sus.2020.13
183. Lapierre LM, Li Y, Kwan HK, Greenhaus JH, DiRenzo MS, Shao P (2018) A meta-analysis of the antecedents of work–family enrichment. J Organ Behav 39(4):385–401. https://doi.org/10.1002/job.2234
184. Larrán Jorge M, Andrades Peña FJ, Herrera Madueño J (2019) An analysis of university sustainability reports from the GRI database: an examination of influential variables. J Environ Plann Manage 62(6):1019–1044. https://doi.org/10.1080/09640568.2018.1457952
185. Laukkanen M, Tura N (2020) The potential of sharing economy business models for sustainable value creation. J Clean Prod 253:120004. https://doi.org/10.1016/j.jclepro.2020.120004
186. Leiva-Martinez MA, Anderson-Seminario MLM, Alvarez-Risco A, Estrada-Merino A, Mlodzianowska S (2021) Price variation in lower goods as of previous economic crisis and the contrast of the current price situation in the context of covid-19 in peru. In: Advances in business and management forecasting, vol 14, pp 161–166
187. Li S, Wu D, Sun Y (2021) The impact of entrepreneurial optimism and labor law on business performance of new ventures. Front Psychol 12:697002. https://doi.org/10.3389/fpsyg.2021.697002
188. Lin T-L, Lu T-Y, Hsieh M-C, Liu H-Y (2018) From conception to start-up: Who and what affect female entrepreneurship. Contemp Manage Res 14(4):253–276. https://doi.org/10.7903/cmr.17957
189. Llinares-Insa LI, González-Navarro P, Córdoba-Iñesta AI, Zacarés-González JJ (2018) Women's job search competence: a question of motivation, behavior, or gender. Front Psychol 9:137. https://doi.org/10.3389/fpsyg.2018.00137
190. Lopez-Odar D, Alvarez-Risco A, Vara-Horna A, Chafloque-Cespedes R, Sekar MC (2020) Validity and reliability of the questionnaire that evaluates factors associated with perceived environmental behavior and perceived ecological purchasing behavior in Peruvian consumers [Article]. Soc Responsib J 16(3):403–417. https://doi.org/10.1108/SRJ-08-2018-0201
191. Lozano R, Barreiro-Gen M, Lozano FJ, Sammalisto K (2019) Teaching sustainability in European higher education institutions: assessing the connections between competences and pedagogical approaches. Sustainability 11(6):1602. https://doi.org/10.3390/su11061602
192. Manavalan E, Jayakrishna K (2019) An analysis on sustainable supply chain for circular economy. Procedia Manuf 33:477–484. https://doi.org/10.1016/j.promfg.2019.04.059
193. Mani V, Gunasekaran A, Delgado C (2018) Supply chain social sustainability: standard adoption practices in Portuguese manufacturing firms. Int J Prod Econ 198:149–164. https://doi.org/10.1016/j.ijpe.2018.01.032
194. Manolova TS, Eunni RV, Gyoshev BS (2008) Institutional environments for entrepreneurship: evidence from emerging economies in Eastern Europe. Entrepreneurship Theory Pract 32(1):203–218. https://doi.org/10.1111/j.1540-6520.2007.00222.x
195. Manzoor SR, Ho JSY, Al Mahmud A (2021) Revisiting the 'university image model' for higher education institutions' sustainability. J Mark High Educ 31(2):220–239. https://doi.org/10.1080/08841241.2020.1781736
196. Marlow S (2020) Gender and entrepreneurship: past achievements and future possibilities. Int J Gend Entrepreneurship 12(1):39–52. https://doi.org/10.1108/IJGE-05-2019-0090
197. Martinez-Martinez A, Cegarra-Navarro J-G, Garcia-Perez A, Wensley A (2019) Knowledge agents as drivers of environmental sustainability and business performance in the hospitality sector. Tour Manage 70:381–389. https://doi.org/10.1016/j.tourman.2018.08.030

198. Mehtap S, Ozmenekse L, Caputo A (2019) "I'ma stay at home businesswoman": an insight into informal entrepreneurship in Jordan. J Entrepreneurship Emerg Econ 11(1):44–65. https://doi.org/10.1108/JEEE-10-2017-0080
199. Mejía-Acosta N, Alvarez-Risco A, Solís-Tarazona Z, Matos-Valerio E, Zegarra-Arellano E, Del-Aguila-Arcentales S (2016) Adverse drug reactions reported as a result of the implementation of pharmaceutical care in the Institutional Pharmacy DIGEMID—Ministry of Health. Pharm Care Espana 18(2):67–74
200. Michelino F, Cammarano A, Celone A, Caputo M (2019) The linkage between sustainability and innovation performance in IT hardware sector. Sustainability 11(16):4275. https://doi.org/10.3390/su11164275
201. Modarresi M, Arasti Z, Talebi K, Farasatkhah M (2017) Growth barriers of women-owned home-based businesses in Iran: an exploratory study. Gend Manag Int J 32(4):244–267. https://doi.org/10.1108/GM-03-2016-0069
202. Mohamued EA, Ahmed M, Pypłacz P, Liczmańska-Kopcewicz K, Khan MA (2021) Global oil price and innovation for sustainability: the impact of R&D spending, oil price and oil price volatility on GHG emissions. Energies 14(6). https://doi.org/10.3390/en14061757
203. Morales-Ríos F, Alvarez-Risco A, Castillo-Benancio S, de las Mercedes Anderson-Seminario M, Del-Aguila-Arcentales S (2022) Material selection for circularity and footprints. In: Alvarez-Risco A, Muthu SS, Del-Aguila-Arcentales S (eds) Circular economy: impact on carbon and water footprint. Springer Singapore, pp 205–221. https://doi.org/10.1007/978-981-19-0549-0_10
204. Mousa SK, Othman M (2020) The impact of green human resource management practices on sustainable performance in healthcare organisations: a conceptual framework. J Clean Prod 243:118595. https://doi.org/10.1016/j.jclepro.2019.118595
205. Moya-Clemente I, Ribes-Giner G, Pantoja-Díaz O (2020) Configurations of sustainable development goals that promote sustainable entrepreneurship over time. Sustain Dev 28(4):572–584. https://doi.org/10.1002/sd.2009
206. Muñoz P, Cohen B (2018) Sustainable entrepreneurship research: taking stock and looking ahead [https://doi.org/10.1002/bse.2000]. Bus Strategy Environ 27(3):300–322. https://doi.org/10.1002/bse.2000
207. Mzembe AN, Lindgreen A, Idemudia U, Melissen F (2020) A club perspective of sustainability certification schemes in the tourism and hospitality industry. J Sustain Tour 28(9):1332–1350. https://doi.org/10.1080/09669582.2020.1737092
208. Nair SR (2020) The link between women entrepreneurship, innovation and stakeholder engagement: a review. J Bus Res 119:283–290. https://doi.org/10.1016/j.jbusres.2019.06.038
209. Nakamura H (2019) Relationship among land price, entrepreneurship, the environment, economics, and social factors in the value assessment of Japanese cities. J Clean Prod 217:144–152. https://doi.org/10.1016/j.jclepro.2019.01.201
210. Neumeyer X, Santos SC, Caetano A, Kalbfleisch P (2019) Entrepreneurship ecosystems and women entrepreneurs: a social capital and network approach. Small Bus Econ 53(2):475–489. https://doi.org/10.1007/s11187-018-9996-5
211. Niemann CC, Dickel P, Eckardt G (2020) The interplay of corporate entrepreneurship, environmental orientation, and performance in clean-tech firms—a double-edged sword [https://doi.org/10.1002/bse.2357]. Bus Strategy Environ 29(1):180–196. https://doi.org/10.1002/bse.2357
212. Norström AV, Cvitanovic C, Löf MF, West S, Wyborn C, Balvanera P, … Österblom H (2020) Principles for knowledge co-production in sustainability research. Nat Sustain 3(3):182-190. https://doi.org/10.1038/s41893-019-0448-2
213. Nouri P, AhmadiKafeshani A (2019) Do female and male entrepreneurs differ in their proneness to heuristics and biases? J Entrep Emerg Economies 12(3):357–375. https://doi.org/10.1108/JEEE-05-2019-0062
214. Olawumi TO, Chan DWM (2018) A scientometric review of global research on sustainability and sustainable development. J Cleaner Prod 183:231–250. https://doi.org/10.1016/j.jclepro.2018.02.162

215. Omoloso O, Mortimer K, Wise WR, Jraisat L (2021) Sustainability research in the leather industry: a critical review of progress and opportunities for future research. J Cleaner Prod 285:125441. https://doi.org/10.1016/j.jclepro.2020.125441
216. Ozturkoglu Y, Sari FO, Saygili E (2021) A new holistic conceptual framework for sustainability oriented hospitality innovation with triple bottom line perspective. J Hosp Tour Technol 12(1):39–57. https://doi.org/10.1108/JHTT-02-2019-0022
217. Panda S (2018) Constraints faced by women entrepreneurs in developing countries: review and ranking. Gend Manage Int J 33(4):315–331. https://doi.org/10.1108/GM-01-2017-0003
218. Patterson L, Varadarajan DS, Salim BS (2020) Women in STEM/SET: gender gap research review of the United Arab Emirates (UAE)–a meta-analysis. Gend Manage Int J 36(8):881–911. https://doi.org/10.1108/GM-11-2019-0201
219. Paulin M, Lachance-Grzela M, McGee S (2017) Bringing work home or bringing family to work: personal and relational consequences for working parents. J Fam Econ Issues 38(4):463–476. https://doi.org/10.1007/s10834-017-9524-9
220. Peçanha Enqvist J, West S, Masterson VA, Haider LJ, Svedin U, Tengö M (2018) Stewardship as a boundary object for sustainability research: linking care, knowledge and agency. Landscape Urban Plan 179:17–37. https://doi.org/10.1016/j.landurbplan.2018.07.005
221. Peeters P (2018) Why space tourism will not be part of sustainable tourism. Tour Recreat Res 43(4):540–543. https://doi.org/10.1080/02508281.2018.1511942
222. Peña Miguel N, Corral Lage J, Mata Galindez A (2020) Assessment of the development of professional skills in university students: sustainability and serious games. Sustainability 12(3). https://doi.org/10.3390/su12031014
223. Pham H, Kim S-Y (2019) The effects of sustainable practices and managers' leadership competences on sustainability performance of construction firms. Sustain Prod Consumption 20:1–14. https://doi.org/10.1016/j.spc.2019.05.003
224. Plewnia F, Guenther E (2018) Mapping the sharing economy for sustainability research. Manag Decis 56(3):570–583. https://doi.org/10.1108/MD-11-2016-0766
225. Poortinga W, Whitmarsh L, Steg L, Böhm G, Fisher S (2019) Climate change perceptions and their individual-level determinants: a cross-European analysis. Glob Environ Chang 55:25–35. https://doi.org/10.1016/j.gloenvcha.2019.01.007
226. Quagrainie FA, Adams S, Kabalan AAM, Dankwa AD (2020) Micro-entrepreneurship, sustainable development goal one and cultural expectations of Ghanaian women. J Entrepreneurship Emerg Economies 13(1):86–106. https://doi.org/10.1108/JEEE-11-2019-0174
227. Quispe-Cañari JF, Fidel-Rosales E, Manrique D, Mascaró-Zan, J, Huamán-Castillón KM, Chamorro–Espinoza SE, … Mejia CR (2021) Self-medication practices during the COVID-19 pandemic among the adult population in Peru: a cross-sectional survey. Saudi Pharm J 29(1):1–11. https://doi.org/10.1016/j.jsps.2020.12.001
228. Raghuvanshi J, Agrawal R, Ghosh P (2017) Analysis of barriers to women entrepreneurship: the DEMATEL approach. J Entrepreneurship 26(2):220–238. https://doi.org/10.1177/0971355717708848
229. Rajesh R (2020) Exploring the sustainability performances of firms using environmental, social, and governance scores. J Clean Prod 247:119600. https://doi.org/10.1016/j.jclepro.2019.119600
230. Ramísio PJ, Pinto LMC, Gouveia N, Costa H, Arezes D (2019) Sustainability strategy in higher education institutions: lessons learned from a nine-year case study. J Clean Prod 222:300–309. https://doi.org/10.1016/j.jclepro.2019.02.257
231. Ransfield AK, Reichenberger I (2021) Māori Indigenous values and tourism business sustainability. AlterNative: Int J Indigenous Peoples 17(1):49–60. https://doi.org/10.1177/1177180121994680
232. Rau H, Goggins G, Fahy F (2018) From invisibility to impact: Recognising the scientific and societal relevance of interdisciplinary sustainability research. Res Policy 47(1):266–276. https://doi.org/10.1016/j.respol.2017.11.005

233. Rojas-Osorio M, Alvarez-Risco A (2019) Intention to use smartphones among Peruvian university students. Int J Interact Mobile Technol 13(3):40–52. https://doi.org/10.3991/ijim.v13i03.9356
234. Román BR, Moscoso S, Chung SA, Terceros BL, Álvarez-Risco A, Yáñez JA (2020) Treatment of COVID-19 in peru and bolivia, and self-medication risks. Revista Cubana de Farmacia 53(2):1–20, Article e435
235. Sakao T, Brambila-Macias SA (2018) Do we share an understanding of transdisciplinarity in environmental sustainability research? J Cleaner Prod 170:1399–1403. https://doi.org/10.1016/j.jclepro.2017.09.226
236. Salmerón-Manzano E, Manzano-Agugliaro F (2018) The higher education sustainability through virtual laboratories: the Spanish university as case of study. Sustainability 10(11). https://doi.org/10.3390/su10114040
237. Samantroy E, Tomar J (2018) Women entrepreneurship in india: evidence from economic censuses. Social Change 48(2):188–207. https://doi.org/10.1177/0049085718768898
238. Sarango-Lalangui P, Santos JLS, Hormiga E (2018) The development of sustainable entrepreneurship research field. Sustainability 10(6):2005. https://doi.org/10.3390/su10062005
239. Schlamp S, Gerpott FH, Voelpel SC (2020) Same talk, different reaction? Communication, emergent leadership and gender. J Manage Psychol 36(1):51–74. https://doi.org/10.1108/JMP-01-2019-0062
240. Sedláček P, Sterk V (2017) The growth potential of startups over the business cycle. Am Econ Rev 107(10):3182–3210. https://doi.org/10.1257/aer.20141280
241. Serafeim G (2020) Public sentiment and the price of corporate sustainability. Financ Anal J 76(2):26–46. https://doi.org/10.1080/0015198X.2020.1723390
242. Shuqin C, Minyan L, Hongwei T, Xiaoyu L, Jian G (2019) Assessing sustainability on Chinese university campuses: development of a campus sustainability evaluation system and its application with a case study. J Build Eng 24:100747. https://doi.org/10.1016/j.jobe.2019.100747
243. Sroufe R, Gopalakrishna-Remani V (2018) Management, social sustainability, reputation, and financial performance relationships: an empirical examination of U.S. Firms. Organ Environ 32(3):331–362. https://doi.org/10.1177/1086026618756611
244. Stead V (2017) Belonging and women entrepreneurs: women's navigation of gendered assumptions in entrepreneurial practice. Int Small Bus J 35(1):61–77. https://doi.org/10.1177/0266242615594413
245. Stefan D, Vasile V, Oltean A, Comes C-A, Stefan A-B, Ciucan-Rusu L, … Timus M (2021) Women entrepreneurship and sustainable business development: key findings from a SWOT–AHP analysis. Sustainability 13(9):5298. https://doi.org/10.3390/su13095298
246. Stubbs W (2017) Sustainable entrepreneurship and B corps. Bus Strateg Environ 26(3):331–344. https://doi.org/10.1002/bse.1920
247. Sung E, Kim H, Lee D (2018) Why do people consume and provide sharing economy accommodation?—a sustainability perspective. Sustainability 10(6):2072. https://doi.org/10.3390/su10062072
248. Swail J, Marlow S (2018) 'Embrace the masculine; attenuate the feminine'–gender, identity work and entrepreneurial legitimation in the nascent context. Entrep Reg Dev 30(1–2):256–282. https://doi.org/10.1080/08985626.2017.1406539
249. Swim JK, Geiger N (2018) The gendered nature of stereotypes about climate change opinion groups. Group Process Intergroup Relat 21(3):438–456. https://doi.org/10.1177/1368430217747406
250. Tahir MW, Kauser R, Bury M, Bhatti JS (2018) 'Individually-led'or 'female-male partnership'models for entrepreneurship with the BISP support: the story of women's financial and social empowerment from Pakistan. Women's Stud Int Forum 68:1–10. https://doi.org/10.1016/j.wsif.2018.01.011
251. Terán-Yépez E, Marín-Carrillo GM, del Pilar Casado-Belmonte M, de las Mercedes Capobianco-Uriarte M (2020) Sustainable entrepreneurship: Review of its evolution and new trends. J Cleaner Prod 252:119742. https://doi.org/10.1016/j.jclepro.2019.119742

252. Thirumalesh Madanaguli A, Kaur P, Bresciani S, Dhir A (2021) Entrepreneurship in rural hospitality and tourism. A systematic literature review of past achievements and future promises. Int J Contemp Hospitality Manage 33(8):2521–2558. https://doi.org/10.1108/IJCHM-09-2020-1121
253. Thompson-Whiteside H, Turnbull S, Fletcher-Brown J (2021) How women in the UAE enact entrepreneurial identities to build legitimacy. Int Small Bus J 39(7):643–661. https://doi.org/10.1177/0266242620979138
254. Treanor L, Marlow S, Swail J (2021) Rationalizing the postfeminist paradox: the case of UK women veterinary professionals. Gend Work Organ 28(1):337–360. https://doi.org/10.1111/gwao.12568
255. Tremblay D, Fortier F, Boucher JF, Riffon O, Villeneuve C (2020) Sustainable development goal interactions: an analysis based on the five pillars of the 2030 agenda. Sustain Devlop 28(6):1584–1596. https://doi.org/10.1002/sd.2107
256. Trosper RL (2002) Northwest coast indigenous institutions that supported resilience and sustainability. Ecol Econ 41(2):329–344. https://doi.org/10.1016/S0921-8009(02)00041-1
257. ul Haq MA, Victor S, Akram F (2021) Exploring the motives and success factors behind female entrepreneurs in India. Qual Quant 55(3):1105–1132. https://doi.org/10.1007/s11135-020-01046-x
258. Urbaniec M (2018) Sustainable entrepreneurship: Innovation-related activities in European enterprises. Pol J Environ Stud 27(4):1773–1779. https://doi.org/10.15244/pjoes/78155
259. van Zanten JA, van Tulder R (2021) Analyzing companies' interactions with the sustainable development goals through network analysis: four corporate sustainability imperatives. Bus Strateg Environ 30:2396–2420. https://doi.org/10.1002/bse.2753
260. Vázquez A, Larzabal-Fernández A, Lois D (2021) Situational materialism increases climate change scepticism in men compared to women. J Exp Soc Psychol 96:104163. https://doi.org/10.1016/j.jesp.2021.104163
261. Veleva V, Bodkin G (2018) Corporate-entrepreneur collaborations to advance a circular economy. J Cleaner Prod 188:20–37. https://doi.org/10.1016/j.jclepro.2018.03.196
262. Venugopalan M, Bastian BL, Viswanathan P (2021) The role of multi-actor engagement for women's empowerment and entrepreneurship in Kerala India. Adm Sci 11(1):31. https://doi.org/10.3390/admsci11010031
263. Villanueva-Moya L, Expósito F (2021) Women's evaluation of costs and benefits in public versus private spheres: the role of gender ideology. Analyses Soc Issues Public Policy:1–29. https://doi.org/10.1111/asap.12238
264. Vismara S (2019) Sustainability in equity crowdfunding. Technol Forecast Soc Change 141:98–106. https://doi.org/10.1016/j.techfore.2018.07.014
265. Vizcardo D, Salvador LF, Nole-Vara A, Dávila KP, Alvarez-Risco A, Yáñez JA, Mejia CR (2022) Sociodemographic predictors associated with the willingness to get vaccinated against COVID-19 in Peru: a cross-sectional survey. Vaccines 10(1), Article 48. https://doi.org/10.3390/vaccines10010048
266. Vuorio AM, Puumalainen K, Fellnhofer K (2018) Drivers of entrepreneurial intentions in sustainable entrepreneurship. Int J Entrepreneurship Behav Res 24(2):359–381. https://doi.org/10.1108/IJEBR-03-2016-0097
267. Weidhaas AD (2018) Female business owners hiding in plain sight. Int J Gend Entrep 10(1):2–18. https://doi.org/10.1108/IJGE-07-2017-0032
268. Welsh DH, Kaciak E, Mehtap S, Pellegrini MM, Caputo A, Ahmed S (2021) The door swings in and out: the impact of family support and country stability on success of women entrepreneurs in the Arab world. Int Small Bus J 39(7):619–642. https://doi.org/10.1177/0266242620952356
269. Wut T-M, Chan W-T, Lee SW (2021) Unconventional entrepreneurship: women handicraft entrepreneurs in a market-driven economy. Sustainability 13(13):7261. https://doi.org/10.3390/su13137261
270. Xheneti M, Karki ST, Madden A (2019) Negotiating business and family demands within a patriarchal society–the case of women entrepreneurs in the Nepalese context. Entrepreneurship Reg Dev 31(3–4):259–278

271. Yan J, Kim S, Zhang SX, Foo MD, Alvarez-Risco A, Del-Aguila-Arcentales S, Yáñez JA (2021) Hospitality workers' COVID-19 risk perception and depression: a contingent model based on transactional theory of stress model. Int J Hospitality Manage 95, Article 102935. https://doi.org/10.1016/j.ijhm.2021.102935
272. Yáñez JA, Alvarez-Risco A, Delgado-Zegarra J (2020) Covid-19 in Peru: from supervised walks for children to the first case of Kawasaki-like syndrome. The BMJ, 369, Article m2418. https://doi.org/10.1136/bmj.m2418
273. Yáñez JA, Jahanshahi AA, Alvarez-Risco A, Li J, Zhang SX (2020) Anxiety, distress, and turnover intention of healthcare workers in Peru by their distance to the epicenter during the COVID-19 crisis. Am J Trop Med Hyg 103(4):1614–1620. https://doi.org/10.4269/ajtmh.20-0800
274. Yáñez S, Uruburu Á, Moreno A, Lumbreras J (2019) The sustainability report as an essential tool for the holistic and strategic vision of higher education institutions. J Clean Prod 207:57–66. https://doi.org/10.1016/j.jclepro.2018.09.171
275. Yarime M, Tanaka Y (2012) The issues and methodologies in sustainability assessment tools for higher education institutions: a review of recent trends and future challenges. J Educ Sustain Devlop 6(1):63–77. https://doi.org/10.1177/097340821100600113
276. Zamora-Polo F, Sánchez-Martín J (2019) Teaching for a better world. sustainability and sustainable development goals in the construction of a change-maker university. Sustainability 11(15). https://doi.org/10.3390/su11154224
277. Zhang SX, Chen J, Afshar Jahanshahi A, Alvarez-Risco A, Dai H, Li J, Patty-Tito RM (2021) Succumbing to the COVID-19 pandemic—healthcare workers not satisfied and intend to leave their jobs. Int J Ment Heal Addict. https://doi.org/10.1007/s11469-020-00418-6
278. Zhang SX, Chen J, Jahanshahi AA, Alvarez-Risco A, Dai H, Li J, Patty-Tito RM (2021) Correction to: succumbing to the COVID-19 pandemic—healthcare workers not satisfied and intend to leave their jobs (international journal of mental health and addiction. Int J Ment Heal Addict. https://doi.org/10.1007/s11469-021-00502-5
279. Zhang SX, Sun S, Afshar Jahanshahi A, Alvarez-Risco A, Ibarra VG, Li J, Patty-Tito RM (2020) Developing and testing a measure of COVID-19 organizational support of healthcare workers—results from Peru, Ecuador, and Bolivia. Psychiatry Res 291. https://doi.org/10.1016/j.psychres.2020.113174
280. Zhou Q (2021) The impact of cross-cultural adaptation on entrepreneurial psychological factors and innovation ability for new entrepreneurs. Front Psychol 12:724544. https://doi.org/10.3389/fpsyg.2021.724544
281. Zhu L, Kara O, Zhu X (2018) A comparative study of women entrepreneurship in transitional economies: the case of China and Vietnam. J Entrepreneurship Emerg Econ 11(6):66–80. https://doi.org/10.1108/JEEE-04-2017-0027

Growing the Green Entrepreneurial Intention Among Youth—A Worldwide Comparative Analysis

Luigi Leclercq-Machado, Aldo Alvarez-Risco, María de las Mercedes Anderson-Seminario, and Shyla Del-Aguila-Arcentales

Abstract Entrepreneurship is recently viewed as a motor of economic, social, and environmental development. As a result, countries and educational institutions are looking toward implementing entrepreneurial education among young people, such as scholars. Nonetheless, it remains weak since students' intention to become entrepreneurs is not substantial. This chapter aims to understand how scholars' educational systems, and government can increase green entrepreneurial intention. A worldwide comparative analysis review was made to determine the current practices in different parts of the globe. National, private, and educational institutions play a crucial role in developing green entrepreneurial intention. Students look for the necessary knowledge, skills, abilities, and an excellent ecosystem to develop a new venture.

Keywords Green · Youth · Young consumer · Manager · Ecology · Sustainable · Circular · SDG · Entrepreneur · Entrepreneurship · Business · International business

1 Introduction

Entrepreneurship is nowadays viewed as a motor of economic development, job creation, and resilience [181, 200, 222]. Consequently, this last contribution to the country's well-being is undeniable [146]. Focusing on sustainability, eco-entrepreneurship is becoming one of the targets for students' learning programs [173]. That is one of the main reasons for the growing trend of implementing green entrepreneurship programs in educational institutions worldwide [142, 146]. Though the government and institutions developed programs for entrepreneurial activities, it remains difficult for students to take this road [142]. For instance, individual and

L. Leclercq-Machado · A. Alvarez-Risco · M. de las Mercedes Anderson-Seminario
Universidad de Lima, Lima, Perú

S. Del-Aguila-Arcentales (✉)
Escuela de Posgrado, Universidad San Ignacio de Loyola, Lima, Peru
e-mail: sdelaguila@usil.edu.pe

A. Alvarez-Risco et al. (eds.), *Footprint and Entrepreneurship*, Environmental Footprints and Eco-design of Products and Processes,
https://doi.org/10.1007/978-981-19-8895-0_8

external factors are crucial for long-term green entrepreneurship [133]. Also, most scholars believe that a specific group of people can only do entrepreneurship rather than develop their competencies to engage in entrepreneurial activities [226]. Therefore, entrepreneurial education is not directly related to students' intention to start entrepreneurial activities [9].

As educational institutions do not fulfill students' needs, entrepreneurial activities remain challenging to achieve [210]. This chapter aims to outline the current practices of entrepreneurial education implementation in educational institutions worldwide. Also, the author discusses how vital are institutions for the enhancement of entrepreneurial intention among students. This chapter is divided into the following sections. The first one consisted of a brief introduction to green entrepreneurship. Section 2 focuses on green entrepreneurial intentions among young people, especially students, and the role of educational institutions and external ones. Finally, the author provided examples of institutions worldwide that implemented entrepreneurial education and developed a critical aspect of the importance of entrepreneurial education for countries' economic, social, and environmental development.

2 From Traditional Entrepreneurship to Green Entrepreneurship

Entrepreneurship has been defined as a business development through resources investment aimed at reaching an objective or opportunity [170, 206]. Recent studies argued that entrepreneurship was a possible solution to environmental degradation [167]. Nonetheless, the traditional approach only considered economic benefits [213]. As a result, the entrepreneurship concept was extended to include non-economic gains [213]. Green entrepreneurship, also known as sustainable and eco-entrepreneurship, focuses on ecosystem preservation by pursuing business opportunities that benefit economically, socially, and environmentally [205].

Green entrepreneurs can be characterized as developing businesses that include environmentally friendly practices and technology [57]. It offers products and services with environmental benefits [188]. This individual is concerned with environmental issues and looks forward to ecosystem preservation [57]. Green entrepreneurship has been advantageous since it reduces deforestation, preserves the ecosystem, and improves quality [205]. Moreover, it is interesting to mention that green entrepreneurs combine business and environmental objectives to achieve social, sustainable, and economic development [125, 188]. As public and private sectors collaborate to overcome environmental issues, entrepreneurs become critical players in this battle [23].

3 Green Entrepreneurship Intention among Students

3.1 Students and Entrepreneurship

Entrepreneurship education's importance has increased since students must be prepared to face the competitive market [147]. As a result, universities play an essential role in the ecosystem [214]. Higher education institutions, for instance, provide training facilities to prepare students to develop new ventures [212]. Indeed, entrepreneurship educational programs offer students the possibility of developing a skill set and knowledge to establish ventures nowadays [173, 190, 203]. Entrepreneurship education fosters entrepreneurial behaviors consisting of ideas and resources [82]. Developing this last and incorporating a sustainable approach accelerates the transition toward a greener economy [56]. Therefore, entrepreneurial education programs provide skills, knowledge, and abilities to inspire future entrepreneurs [131].

Entrepreneurial education encourages young people such as scholars to begin a business, considering sustainable economic development to focusing on a motivational stage [194]. It is necessary to develop a suitable environment for learning and work to increase entrepreneurial intention from students [197]. For instance, a study made in higher education in India evidenced that institutional culture, lack of academic rigor, regional differences, and economic and gender gaps affect the entrepreneurial intention of students, which confirms the importance of an entrepreneurial education ecosystem [145].

An entrepreneurial ecosystem may include several factors such as universities, industries, and governments [159]. The principles to achieve an excellent entrepreneurial ecosystem are easy to access to the market, enough funding, networking, professional services such as mentoring and coaching, entrepreneurial education programs, and entrepreneurial orientation culture [29]. This environment can stimulate students and discard the idea that only a few people can determine business opportunities and obtain the resources to develop a new venture [79].

3.2 Green Entrepreneurship Intention

Green entrepreneurship intention (GEI), according to the current literature, can be defined as the predisposition to begin a sustainable entrepreneurial activity based on factors such as interests, decisions, concerns, and support, as stated by Alvarez-Risco, Mlodzianowska, Zamora-Ramos, et al. [42]. Beginning a venture is not oblivious to external factors since the environmental context, such as institutional framework and external interactions, influences decisions [118]. Analyzing the entrepreneurial orientation from students is, as a result, crucial as it explains the actual process and decision-making from an individual to commence a new business [145] (Fig. 1).

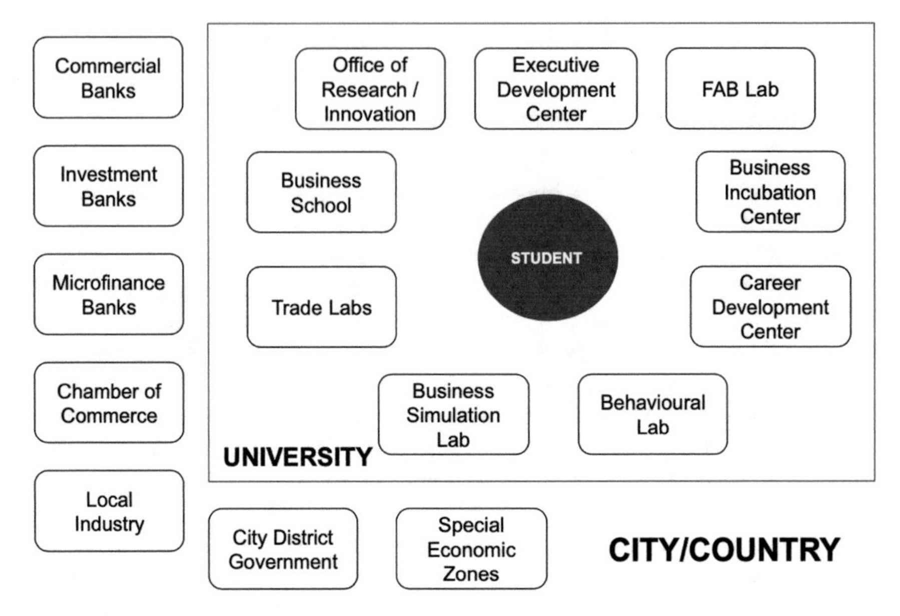

Fig. 1 Entrepreneurial ecosystem of scholars. *Source* Adapted from the role of universities and entrepreneurial ecosystem in producing entrepreneurs for industry 4.0 [203]

Self-efficacy can be defined as the individual's confidence to begin a new activity such as a venture [146]. Furthermore, it measures the individual's competences to perform a specific action [65, 165]. People perceive their skills and abilities differently [23]. As a result, only, a group of students may consider having enough resources, knowledge, skills, among others, to develop a new business [78]. Previous research found that self-efficacy in entrepreneurship is directly related to entrepreneurial intention [65]. That is because efficacy is significant for business start-ups and development [90]. According to Alvarez-Risco et al. [52], several factors can influence self-efficacy.

- **Education Development Support**

This first variable is understood as the training activities implemented by the educational system, such as universities, to develop businesses [76]. In this construct, courses, and programs focused on establishing business plans, ventures, and others are present [53]. By learning and inspiration, these initiatives are aimed to arouse interest and curiosity from students, which results in an increasingly entrepreneurial intention [23]. To increase students' well-being, universities, for instance, are required to invest in educational courses such as entrepreneurial programs [223]. As a result, it has been argued that skills, knowledge and abilities developed in this last will influence and improve entrepreneurial intention from scholars [146, 183].

As the growing awareness of environmental problems, sustainable approaches and practices are being implemented in all educational programs though it remains slow [23, 216]. Nonetheless, students who participated in green entrepreneurial courses are more likely to engage in entrepreneurial activities [76, 147]. A high degree of support influences the climate conditions in which scholars learn, as green entrepreneurship programs can offer knowledge, skills, and abilities to establish sustainable ventures [122, 173]. As entrepreneurship is a motor of economic, social, and environmental development, educational programs must support and promote entrepreneurial initiatives [169].

- **Institutional Support**

Institutional support is also an effort made by educational institutions but centered on promoting and motivating entrepreneurial activities among students [53]. Universities act as a vital source of motivation for students to promote innovation and new business ideas [81]. Indeed, students who take entrepreneurial courses feel that universities, for instance, motivate and encourage them to become entrepreneurs [76]. That is due to mentoring support from higher education institutions where scholars are assessed and discuss their business ideas [214]. Finally, it has been evidenced that a higher degree of institutionalization of entrepreneurship increases climate conditions and enhances entrepreneurial intention from students [76]. As education support exists, students are backed up by universities and other institutions as they provide adequate conditions to foster entrepreneurship [57].

- **Country Support**

Country support is the actual effort of the nation to promote new ventures [118]. Previous authors also refer to structural supports, combining private and public institutions, as key players in establishing a good ecosystem that promotes entrepreneurship [146]. Universities foment entrepreneurship, and governments are implementing a more significant amount of programs to support entrepreneurs [179]. For instance, government policies can enhance the existing conditions and improve the entrepreneurial ecosystem of students [56, 101].

To sum up, pro-environmental behavior development among students is crucial for the transition toward green economies [216]. Education, awareness-raising, and support remain critical players to increase green entrepreneurial intention in scholars [111, 216]. [220] mentioned that entrepreneurial education sensitizes students to entrepreneurship opportunities (Fig. 2).

4 Green Entrepreneurship Intention Promotion Worldwide

4.1 Higher Education Cases

Entrepreneurship education fosters students' intention to perform entrepreneurial activities, providing the necessary knowledge regarding this last [58]. In Australia, the Swinburne University of Technology, University of Queensland, University of

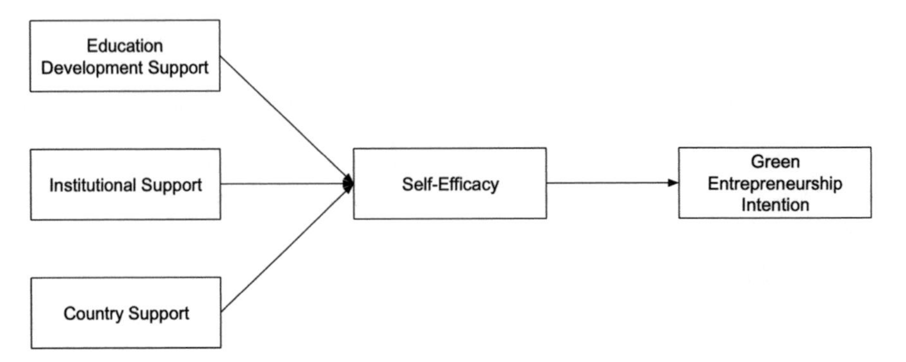

Fig. 2 Green entrepreneurship intention framework. *Source* Alvarez-Risco et al. [53]

Melbourne, and other 38 higher education institutions offer around 307 subjects, specifically entrepreneurship [157]. A sustainability entrepreneurship program called The Business Oriented Technological System Analysis (BOTSA) was implemented by the Eindhoven University of Technology based in the Netherlands [221]. This last focuses on providing the necessary knowledge to students of technological opportunities as tools for ventures creation. Focusing on the University of Queensland provides a Startup Academy to assess entrepreneurial projects, the ILab Accelerator for mentorship, and Idea Hub Unlimited programs such as Venture Curiosity for brainstorming [112]. The primary objectives of these lasts are to develop students' critical thinking and teach them how to develop an effective business that generates economic and environmental benefits [71].

In the United States, Illinois registered around 978 start-up creations between 2014 and 2018, supported by universities [135]. Among the 13 existing incubators and research parks in this state, the University of Illinois implemented the Enterprise-Works technology incubator as a facility to help students build and grow their future ventures (Illinois Science & Technology [135]. Despite the growing assistance from universities, programs are not the only tools to promote green entrepreneurship. The University of Washington, as an example, created the Global Social Entrepreneurship Competition [134]. In Europe, 15 universities offered different approaches to entrepreneurial education. For instance, Luiss Business School in Italy created the Executive Master In Circular Economy and incorporated seminars and workshops as learning methods [108]. Finally, China presents three types of entrepreneurial education tools. The first one consists basically of an entrepreneurial lectures program, the second is focused on testing centers and university science parks as incubators, and the last one is related to work internships designed to increase students' awareness of entrepreneurial opportunities [99, 129, 150, 238].

Educational institutions play an essential role in developing entrepreneurial individuals and fomenting a business culture involving economic and environmental benefits [77]. As it is being evidenced, entrepreneurial programs foster and enhance entrepreneurship intentions from students [173]. By improving skills and competencies, entrepreneurial intention increases [74]. Indeed, it develops entrepreneurial spirit and aspiration, resulting in a higher probability of developing a new venture. That is one of the main reasons why China implemented as a mandatory requirement entrepreneurial education in higher education institutions [150]. Environmental aspects are also starting to be incorporated in education to raise concerns of students [197]. Nonetheless, programs are not enough. Universities adopted different tools such as sessions, mentorship, coaching, award ceremonies, among others, to help scholars begin their entrepreneurial activities [102].

The green entrepreneurship is part of the efforts of sustainability which include the citizens [10, 13–15, 17, 24, 35, 40, 53, 54, 62, 69, 92, 110, 121, 139, 156, 167, 198, 209, 219, 230, 234], organizations [4, 25, 27, 31, 38, 75, 80, 83, 85, 148, 152, 185, 202, 207, 215, 232, 231, 227] and companies [6, 16, 20, 48, 67, 73, 98, 106, 155, 171, 186, 191, 208, 218] and activities as sports [95, 127, 137].

Then, it is considered in sustainability the activities of tourism [11, 14, 21, 84–86, 119, 168, 180, 193], education [5, 17, 25, 47, 49, 50, 51, 91, 132, 149, 192].

Then, green entrepreneurship includes circular economy approach [2, 12, 28, 36, 39, 72, 109, 141, 151, 154, 217], prices [60, 61, 64, 66, 128, 130, 144, 149, 163, 204], hospitality [3, 22, 116, 124, 136, 158, 178, 153, 225].

Also, the creation of new products and services in green entrepreneurship based in sustainability needs analysis of intellectual property [7, 30, 34, 59, 68, 70, 115, 161], health efforts [1, 8, 18, 19, 26, 32–33, 49, 37, 51–41, 84, 93, 97, 109, 114, 160, 166, 189, 199, 230, 222–237], and research [94, 143, 172, 176, 177, 184, 187, 196, 201].

4.2 Institutions Cases

Institutions, private and public, can also raise awareness and incent scholars to begin entrepreneurial actions. An example is the *Merdeka Belajar Kampus Merdeka* (MBKM), developed by the Indonesian Ministry of Education and Culture in 2021, allowing students to study at different universities and deepen their knowledge in entrepreneurship [174]. In Europe, for instance, the Contamination Lab was financed and formulated by the Italian Ministry of Education, University, and Research [120]. Another one is the Climate Launchpad developed by the European Institute of Innovation and Technology [102]. Those initiatives are being promoted by the sense of entrepreneurship principle formulated by the European Commission as a driver of personal fulfillment, social, environmental, and economic development [120].

In South Africa, the National Development Plan and The Vision 2030 promote youth training and entrepreneurial involvement [175]. Moreover, several entrepreneurial programs and funds were created in South Africa to increase entrepreneurship intentions, such as The Uganda Youth Venture Capital Fund [123]. Despite the growing entrepreneurship support practices worldwide, it is not standardized. In Saudi Arabia, for instance, incubators remain weak, limiting the adequate assessment for new entrepreneurial activities [140]. In 2008, the Ministry of Education in China launched a pilot entrepreneurship program collaborating with Tsinghua

University and Renmin University [162]. Promoting and raising entrepreneurial possibilities awareness among university students in China was crucial [150]. It launched the "Mass Entrepreneurship and Innovation" campaign to invest in universities and contribute to implementing entrepreneurship programs [229]. Encouraging scholars to become entrepreneurs became one of the scope of countries [224]. As a result, entrepreneurship is considered helpful for the government to achieve sustainable growth [195]. Nonetheless, several barriers are present during the process, and institutions can intervene as support [233]. Indeed, these lasts can offer incentives to promote responsible entrepreneurship [55].

Recent Suggested Lectures About Green Approach

- Waste reduction and carbon footprint [126]
- Material selection for circularity and footprints [164]
- 3D print, circularity, and footprints [103]
- Virtual tourism [104]
- Leadership for sustainability in crisis time [45]
- Virtual education and circularity [96]
- Circular economy for packaging [89]
- Students oriented to circular learning [105]
- Food and circular economy [100]
- Water footprint and food supply chain management [138]
- Waste footprint [63]
- Measuring circular economy [107]
- Carbon footprint [117]

5 Closing Remarks

Entrepreneurial education was born to raise entrepreneurship behaviors [197]. The amount of entrepreneurship educational programs is growing worldwide [77]. Several universities are adopting and partnering with overseas organizations to increase entrepreneurial programs portfolio [211]. Moreover, despite being different, all programs are focused on economic development stimulation [29, 182]. Indeed, one of the reasons that universities implemented these programs is for the positive impact which entrepreneurship generates to achieve sustainable growth [77]. Knowledge, skills, and abilities to become entrepreneurs are imperative, meaning that education is vital for creating new ventures [77]. Institutional and country support (education, external organizations, and government) are necessary for green entrepreneurship development [113]. It is recommended to facilitate access to financing and other resources to contribute to students' entrepreneurship creation [175]. On the other hand, educational institutions can deepen their modules by incorporating the current legislation and regulations existing on entrepreneurial activities and helping students raise their self-efficacy [211]. Green entrepreneurial intention can only be achieved if students are aware of universities and external institutions [228].

References

1. Aaron L (2020) The Oft forgotten part of nutrition: lessons from an integrated approach to understand consumer food safety behaviors. Curr Develop Nutrition, 4(Supplement_2):1322–1322. https://doi.org/10.1093/cdn/nzaa059_039
2. Abad-Segura E, Batlles-delaFuente A, González-Zamar M-D, Belmonte-Ureña LJ (2021) Implications for Sustainability of the joint application of bioeconomy and circular economy: a worldwide trend study. Sustainability 13(13):7182. https://doi.org/10.3390/su13137182
3. Abad-Segura E, Cortés-García FJ, Belmonte-Ureña LJ (2019) The sustainable approach to corporate social responsibility: a global analysis and future trends. Sustainability 11(19):5382. https://doi.org/10.3390/su11195382
4. Abad-Segura E, González-Zamar M-D, Infante-Moro JC, Ruipérez García G (2020) Sustainable management of digital transformation in higher education: global research trends. Sustainability 12(5):2107. https://doi.org/10.3390/su12052107
5. Abad-Segura E, González-Zamar MD (2021) Sustainable economic development in higher education institutions: a global analysis within the SDGs framework. J Cleaner Prod 294:126133. https://doi.org/10.1016/j.jclepro.2021.126133
6. Abad-Segura E, Morales ME, Cortés-García FJ, Belmonte-Ureña LJ (2020) Industrial processes management for a sustainable society: global research analysis. Processes, 8(5). https://doi.org/10.3390/pr8050631
7. Abdaljaleel M, Singer EJ, Yong WH (2019) Sustainability in biobanking. In: Yong WH (ed) Biobanking: methods and protocols. Springer, New York, pp 1–6. https://doi.org/10.1007/978-1-4939-8935-5_1
8. Aberilla JM, Gallego-Schmid A, Stamford L, Azapagic A (2020) Environmental sustainability of cooking fuels in remote communities: Life cycle and local impacts. Sci Total Environ 713:136445. https://doi.org/10.1016/j.scitotenv.2019.136445
9. Abina M, Oyeniran I, Onikosi-Alliyu S (2015) Determinants of eco entrepreneurial intention among students: a case study of University students in Ilorin and Malete. Ethiopian J Environ Stud Manage 8:107–112–107–112. https://webofproceedings.org/proceedings_series/ESSP/IWEDSS%202019/IWEDSS19220.pdf
10. Acevedo-Duque Á, Gonzalez-Diaz R, Vargas EC, Paz-Marcano A, Muller-Pérez S, Salazar-Sepúlveda G, D'Adamo I et al (2021) Resilience, leadership and female entrepreneurship within the context of SMEs: evidence from Latin America. Sustainability 13(15):8129. https://doi.org/10.3390/su13158129
11. Acevedo-Duque Á, Vega-Muñoz A, Salazar-Sepúlveda G (2020) Analysis of hospitality, leisure, and tourism studies in Chile. Sustainability 12(18):7238. https://doi.org/10.3390/su12187238
12. Ada E, Sagnak M, Mangla SK, Kazancoglu Y (2021) A circular business cluster model for sustainable operations management. Int J Logistics Res Appl 1–19. https://doi.org/10.1080/13675567.2021.2008335
13. Adam A (2015) Challenges of public finance sustainability in the European Union. Proc Econ Finance 23:298–302. https://doi.org/10.1016/S2212-5671(15)00507-9
14. Adam I, Agyeiwaah E, Dayour F (2021) Understanding the social identity, motivations, and sustainable behaviour among backpackers: a clustering approach. J Travel Tour Mark 38(2):139–154. https://doi.org/10.1080/10548408.2021.1887053
15. Adam JN, Adams T, Gerber JD, Haller T (2021) Decentralization for increased sustainability in natural resource management? two cautionary cases from Ghana. Sustainability, 13(12). https://doi.org/10.3390/su13126885
16. Adam M (2018) The role of human resource management (HRM) for the implementation of sustainable product-service systems (PSS)—an analysis of fashion retailers. Sustainability 10(7). https://doi.org/10.3390/su10072518
17. Adams R, Martin S, Boom K (2018) University culture and sustainability: designing and implementing an enabling framework. J Clean Prod 171:434–445. https://doi.org/10.1016/j.jclepro.2017.10.032

18. Adesogan AT, Havelaar AH, McKune SL, Eilittä M, Dahl GE (2020) Animal source foods: Sustainability problem or malnutrition and sustainability solution? Perspective matters. Glob Food Secur 25:100325. https://doi.org/10.1016/j.gfs.2019.100325
19. Adomah-Afari A, Chandler JA (2018) The role of government and community in the scaling up and sustainability of mutual health organisations: an exploratory study in Ghana. Soc Sci Med 207:25–37. https://doi.org/10.1016/j.socscimed.2018.04.044
20. Adomako S, Amankwah-Amoah J, Danso A, Konadu R, Owusu-Agyei S (2019) Environmental sustainability orientation and performance of family and nonfamily firms. Bus Strateg Environ 28(6):1250–1259. https://doi.org/10.1002/bse.2314
21. Adongo CA, Taale F, Adam I (2018) Tourists' values and empathic attitude toward sustainable development in tourism. Ecol Econ 150:251–263. https://doi.org/10.1016/j.ecolecon.2018.04.013
22. Agyeiwaah E (2019) Exploring the relevance of sustainability to micro tourism and hospitality accommodation enterprises (MTHAEs): evidence from home-stay owners. J Clean Prod 226:159–171. https://doi.org/10.1016/j.jclepro.2019.04.089
23. Ahmad NH, Halim HA, Ramayah T, Rahman SA (2015) Green entrepreneurship inclination among generation Y: the road towards a green economy. Prob Perspect Manage 13(2):211–218
24. Al-Naqbi AK, Alshannag Q (2018) The status of education for sustainable development and sustainability knowledge, attitudes, and behaviors of UAE University students. Int J Sustain High Educ 19(3):566–588. https://doi.org/10.1108/IJSHE-06-2017-0091
25. Aleixo AM, Leal S, Azeiteiro UM (2018) Conceptualization of sustainable higher education institutions, roles, barriers, and challenges for sustainability: an exploratory study in Portugal. J Clean Prod 172:1664–1673. https://doi.org/10.1016/j.jclepro.2016.11.010
26. Ali M, de Azevedo ARG, Marvila MT, Khan MI, Memon AM, Masood F, Haq IU et al (2021) The influence of COVID-19-induced daily activities on health parameters—a case study in Malaysia. Sustainability 13(13):7465. https://doi.org/10.3390/su13137465
27. Aliabadi V, Ataei P, Gholamrezai S, Aazami M (2019) Components of sustainability of entrepreneurial ecosystems in knowledge-intensive enterprises: the application of fuzzy analytic hierarchy process. Small Enterp Res 26(3):288–306. https://doi.org/10.1080/13215906.2019.1671215
28. Alkhuzaim L, Zhu Q, Sarkis J (2021) Evaluating emergy analysis at the nexus of circular economy and sustainable supply chain management. Sustain Prod Consumption 25:413–424. https://doi.org/10.1016/j.spc.2020.11.022
29. Allahar H, Sookram R (2019) A university business school as an entrepreneurial ecosystem hub. Technol Innovation Manage Rev 9(11)
30. Alonso-Fradejas A (2021) The resource property question in climate stewardship and sustainability transitions. Land Use Policy 108:105529. https://doi.org/10.1016/j.landusepol.2021.105529
31. Alshuwaikhat HM, Adenle YA, Saghir B (2016) Sustainability assessment of higher education institutions in Saudi Arabia. Sustainability, 8(8). https://doi.org/10.3390/su8080750
32. Álvarez-Risco A, Arellano EZ, Valerio EM, Acosta NM, Tarazona ZS (2013) Pharmaceutical care campaign as a strategy for implementation of pharmaceutical services: experience Peru. Pharmaceutical Care Espana 15(1):35–37
33. Alvarez-Risco A, Del-Aguila-Arcentales S (2015) Prescription errors as a barrier to pharmaceutical care in public health facilities: experience Peru. Pharmaceutical Care Espana 17(6):725–731
34. Alvarez-Risco A, Del-Aguila-Arcentales S (2021) A note on changing regulation in international business: the world intellectual property organization (wipo) and artificial intelligence. Progress Int Bus Res 15:363–371. https://doi.org/10.1108/S1745-886220210000015020
35. Alvarez-Risco A, Del-Aguila-Arcentales S, Delgado-Zegarra J, Yáñez JA, Diaz-Risco S (2019) Doping in sports: findings of the analytical test and its interpretation by the public. Sport Sci Health 15(1):255–257. https://doi.org/10.1007/s11332-018-0484-8
36. Alvarez-Risco A, Del-Aguila-Arcentales S, Diaz-Risco S (2018) Pharmacovigilance as a tool for sustainable development of healthcare in Peru. PharmacoVigilance Rev 10(2):4–6

37. Alvarez-Risco A, Del-Aguila-Arcentales S, Stevenson JG (2015) Pharmacists and mass communication for implementing pharmaceutical care. Am J Pharm Benefits 7(3):e125–e126
38. Alvarez-Risco A, Del-Aguila-Arcentales S, Yáñez JA, Alvarez-Risco A (2021) Telemedicine in Peru as a result of the COVID-19 pandemic: perspective from a country with limited internet access. Am J Trop Med Hyg 105(1):6–11. https://doi.org/10.4269/ajtmh.21-0255
39. Alvarez-Risco A, Estrada-Merino A, Anderson-Seminario MM, Mlodzianowska S, García-Ibarra V, Villagomez-Buele C, Carvache-Franco M (2021) Multitasking behavior in online classrooms and academic performance: case of university students in Ecuador during COVID-19 outbreak. Interactive Technol Smart Educ 18(3):422–434. https://doi.org/10.1108/ITSE-08-2020-0160
40. Alvarez-Risco A, Mejia CR, Delgado-Zegarra J, Del-Aguila-Arcentales S, Arce-Esquivel AA, Valladares-Garrido MJ, Yáñez JA et al (2020) The Peru approach against the COVID-19 infodemic: insights and strategies. Am J Trop Med Hyg 103(2):583–586. https://doi.org/10.4269/ajtmh.20-0536
41. Alvarez-Risco A, Van Mil JWF (2007) Pharmaceutical care in community pharmacies: practice and research in Peru. Ann Pharmacother 41(12):2032–2037. https://doi.org/10.1345/aph.1K117
42. Alvarez-Risco A, Mlodzianowska S, Zamora-Ramos U, Del-Aguila S (2021) Green entrepreneurship intention in university students: the case of Peru. Entrepreneurial Bus Econ Rev 9(4):85–100
43. Alvarez-Risco A, Quiroz-Delgado D, Del-Aguila-Arcentales S (2016) Pharmaceutical care in hypertension patients in a peruvian hospital [Article]. Indian J Public Health Res Develop 7(3):183–188. https://doi.org/10.5958/0976-5506.2016.00153.4
44. Alvarez-Risco A, Turpo-Cama A, Ortiz-Palomino L, Gongora-Amaut N, Del-Aguila-Arcentales S (2016) Barriers to the implementation of pharmaceutical care in pharmacies in Cusco Peru. Pharmaceutical Care Espana 18(5):194–205
45. Alvarez-Risco A, Del-Aguila-Arcentales S, Villalobos-Alvarez D, Diaz-Risco S (2022) Leadership for sustainability in crisis time. In: Alvarez-Risco A, Muthu SS, Del-Aguila-Arcentales S (eds) Circular economy. Environmental footprints and eco-design of products and processes. Springer, Singapore. https://doi.org/10.1007/978-981-19-0549-0_3
46. Alvarez-Risco A, Dawson J, Johnson W, Conteh-Barrat M, Aslani P, Del-Aguila-Arcentales S, Diaz-Risco S (2020) Ebola virus disease outbreak: a global approach for health systems. Revista Cubana de Farmacia 53(4):1–13, Article e491
47. Alvarez-Risco A, Del-Aguila-Arcentales S (2021b) Public policies and private efforts to increase women entrepreneurship based on STEM background. In: Contributions to management science, pp 75–87. https://doi.org/10.1007/978-3-030-83792-1_5
48. Alvarez-Risco A, Del-Aguila-Arcentales S (2022) Sustainable initiatives in international markets. In: Contributions to management science, pp 181–191. https://doi.org/10.1007/978-3-030-85950-3_10
49. Alvarez-Risco A, Del-Aguila-Arcentales S, Rosen MA, García-Ibarra V, Maycotte-Felkel S, Martínez-Toro GM (2021) Expectations and interests of university students in covid-19 times about sustainable development goals: evidence from Colombia, Ecuador, Mexico, and Peru. Sustainability (Switzerland) 13(6), Article 3306. https://doi.org/10.3390/su13063306
50. Alvarez-Risco A, Del-Aguila-Arcentales S, Yáñez JA, Rosen MA, Mejia CR (2021) Influence of technostress on academic performance of university medicine students in Peru during the covid-19 pandemic. Sustainability (Switzerland), 13(16), Article 8949. https://doi.org/10.3390/su13168949
51. Alvarez-Risco A, Delgado-Zegarra J, Yáñez JA, Diaz-Risco S, Del-Aguila-Arcentales S (2018) Predation risk by gastronomic boom—case Peru. J Landscape Ecol (Czech Republic) 11(1):100–103. https://doi.org/10.2478/jlecol-2018-0003
52. Alvarez-Risco A, Mlodzianowska S, García-Ibarra V, Rosen MA, Del-Aguila-Arcentales S (2021) Factors affecting green entrepreneurship intentions in business university students in COVID-19 pandemic times: case of ecuador. Sustainability, 13(11). https://doi.org/10.3390/su13116447

53. Alvarez-Risco A, Mlodzianowska S, García-Ibarra V, Rosen MA, Del-Aguila-Arcentales S (2021) Factors affecting green entrepreneurship intentions in business university students in covid-19 pandemic times: case of Ecuador. Sustainability (Switzerland), 13(11), Article 6447. https://doi.org/10.3390/su13116447
54. Alvarez-Risco A, Mlodzianowska S, Zamora-Ramos U, Del-Aguila-Arcentales S (2021) Green entrepreneurship intention in university students: the case of Peru. Entrepreneurial Bus Econ Rev 9(4):85–100. https://doi.org/10.15678/EBER.2021.090406
55. Alwakid W, Aparicio S, Urbano D (2021) The influence of green entrepreneurship on sustainable development in Saudi Arabia: the role of formal institutions. Int J Environ Res Public Health 18(10):5433
56. Alwakid W, Aparicio S, Urbano D (2020) Cultural antecedents of green entrepreneurship in Saudi Arabia: an institutional approach. Sustainability, 12(9). https://doi.org/10.3390/su12093673
57. Amankwah J, Sesen H (2021) On the relation between green entrepreneurship intention and behavior. Sustainability, 13(13). https://doi.org/10.3390/su13137474
58. Ambad SNA, Damit DHDA (2016) Determinants of entrepreneurial intention among undergraduate students in Malaysia. Procedia Econ Finance 37:108–114
59. Ang KL, Saw ET, He W, Dong X, Ramakrishna S (2021) Sustainability framework for pharmaceutical manufacturing (PM): a review of research landscape and implementation barriers for circular economy transition. J Clean Prod 280:124264. https://doi.org/10.1016/j.jclepro.2020.124264
60. Angulo-Mosquera LS, Alvarado-Alvarado AA, Rivas-Arrieta MJ, Cattaneo CR, Rene ER, García-Depraect O (2021) Production of solid biofuels from organic waste in developing countries: a review from sustainability and economic feasibility perspectives. Sci Total Environ 795:148816. https://doi.org/10.1016/j.scitotenv.2021.148816
61. Apcho-Ccencho LV, Cuya-Velásquez BB, Rodríguez DA, Anderson-Seminario MLM, Alvarez-Risco A, Estrada-Merino A, Mlodzianowska S (2021) The impact of international price on the technological industry in the united states and china during times of crisis: commercial war and covid-19. Adv Bus Manage Forecasting 14:149–160
62. Aragon-Correa JA, Marcus AA, Rivera JE, Kenworthy AL (2017) Sustainability management teaching resources and the challenge of balancing planet, people, and profits. Acad Manage Learn Educ 16(3):469–483. https://doi.org/10.5465/amle.2017.0180
63. Arias-Meza M, Alvarez-Risco A, Cuya-Velásquez BB, de las Mercedes Anderson-Seminario M, Del-Aguila-Arcentales S (2022) Fashion and textile circularity and waste footprint. In: Alvarez-Risco A, Muthu SS, Del-Aguila-Arcentales S (eds) Circular economy. Environmental footprints and eco-design of products and processes. Springer, Singapore. https://doi.org/10.1007/978-981-19-0549-0_9
64. Aschemann-Witzel J, Giménez A, Ares G (2018) Convenience or price orientation? consumer characteristics influencing food waste behaviour in the context of an emerging country and the impact on future sustainability of the global food sector. Glob Environ Chang 49:85–94. https://doi.org/10.1016/j.gloenvcha.2018.02.002
65. Awotunde OM, Westhuizen TVD (2021) Entrepreneurial self-efficacy development: an effective intervention for sustainable student entrepreneurial intentions. Int J Innovation Sustain Develop 15(4):475–495
66. Aydın B, Alvarez MD (2020) Understanding the tourists' perspective of sustainability in cultural tourist destinations. Sustainability 12(21):8846. https://doi.org/10.3390/su12218846
67. Bamgbade JA, Kamaruddeen AM, Nawi MNM, Adeleke AQ, Salimon MG, Ajibike WA (2019) Analysis of some factors driving ecological sustainability in construction firms. J Clean Prod 208:1537–1545. https://doi.org/10.1016/j.jclepro.2018.10.229
68. Bannerman S (2020) The World intellectual property organization and the sustainable development agenda. Futures 122:102586. https://doi.org/10.1016/j.futures.2020.102586
69. Barr S, Gilg A, Shaw G (2011) 'Helping people make better choices': exploring the behaviour change agenda for environmental sustainability. Appl Geogr 31(2):712–720. https://doi.org/10.1016/j.apgeog.2010.12.003

70. Barragán-Ocaña A, Silva-Borjas P, Olmos-Peña S, Polanco-Olguín M (2020) Biotechnology and bioprocesses: their contribution to sustainability. Processes, 8(4). https://doi.org/10.3390/pr8040436
71. Barron E, Ruiz LE (2021) Evaluating the effect of entrepreneurial programs elements on students: a scale development. BAR-Brazilian Adminis Rev 18
72. Barros MV, Salvador R, do Prado GF, de Francisco AC, Piekarski CM (2021) Circular economy as a driver to sustainable businesses. Cleaner Environ Syst 2:100006. https://doi.org/10.1016/j.cesys.2020.100006
73. Batista AA, Francisco AC (2018) Organizational sustainability practices: a study of the firms listed by the corporate sustainability index. Sustainability, 10(1). https://doi.org/10.3390/su10010226
74. Belas J, Belás Ľ, Čepel M, Rozsa Z (2019) The impact of the public sector on the quality of the business environment in the SME segment. Administratie si Management Public. https://doi.org/10.24818/amp/2019.32-02
75. Ben Youssef A, Boubaker S, Omri A (2018) Entrepreneurship and sustainability: the need for innovative and institutional solutions. Technol Forecast Soc Chang 129:232–241. https://doi.org/10.1016/j.techfore.2017.11.003
76. Bergmann H, Geissler M, Hundt C, Grave B (2018) The climate for entrepreneurship at higher education institutions. Res Policy 47(4):700–716. https://doi.org/10.1016/j.respol.2018.01.018
77. Betáková J, Havierniková K, Okręglicka M, Mynarzova M, Magda R (2020) The role of universities in supporting entrepreneurial intentions of students toward sustainable entrepreneurship
78. Bhaskar AU, Garimella S (2017) A study of predictors of entrepreneurial intentions: development of comprehensive measures. Glob Bus Rev 18(3):629–651
79. Blenker P, Frederiksen SH, Korsgaard S, Müller S, Neergaard H, Thrane C (2012) Entrepreneurship as everyday practice: towards a personalized pedagogy of enterprise education. Ind High Educ 26(6):417–430. https://doi.org/10.5367/ihe.2012.0126
80. Bokpin GA (2017) Foreign direct investment and environmental sustainability in Africa: the role of institutions and governance. Res Int Bus Finance 39:239–247. https://doi.org/10.1016/j.ribaf.2016.07.038
81. Bonaccorsi A, Colombo MG, Guerini M, Rossi-Lamastra C (2013) University specialization and new firm creation across industries. Small Bus Econ 41(4):837–863. https://doi.org/10.1007/s11187-013-9509-5
82. Borasi R, Finnigan K (2010) Entrepreneurial attitudes and behaviors that can help prepare successful change-agents in education. The New Educator 6(1):1–29. https://doi.org/10.1080/1547688X.2010.10399586
83. Brito RM, Rodríguez C, Aparicio JL (2018) Sustainability in teaching: an evaluation of university teachers and students. Sustainability 10(2):439. https://doi.org/10.3390/su10020439
84. Brown KA, Harris F, Potter C, Knai C (2020) The future of environmental sustainability labelling on food products. The Lancet Planetary Health 4(4):e137–e138. https://doi.org/10.1016/S2542-5196(20)30074-7
85. Brown HS, de Jong M, Levy DL (2009) Building institutions based on information disclosure: lessons from GRI's sustainability reporting. J Clean Prod 17(6):571–580. https://doi.org/10.1016/j.jclepro.2008.12.009
86. Carvache-Franco M, Carvache-Franco O, Carvache-Franco W, Alvarez-Risco A, Estrada-Merino A (2021) Motivations and segmentation of the demand for coastal cities: a study in Lima Peru. Int J Tourism Res 23(4):517–531. https://doi.org/10.1002/jtr.2423
87. Carvache-Franco M, Alvarez-Risco A, Carvache-Franco O, Carvache-Franco W, Estrada-Merino A, Villalobos-Alvarez D (2021) Perceived value and its influence on satisfaction and loyalty in a coastal city: a study from Lima, Peru. J Policy Res Tourism Leisure Events. https://doi.org/10.1080/19407963.2021.1883634

88. Carvache-Franco M, Alvarez-Risco A, Carvache-Franco W, Carvache-Franco O, Estrada-Merino A, Rosen MA (2021) Coastal cities seen from loyalty and their tourist motivations: a study in Lima, Peru. Sustainability (Switzerland) 13(21), Article 11575. https://doi.org/10.3390/su132111575
89. Castillo-Benancio S et al (2022) Circular economy for packaging and carbon footprint. In: Alvarez-Risco A, Muthu SS, Del-Aguila-Arcentales S (eds) Circular economy. Environmental footprints and eco-design of products and processes. Springer, Singapore. https://doi.org/10.1007/978-981-19-0549-0_6
90. Ceresia F, Mendola C (2019) Entrepreneurial self-identity, perceived corruption, exogenous and endogenous obstacles as antecedents of entrepreneurial intention in Italy. Soc Sci 8(2). https://doi.org/10.3390/socsci8020054
91. Chafloque-Cespedes R, Alvarez-Risco A, Robayo-Acuña PV, Gamarra-Chavez CA, Martinez-Toro GM, Vicente-Ramos W (2021) Effect of sociodemographic factors in entrepreneurial orientation and entrepreneurial intention in university students of Latin American business schools. Contemp Issues Entrepreneurship Res 11:151–165. https://doi.org/10.1108/S2040-724620210000011010
92. Chafloque-Céspedes R, Vara-Horna A, Asencios-Gonzales Z, López-Odar D, Alvarez-Risco A, Quipuzco-Chicata L, Sánchez-Villagomez M et al (2020) Academic presenteeism and violence against women in schools of business and engineering in Peruvian universities. Lecturas de Economia, 93, 127–153. https://doi.org/10.17533/udea.le.n93a340726
93. Chen X, Zhang SX, Jahanshahi AA, Alvarez-Risco A, Dai H, Li J, Ibarra VG (2020) Belief in a COVID-19 conspiracy theory as a predictor of mental health and well-being of health care workers in ecuador: cross-sectional survey study. JMIR Public Health Surveillance 6(3), Article e20737. https://doi.org/10.2196/20737
94. Chung SA, Olivera S, Román BR, Alanoca E, Moscoso S, Terceros BL, Yáñez JA (2021) Themes of scientific production of the Cuban journal of pharmacy indexed in scopus (1967–2020). Revista Cubana de Farmacia, 54(1), Article e511.
95. Constantin P-N, Stanescu R, Stanescu M (2020) Social entrepreneurship and sport in Romania: how can former athletes contribute to sustainable social change? Sustainability 12(11):4688. https://doi.org/10.3390/su12114688
96. Contreras-Taica A et al (2022) Virtual education: carbon footprint and circularity. In: Alvarez-Risco A, Muthu SS, Del-Aguila-Arcentales S (eds) Circular economy. Environmental footprints and eco-design of products and processes. Springer, Singapore. https://doi.org/10.1007/978-981-19-0549-0_13
97. Cruz JP, Alshammari F, Felicilda-Reynaldo RFD (2018) Predictors of Saudi nursing students' attitudes towards environment and sustainability in health care. Int Nurs Rev 65(3):408–416. https://doi.org/10.1111/inr.12432
98. Cruz-Torres W, Alvarez-Risco A, Del-Aguila-Arcentales S (2021) Impact of enterprise resource planning (ERP) implementation on performance of an education enterprise: a structural equation modeling (SEM) [Article]. Stud Bus Econ 16(2):37–52. https://doi.org/10.2478/sbe-2021-0023
99. Cui J, Sun J, Bell R (2021) The impact of entrepreneurship education on the entrepreneurial mindset of college students in China: the mediating role of inspiration and the role of educational attributes. Int J Manage Educ 19(1):100296
100. Cuya-Velásquez BB et al (2022) Circular economy for food loss reduction and water footprint. In: Alvarez-Risco A, Muthu SS, Del-Aguila-Arcentales S (eds) Circular economy. Environmental footprints and eco-design of products and processes. Springer, Singapore. https://doi.org/10.1007/978-981-19-0549-0_4
101. Dai W, Si S (2018) Government policies and firms' entrepreneurial orientation: strategic choice and institutional perspectives. J Bus Res 93:23–36. https://doi.org/10.1016/j.jbusres.2018.08.026
102. Daub CH, Hasler M, Verkuil AH, Milow U (2020) Universities talk, students walk: promoting innovative sustainability projects. Int J Sustain Higher Educ

103. De-la-Cruz-Diaz M, Alvarez-Risco A, Jaramillo-Arévalo M, de las Mercedes Anderson-Seminario M, Del-Aguila-Arcentales S (2022) 3D print, circularity, and footprints. In: Alvarez-Risco A, Muthu SS, Del-Aguila-Arcentales S (eds) Circular economy. Environmental footprints and eco-design of products and processes. Springer, Singapore. https://doi.org/10.1007/978-981-19-0549-0_5
104. De-la-Cruz-Diaz M et al (2022) Virtual tourism, carbon footprint, and circularity. In: Alvarez-Risco A, Muthu SS, Del-Aguila-Arcentales S (eds) Circular economy. Environmental footprints and eco-design of products and processes. Springer, Singapore. https://doi.org/10.1007/978-981-19-0549-0_12
105. De las Mercedes Anderson-Seminario M, Alvarez-Risco A (2022) Better students, better companies, better life: circular learning. In: Alvarez-Risco A, Muthu SS, Del-Aguila-Arcentales S (eds) Circular economy. Environmental footprints and eco-design of products and processes. Springer, Singapore. https://doi.org/10.1007/978-981-19-0549-0_2
106. Del-Aguila-Arcentales S, Alvarez-Risco A (2013) Human error or burnout as explanation for mistakes in pharmaceutical laboratories. Accred Qual Assur 18(5):447–448. https://doi.org/10.1007/s00769-013-1000-0
107. Del-Aguila-Arcentales S, Alvarez-Risco A, Muthu SS (2022) Measuring circular economy. In: Alvarez-Risco A, Muthu SS, Del-Aguila-Arcentales S (eds) Circular economy. Environmental footprints and eco-design of products and processes. Springer, Singapore. https://doi.org/10.1007/978-981-19-0549-0_1
108. Del Vecchio P, Secundo G, Mele G, Passiante G (2021) Sustainable entrepreneurship education for circular economy: emerging perspectives in Europe. Int J Entrepreneurial Behavior Res
109. Delgado-Zegarra J, Alvarez-Risco A, Yáñez JA (2018) Indiscriminate use of pesticides and lack of sanitary control in the domestic market in Peru. Revista Panamericana de Salud Publica/Pan American Journal of Public Health 42, Article e3. https://doi.org/10.26633/RPSP.2018.3
110. Deslatte A, Swann WL (2019) Elucidating the linkages between entrepreneurial orientation and local government sustainability performance. Am Rev Public Adminis 50(1):92–109. https://doi.org/10.1177/0275074019869376
111. Dimante D, Tambovceva T, Atstaja D (2016) Raising environmental awareness through education. Int J Continuing Eng Educ Life Long Learn 26(3):259–272
112. Dodgson M, Gann D (2020). Universities should support more student entrepreneurs. Here's why–and how. World Economic Forum
113. Domańska A, Żukowska B, Zajkowski R (2018) Green entrepreneurship as a connector among social, environmental and economic pillars of sustainable development. Why some countries are more agile? Problemy Ekorozwoju, 13(2)
114. Enciso-Zarate A, Guzmán-Oviedo J, Sánchez-Cardona F, Martínez-Rohenes D, Rodríguez-Palomino JC, Alvarez-Risco A, Diaz-Risco S et al (2016) Evaluation of contamination by cytotoxic agents in colombian hospitals. Pharmaceutical Care Espana 18(6):241–250
115. Eppinger E, Jain A, Vimalnath P, Gurtoo A, Tietze F, Hernandez Chea R (2021) Sustainability transitions in manufacturing: the role of intellectual property. Curr Opin Environ Sustain 49:118–126. https://doi.org/10.1016/j.cosust.2021.03.018
116. Ertuna B, Karatas-Ozkan M, Yamak S (2019) Diffusion of sustainability and CSR discourse in hospitality industry. Int J Contemp Hosp Manag 31(6):2564–2581. https://doi.org/10.1108/IJCHM-06-2018-0464
117. Esquerre-Botton S, Alvarez-Risco A, Leclercq-Machado L, de las Mercedes Anderson-Seminario M, Del-Aguila-Arcentales S (2022) Food loss reduction and carbon footprint practices worldwide: a benchmarking approach of circular economy. In: Alvarez-Risco A, Muthu SS, Del-Aguila-Arcentales S (eds) Circular economy. Environmental footprints and eco-design of products and processes. Springer, Singapore. https://doi.org/10.1007/978-981-19-0549-0_8
118. Fichter K, Tiemann I (2018) Factors influencing university support for sustainable entrepreneurship: insights from explorative case studies. J Cleaner Prod 175:512–524. https://doi.org/10.1016/j.jclepro.2017.12.031

119. Figueroa-Domecq C, Kimbu A, de Jong A, Williams AM (2020) Sustainability through the tourism entrepreneurship journey: a gender perspective. J Sustain Tourism 1–24. https://doi.org/10.1080/09669582.2020.1831001
120. Fiore E, Sansone G, Paolucci E (2019) Entrepreneurship education in a multidisciplinary environment: evidence from an entrepreneurship programme held in Turin. Adminis Sci 9(1):28
121. Galleli B, Teles NEB, Santos JARD, Freitas-Martins MS, Hourneaux Junior F (2021) Sustainability university rankings: a comparative analysis of UI green metric and the times higher education world university rankings. Int J Sustain Higher Educ, Ahead-of-print(ahead-of-print). https://doi.org/10.1108/IJSHE-12-2020-0475
122. Geissler M, Jahn S, Haefner P (2010) The entrepreneurial climate at universities: the impact of organizational factors. The theory and practice of entrepreneurship 12–31
123. Gemma A, Ibrahim K (2015) Creating youth employment through entrepreneurship financing: the Uganda youth venture capital fund
124. Gerdt S-O, Wagner E, Schewe G (2019) The relationship between sustainability and customer satisfaction in hospitality: an explorative investigation using eWOM as a data source. Tour Manage 74:155–172. https://doi.org/10.1016/j.tourman.2019.02.010
125. Gibbs D, O'Neill K (2012) Green entrepreneurship: building a green economy?—evidence from the UK. In: Underwood S, Blundel R, Lyon F, Schaefer A (eds) Social and sustainable enterprise: changing the nature of business, vol 2. Emerald Group Publishing Limited, pp 75–96. https://doi.org/10.1108/S2040-7246(2012)0000002008
126. Gómez-Prado R, Alvarez-Risco A, Sánchez-Palomino J, de las Mercedes Anderson-Seminario M, Del-Aguila-Arcentales S (2022) Circular economy for waste reduction and carbon footprint. In: Alvarez-Risco A, Muthu SS, Del-Aguila-Arcentales S (eds) Circular economy. Environmental footprints and eco-design of products and processes. Springer, Singapore. https://doi.org/10.1007/978-981-19-0549-0_7
127. González-Serrano MH, Añó Sanz V, González-García RJ (2020) Sustainable sport entrepreneurship and innovation: a bibliometric analysis of this emerging field of research. Sustainability 12(12):5209. https://doi.org/10.3390/su12125209
128. Grewal J, Hauptmann C, Serafeim G (2021) Material sustainability information and stock price informativeness. J Bus Ethics 171(3):513–544. https://doi.org/10.1007/s10551-020-04451-2
129. Guo M, Nowakowska-Grunt J, Gorbanyov V, Egorova M (2020) Green technology and sustainable development: assessment and green growth frameworks. Sustainability, 12(16). https://doi.org/10.3390/su12166571
130. Hall MR (2019) The sustainability price: expanding environmental life cycle costing to include the costs of poverty and climate change. Int J Life Cycle Assess 24(2):223–236. https://doi.org/10.1007/s11367-018-1520-2
131. Hameed I, Zaman U, Waris I, Shafique O (2021) A serial-mediation model to link entrepreneurship education and green entrepreneurial behavior: application of resource-based view and flow theory. Int J Environ Res Public Health 18(2). https://doi.org/10.3390/ijerph18020550
132. Hermann RR, Bossle MB (2020) Bringing an entrepreneurial focus to sustainability education: a teaching framework based on content analysis. J Cleaner Prod 246:119038. https://doi.org/10.1016/j.jclepro.2019.119038
133. Hussain I, Nazir M, Hashmi SB, Di Vaio A, Shaheen I, Waseem MA, Arshad A (2021) Green and sustainable entrepreneurial intentions: a mediation-moderation perspective. Sustainability 13(15). https://doi.org/10.3390/su13158627
134. Huster K, Petrillo C, O'Malley G, Glassman D, Rush J, Wasserheit J (2017) Global social entrepreneurship competitions: incubators for innovations in global health? J Manag Educ 41(2):249–271
135. Illinois Science and Technology Coalition (2019) Illinois innovation index. https://www.istcoalition.org/wp-content/uploads/UE_Index_19_5.22.pdf
136. Jones P, Comfort D (2020) The COVID-19 crisis and sustainability in the hospitality industry. Int J Contemp Hosp Manag 32(10):3037–3050. https://doi.org/10.1108/IJCHM-04-2020-0357

137. Jones P, Ratten V, Hayduk T (2020) Sport, fitness, and lifestyle entrepreneurship. Int Entrepreneurship Manage J 16(3):783–793. https://doi.org/10.1007/s11365-020-00666-x
138. Juarez-Rojas L, Alvarez-Risco A, Campos-Dávalos N, de las Mercedes Anderson-Seminario M, Del-Aguila-Arcentales S (2022) Water footprint in the textile and food supply chain management: trends to become circular and sustainable. In: Alvarez-Risco A, Muthu SS, Del-Aguila-Arcentales S (eds) Circular economy. Environmental footprints and eco-design of products and processes. Springer, Singapore. https://doi.org/10.1007/978-981-19-0549-0_11
139. Khan S, Henderson C (2020) How Western Michigan University is approaching its commitment to sustainability through sustainability-focused courses. J Clean Prod 253:119741. https://doi.org/10.1016/j.jclepro.2019.119741
140. Khan MR (2013) Mapping entrepreneurship ecosystem of Saudi Arabia. World J Entrepreneurship Manage Sustain Develop
141. Khan SAR, Razzaq A, Yu Z, Miller S (2021) Industry 4.0 and circular economy practices: a new era business strategies for environmental sustainability. Bus Strat Environ 30(8):4001–4014. https://doi.org/10.1002/bse.2853
142. Kim M, Park MJ (2019) Entrepreneurial education program motivations in shaping engineering students' entrepreneurial intention. J Entrepreneurship Emerg Econ 11(3):328–350. https://doi.org/10.1108/JEEE-08-2018-0082
143. Kong L, Liu Z, Wu J (2020) A systematic review of big data-based urban sustainability research: state-of-the-science and future directions. J Clean Prod 273:123142. https://doi.org/10.1016/j.jclepro.2020.123142
144. Kozlowski A, Searcy C, Bardecki M (2018) The reDesign canvas: fashion design as a tool for sustainability. J Cleaner Prod 183:194–207. https://doi.org/10.1016/j.jclepro.2018.02.014
145. Kumar S, Paray ZA, Dwivedi AK (2021) Student's entrepreneurial orientation and intentions. Higher Educ Skills Work-Based Learn 11(1):78–91. https://doi.org/10.1108/HESWBL-01-2019-0009
146. Kör B, Wakkee I, Mutlutürk M (2020) An investigation of factors influencing entrepreneurial intention amongst university students. J Higher Educ Theory Practice 20(1):70–86
147. Küttim M, Kallaste M, Venesaar U, Kiis A (2014) Entrepreneurship education at university level and students' entrepreneurial intentions. Proc—Soc Behav Sci 110:658–668. https://doi.org/10.1016/j.sbspro.2013.12.910
148. Larrán Jorge M, Andrades Peña FJ, Herrera Madueño J (2019) An analysis of university sustainability reports from the GRI database: an examination of influential variables. J Environ Planning Manage 62(6):1019–1044. https://doi.org/10.1080/09640568.2018.1457952
149. Leiva-Martinez MA, Anderson-Seminario MLM, Alvarez-Risco A, Estrada-Merino A, Mlodzianowska S (2021) Price variation in lower goods as of previous economic crisis and the contrast of the current price situation in the context of covid-19 in Peru. Adv Bus Manage Forecasting 14:161–166
150. Liu T, Walley K, Pugh G, Adkins P (2020) Entrepreneurship education in China. J Entrepreneurship Emerg Econ 12(2):305–326. https://doi.org/10.1108/JEEE-01-2019-0006
151. Lopez-Odar D, Alvarez-Risco A, Vara-Horna A, Chafloque-Cespedes R, Sekar MC (2020) Validity and reliability of the questionnaire that evaluates factors associated with perceived environmental behavior and perceived ecological purchasing behavior in Peruvian consumers [Article]. Soc Respons J 16(3):403–417. https://doi.org/10.1108/SRJ-08-2018-0201
152. Lozano R, Barreiro-Gen M, Lozano FJ, Sammalisto K (2019) Teaching sustainability in European higher education institutions: assessing the connections between competences and pedagogical approaches. Sustainability 11(6):1602. https://doi.org/10.3390/su11061602
153. Madanaguli AT, Kaur P, Bresciani S, Dhir A (2021) Entrepreneurship in rural hospitality and tourism. A systematic literature review of past achievements and future promises. Int J Contemp Hospital Manage 33(8):2521–2558. https://doi.org/10.1108/IJCHM-09-2020-1121
154. Manavalan E, Jayakrishna K (2019) An analysis on sustainable supply chain for circular economy. Procedia Manuf 33:477–484. https://doi.org/10.1016/j.promfg.2019.04.059

155. Mani V, Gunasekaran A, Delgado C (2018) Supply chain social sustainability: standard adoption practices in Portuguese manufacturing firms. Int J Prod Econ 198:149–164. https://doi.org/10.1016/j.ijpe.2018.01.032
156. Manzoor SR, Ho JSY, Al Mahmud A (2021) Revisiting the 'university image model' for higher education institutions' sustainability. J Mark High Educ 31(2):220–239. https://doi.org/10.1080/08841241.2020.1781736
157. Maritz A, Jones C, Shwetzer C (2015) The status of entrepreneurship education in Australian universities. Education+Training. https://doi.org/10.1108/ET-04-2015-0026
158. Martinez-Martinez A, Cegarra-Navarro J-G, Garcia-Perez A, Wensley A (2019) Knowledge agents as drivers of environmental sustainability and business performance in the hospitality sector. Tour Manage 70:381–389. https://doi.org/10.1016/j.tourman.2018.08.030
159. McAdam M, Debackere K (2018) Beyond 'triple helix' toward 'quadruple helix' models in regional innovation systems: implications for theory and practice. Wiley Online Library, vol 48, pp 3–6
160. Mejía-Acosta N, Alvarez-Risco A, Solís-Tarazona Z, Matos-Valerio E, Zegarra-Arellano E, Del-Aguila-Arcentales S (2016) Adverse drug reactions reported as a result of the implementation of pharmaceutical care in the Institutional Pharmacy DIGEMID—ministry of health. Pharmaceutical Care Espana 18(2):67–74
161. Michelino F, Cammarano A, Celone A, Caputo M (2019) The linkage between sustainability and innovation performance in IT hardware sector. Sustainability 11(16):4275. https://doi.org/10.3390/su11164275
162. Millman C, Matlay H, Liu F (2008) Entrepreneurship education in China: a case study approach. J Small Bus Enterp Dev 15(4):802–815. https://doi.org/10.1108/14626000810917870
163. Mohamued EA, Ahmed M, Pypłacz P, Liczmańska-Kopcewicz K, Khan MA (2021) Global oil price and innovation for sustainability: the impact of R&D spending, oil price and oil price volatility on GHG emissions. Energies, 14(6). https://doi.org/10.3390/en14061757
164. Morales-Ríos F, Alvarez-Risco A, Castillo-Benancio S, de las Mercedes Anderson-Seminario M, Del-Aguila-Arcentales S (2022) Material selection for circularity and footprints. In: Alvarez-Risco A, Muthu SS, Del-Aguila-Arcentales S (eds) Circular economy. Environmental footprints and eco-design of products and processes. Springer, Singapore. https://doi.org/10.1007/978-981-19-0549-0_10
165. Morris JE, Lummis GW (2014) Investigating the personal experiences and self-efficacy of Western Australian primary pre-service teachers in the visual arts. Australian Art Educ 36(1):26–47
166. Mousa SK, Othman M (2020) The impact of green human resource management practices on sustainable performance in healthcare organisations: a conceptual framework. J Clean Prod 243:118595. https://doi.org/10.1016/j.jclepro.2019.118595
167. Muñoz P, Cohen B (2018) Sustainable entrepreneurship research: taking stock and looking ahead [https://doi.org/10.1002/bse.2000]. Bus Strat Environ 27(3):300–322. https://doi.org/10.1002/bse.2000
168. Mzembe AN, Lindgreen A, Idemudia U, Melissen F (2020) A club perspective of sustainability certification schemes in the tourism and hospitality industry. J Sustain Tour 28(9):1332–1350. https://doi.org/10.1080/09669582.2020.1737092
169. Nasr KB, Boujelbene Y (2014) Assessing the impact of entrepreneurship education. Proc—Soc Behav Sci 109:712–715. https://doi.org/10.1016/j.sbspro.2013.12.534
170. Nguyen PM, Dinh VT, Luu T-M-N, Choo Y (2020) Sociological and theory of planned behaviour approach to understanding entrepreneurship: comparison of Vietnam and South Korea. Cogent Bus Manage 7(1):1815288. https://doi.org/10.1080/23311975.2020.1815288
171. Niemann CC, Dickel P, Eckardt G (2020) The interplay of corporate entrepreneurship, environmental orientation, and performance in clean-tech firms—a double-edged sword [https://doi.org/10.1002/bse.2357]. Bus Strat Environ 29(1):180–196. https://doi.org/10.1002/bse.2357
172. Norström AV, Cvitanovic C, Löf MF, West S, Wyborn C, Balvanera P, Österblom H et al (2020) Principles for knowledge co-production in sustainability research. Nat Sustain 3(3):182–190. https://doi.org/10.1038/s41893-019-0448-2

173. Nuringsih K, Puspitowati I (2017) Determinants of eco entrepreneurial intention among students: study in the entrepreneurial education practices. Adv Sci Lett 23(8):7281–7284
174. Nuringsih K, Nuryasman M (2021) The role of green entrepreneurship in understanding Indonesia economy development sustainability among young adults. Stud Appl Econ 39(12)
175. Odongo I, Kyei PP (2018) The role of government in promoting youth entrepreneurship: the case of South Africa. J Soc Dev Afr 33(2):11–36
176. Olawumi TO, Chan DWM (2018) A scientometric review of global research on sustainability and sustainable development. J Cleaner Prod 183:231–250. https://doi.org/10.1016/j.jclepro.2018.02.162
177. Omoloso O, Mortimer K, Wise WR, Jraisat L (2021) Sustainability research in the leather industry: a critical review of progress and opportunities for future research. J Cleaner Prod 285:125441. https://doi.org/10.1016/j.jclepro.2020.125441
178. Ozturkoglu Y, Sari FO, Saygili E (2021) A new holistic conceptual framework for sustainability oriented hospitality innovation with triple bottom line perspective. J Hosp Tour Technol 12(1):39–57. https://doi.org/10.1108/JHTT-02-2019-0022
179. Pauceanu AM, Alpenidze O, Edu T, Zaharia RM (2019) What determinants influence students to start their own business? empirical evidence from United Arab Emirates universities. Sustainability, 11(1). https://doi.org/10.3390/su11010092
180. Peeters P (2018) Why space tourism will not be part of sustainable tourism. Tour Recreat Res 43(4):540–543. https://doi.org/10.1080/02508281.2018.1511942
181. Peris-Ortiz M, Devece-Carañana CA, Navarro-Garcia A (2018) Organizational learning capability and open innovation. Manag Decis 56(6):1217–1231. https://doi.org/10.1108/MD-02-2017-0173
182. Peterk SO, Koprivnjak T, Mezulic P (2015) Challenges of evaluation of the influence of entrepreneurship education. Econ Rev: J Econ Bus 13(2):74–86
183. Peterman NE, Kennedy J (2003) Enterprise education: influencing students' perceptions of entrepreneurship. Entrep Theory Pract 28(2):129–144. https://doi.org/10.1046/j.1540-6520.2003.00035.x
184. Peçanha Enqvist J, West S, Masterson VA, Haider LJ, Svedin U, Tengö M (2018) Stewardship as a boundary object for sustainability research: linking care, knowledge and agency. Landsc Urban Plan 179:17–37. https://doi.org/10.1016/j.landurbplan.2018.07.005
185. Peña Miguel N, Corral Lage J, Mata Galindez A (2020) Assessment of the development of professional skills in university students: sustainability and serious games. Sustainability, 12(3). https://doi.org/10.3390/su12031014
186. Pham H, Kim S-Y (2019) The effects of sustainable practices and managers' leadership competences on sustainability performance of construction firms. Sustain Product Consump 20:1–14. https://doi.org/10.1016/j.spc.2019.05.003
187. Plewnia F, Guenther E (2018) Mapping the sharing economy for sustainability research. Manag Decis 56(3):570–583. https://doi.org/10.1108/MD-11-2016-0766
188. Qazi W, Qureshi JA, Raza SA, Khan KA, Qureshi MA (2021) Impact of personality traits and university green entrepreneurial support on students' green entrepreneurial intentions: the moderating role of environmental values. J Appl Res Higher Educ 13(4):1154–1180. https://doi.org/10.1108/JARHE-05-2020-0130
189. Quispe-Cañari JF, Fidel-Rosales E, Manrique D, Mascaró-Zan J, Huamán-Castillón KM, Chamorro-Espinoza SE, Mejia CR et al (2021) Self-medication practices during the COVID-19 pandemic among the adult population in Peru: a cross-sectional survey. Saudi Pharmaceutical J 29(1):1–11. https://doi.org/10.1016/j.jsps.2020.12.001
190. Qureshi S, Mian S (2021) Transfer of entrepreneurship education best practices from business schools to engineering and technology institutions: evidence from Pakistan. J Technol Transf 46(2):366–392
191. Rajesh R (2020) Exploring the sustainability performances of firms using environmental, social, and governance scores. J Clean Prod 247:119600. https://doi.org/10.1016/j.jclepro.2019.119600

192. Ramísio PJ, Pinto LMC, Gouveia N, Costa H, Arezes D (2019) Sustainability strategy in higher education institutions: lessons learned from a nine-year case study. J Clean Prod 222:300–309. https://doi.org/10.1016/j.jclepro.2019.02.257
193. Ransfield AK, Reichenberger I (2021) Māori Indigenous values and tourism business sustainability. AlterNative: Int J Indigenous Peoples 17(1):49–60. https://doi.org/10.1177/1177180121994680
194. Raposo M, Do Paço A (2011) Entrepreneurship education: Relationship between education and entrepreneurial activity. Psicothema 23(3):453–457
195. Ratten V, Jones P (2021) Entrepreneurship and management education: exploring trends and gaps. Int J Manage Educ 19(1):100431
196. Rau H, Goggins G, Fahy F (2018) From invisibility to impact: Recognising the scientific and societal relevance of interdisciplinary sustainability research. Res Policy 47(1):266–276. https://doi.org/10.1016/j.respol.2017.11.005
197. Rauf R, Wijaya H, Tari E (2021) Entrepreneurship education based on environmental insight: opportunities and challenges in the new normal era. Cogent Arts Human 8(1):1945756
198. Rojas-Osorio M, Alvarez-Risco A (2019) Intention to use smartphones among Peruvian university students. Int J Interactive Mobile Technol 13(3):40–52. https://doi.org/10.3991/ijim.v13i03.9356
199. Román BR, Moscoso S, Chung SA, Terceros BL, Álvarez-Risco A, Yáñez JA (2020) Treatment of COVID-19 in Peru and Bolivia, and self-medication risks. Revista Cubana de Farmacia, 53(2):1–20, Article e435.
200. Rusu VD, Roman A (2017) Entrepreneurial activity in the EU: an empirical evaluation of its determinants. Sustainability, 9(10). https://doi.org/10.3390/su9101679
201. Sakao T, Brambila-Macias SA (2018) Do we share an understanding of transdisciplinarity in environmental sustainability research? J Cleaner Prod 170:1399–1403. https://doi.org/10.1016/j.jclepro.2017.09.226
202. Salmerón-Manzano E, Manzano-Agugliaro F (2018) The higher education sustainability through virtual laboratories: the Spanish university as case of study. Sustainability, 10(11). https://doi.org/10.3390/su10114040
203. Samo AH, Channa NA, Qureshi NA (2021) The role of universities and entrepreneurial ecosystem in producing entrepreneurs for industry 4.0. J Higher Educ Theory Pract 21(11):28–40
204. Serafeim G (2020) Public sentiment and the price of corporate sustainability. Financ Anal J 76(2):26–46. https://doi.org/10.1080/0015198X.2020.1723390
205. Shepherd DA, Patzelt H (2011) The new field of sustainable entrepreneurship: studying entrepreneurial action linking "what is to be sustained" with "what is to be developed." Entrep Theory Pract 35(1):137–163. https://doi.org/10.1111/j.1540-6520.2010.00426.x
206. Shepherd DA, Wennberg K, Suddaby R, Wiklund J (2018) What Are we explaining? a review and agenda on initiating, engaging, performing, and contextualizing entrepreneurship. J Manag 45(1):159–196. https://doi.org/10.1177/0149206318799443
207. Shuqin C, Minyan L, Hongwei T, Xiaoyu L, Jian G (2019) Assessing sustainability on Chinese university campuses: development of a campus sustainability evaluation system and its application with a case study. J Build Eng 24:100747. https://doi.org/10.1016/j.jobe.2019.100747
208. Sroufe R, Gopalakrishna-Remani V (2018) Management, social sustainability, reputation, and financial performance relationships: an empirical examination of U.S. firms. Organ Environ 32(3):331–362. https://doi.org/10.1177/1086026618756611
209. Sung E, Kim H, Lee D (2018) Why do people consume and provide sharing economy accommodation?—a sustainability perspective. Sustainability 10(6):2072. https://doi.org/10.3390/su10062072
210. Tang M, Chen X, Li Q, Lu Y (2014) Does Chinese university entrepreneurship education fit students' needs? J Entrepreneurship Emerg Econ 6(2):163–178. https://doi.org/10.1108/JEEE-02-2014-0002

211. Tehseen S, Haider SA (2021) Impact of universities' partnerships on students' sustainable entrepreneurship intentions: a comparative study. Sustainability, 13(9). https://doi.org/10.3390/su13095025
212. Teo T, Zhou M, Fan ACW, Huang F (2019) Factors that influence university students' intention to use Moodle: a study in Macau. Educ Tech Research Dev 67(3):749–766. https://doi.org/10.1007/s11423-019-09650-x
213. Teran-Yepez E, Guerrero-Mora AM (2020) Entrepreneurship theories: critical review of the literature and suggestions for future research [Teorías de emprendimiento: revisión crítica de la literatura y sugerencias para futuras investigaciones]. Revista Espacios, 41(07)
214. Trivedi R (2016) Does university play significant role in shaping entrepreneurial intention? a cross-country comparative analysis. J Small Bus Enterp Dev 23(3):790–811. https://doi.org/10.1108/JSBED-10-2015-0149
215. Trosper RL (2002) Northwest coast indigenous institutions that supported resilience and sustainability. Ecol Econ 41(2):329–344. https://doi.org/10.1016/S0921-8009(02)00041-1
216. Uvarova I, Mavlutova I, Atstaja D (2021) Development of the green entrepreneurial mindset through modern entrepreneurship education. IOP Conf Series: Earth Environ Sci 628:012034. https://doi.org/10.1088/1755-1315/628/1/012034
217. Veleva V, Bodkin G (2018) Corporate-entrepreneur collaborations to advance a circular economy. J Cleaner Prod 188:20–37. https://doi.org/10.1016/j.jclepro.2018.03.196
218. Vismara S (2019) Sustainability in equity crowdfunding. Technol Forecast Soc Change 141:98–106. https://doi.org/10.1016/j.techfore.2018.07.014
219. Vizcardo D, Salvador LF, Nole-Vara A, Dávila KP, Alvarez-Risco A, Yáñez JA, Mejia CR (2022) Sociodemographic predictors associated with the willingness to get vaccinated against COVID-19 in Peru: a cross-sectional survey. Vaccines, 10(1), Article 48. https://doi.org/10.3390/vaccines10010048
220. Walter SG, Block JH (2016) Outcomes of entrepreneurship education: an institutional perspective. J Bus Venturing 31(2):216–233. https://doi.org/10.1016/j.jbusvent.2015.10.003
221. Wijnker M, Van Kasteren H, Romijn H (2015) Fostering sustainable energy entrepreneurship among students: the business oriented technological system analysis (BOTSA) program at eindhoven university of technology. Sustainability, 7(7). https://doi.org/10.3390/su7078205
222. Wong PK, Ho YP, Autio E (2005) Entrepreneurship, innovation and economic growth: evidence from GEM data. Small Bus Econ 24(3):335–350. https://doi.org/10.1007/s11187-005-2000-1
223. Wu S, Wu L (2008) The impact of higher education on entrepreneurial intentions of university students in China. J Small Bus Enterp Dev 15(4):752–774. https://doi.org/10.1108/14626000810917843
224. Xu J, Liu F (2020) Research on entrepreneurship policy of college students based on quantitative analysis of text data mining. J Phys: Conf Ser 1570(1):012071. https://doi.org/10.1088/1742-6596/1570/1/012071
225. Yan J, Kim S, Zhang SX, Foo MD, Alvarez-Risco A, Del-Aguila-Arcentales S, Yáñez JA (2021) Hospitality workers' COVID-19 risk perception and depression: a contingent model based on transactional theory of stress model. Int J Hospital Manage 95, Article 102935. https://doi.org/10.1016/j.ijhm.2021.102935
226. Yao W, Weng M, Ye T (2019) Towards good governance of an entrepreneurial university: the case of Zhejiang university. In: University governance and academic leadership in the EU and China. IGI Global, pp 82–97. https://doi.org/10.4018/978-1-5225-7441-5.ch006
227. Yarime M, Tanaka Y (2012) The issues and methodologies in sustainability assessment tools for higher education institutions: a review of recent trends and future challenges. J Educ Sustain Dev 6(1):63–77. https://doi.org/10.1177/097340821100600113
228. Yi G (2021) From green entrepreneurial intentions to green entrepreneurial behaviors: the role of university entrepreneurial support and external institutional support. Int Entrepreneurship Manage J 17(2):963–979. https://doi.org/10.1007/s11365-020-00649-y
229. Ying-Qiu L (2016) Mass entrepreneurship and innovation: new impetus to development. J Chin Acad Gov. https://en.cnki.com.cn/Article_en/CJFDTotal-LJXZ201606006.htm

230. Yáñez JA, Jahanshahi AA, Alvarez-Risco A, Li J, Zhang SX (2020) Anxiety, distress, and turnover intention of healthcare workers in Peru by their distance to the epicenter during the COVID-19 crisis. Am J Trop Med Hyg 103(4):1614–1620. https://doi.org/10.4269/ajtmh.20-0800
231. Yáñez S, Uruburu Á, Moreno A, Lumbreras J (2019) The sustainability report as an essential tool for the holistic and strategic vision of higher education institutions. J Clean Prod 207:57–66. https://doi.org/10.1016/j.jclepro.2018.09.171
232. Yáñez JA, Alvarez-Risco A, Delgado-Zegarra J (2020) Covid-19 in Peru: from supervised walks for children to the first case of Kawasaki-like syndrome. BMJ 369, Article m2418. https://doi.org/10.1136/bmj.m2418
233. Zahraie B, Everett AM, Walton S, Kirkwood J (2016) Environmental entrepreneurs facilitating change toward sustainability: a case study of the wine industry in New Zealand. Small Enterp Res 23(1):39–57. https://doi.org/10.1080/13215906.2016.1188717
234. Zamora-Polo F, Sánchez-Martín J (2019) Teaching for a better world. Sustainability and sustainable development goals in the construction of a change-maker university. Sustainability 11(15). https://doi.org/10.3390/su11154224
235. Zhang SX, Chen J, Afshar Jahanshahi A, Alvarez-Risco A, Dai H, Li J, Patty-Tito RM (2021) Succumbing to the COVID-19 pandemic—healthcare workers not satisfied and intend to leave their jobs. Int J Mental Health Addict. https://doi.org/10.1007/s11469-020-00418-6
236. Zhang SX, Chen J, Jahanshahi AA, Alvarez-Risco A, Dai H, Li J, Patty-Tito RM (2021) Correction to: succumbing to the COVID-19 pandemic—healthcare workers not satisfied and intend to leave their jobs (International Journal of Mental Health and Addiction, (2021). Int J Mental Health Addict. https://doi.org/10.1007/s11469-021-00502-5
237. Zhang SX, Sun S, Afshar Jahanshahi A, Alvarez-Risco A, Ibarra VG, Li J, Patty-Tito RM (2020) Developing and testing a measure of COVID-19 organizational support of healthcare workers—results from Peru, Ecuador, and Bolivia. Psychiatry Res 291. https://doi.org/10.1016/j.psychres.2020.113174
238. Zhao G (2019) Analysis on the Transitional development of entrepreneurship education in colleges and universities in the age of mass creation In: 5th International workshop on education, development and social sciences (IWEDSS 2019). https://doi.org/10.25236/iwedss.2019.220

Green Marketing and Entrepreneurship

Myreya De-La-Cruz-Diaz, Aldo Alvarez-Risco, Micaela Jaramillo-Arévalo, María de las Mercedes Anderson-Seminario, and Shyla Del-Aguila-Arcentales

Abstract The world's environmental situation is a great source of concern for most people and industries that do nothing but grow over time. People have begun to develop a greater awareness of what they consume and these products' impact, processes, and transport. The concept of green marketing has been generated. Green marketing is a tool used to show the sustainable objectives of the company while meeting the needs and expectations of customers. This concept is being applied more and more among companies, thus causing a particular influence on the behavior of consumers when making purchases. In addition, the decision of its application gives rise to both advantages and challenges, which are related and can be closely observed in entrepreneurship and small businesses. This chapter presents the concept of green marketing, its importance, presence in current markets, the advantages and challenges of its application. In the same way, it is necessary to highlight the approach taken by this work when relating and highlighting the importance and influence of green marketing in entrepreneurship and small companies over large companies.

Keywords Green marketing · Entrepreneurship · Sustainability · Greenwashing · Green marketing mix · Behavior · Green · Brand

1 Introduction

Big companies and entrepreneurship produce goods and services in an environment where they are forced to interact [37]. In recent years, it has been proposed that this interaction needs to be sustainable in the future, and it has become one of the most critical problems on the agendas of both countries and businesses. It has been considered to rethink the traditional forms of marketing to be adapted to their environment

M. De-La-Cruz-Diaz · A. Alvarez-Risco · M. Jaramillo-Arévalo ·
M. de las Mercedes Anderson-Seminario
Universidad de Lima, Lima, Peru

S. Del-Aguila-Arcentales (✉)
Escuela de Posgrado, Universidad San Ignacio de Loyola, Lima, Peru
e-mail: sdelaguila@usil.edu.pe

A. Alvarez-Risco et al. (eds.), *Footprint and Entrepreneurship*, Environmental Footprints and Eco-design of Products and Processes,
https://doi.org/10.1007/978-981-19-8895-0_9

and the new needs of companies because macromarketing, a field of marketing that is responsible for studying the effects it can have on the economy and society in general. At the same time, it focuses on the environment and its problems. It also carries out the vital work of proposing activities that allow businesses to harmonize with the environment [59].

Since 1960, people specializing in marketing have begun to include the term "green" when talking about consumption [52]. However, in recent years, the issue has taken on greater importance as sustainability is one of the most crucial issues in business. In addition to applying what "green" consumption is, other concepts such as environmental responsibility have been introduced [30, 39, 51].

On the other hand, the relationship between the term "green," the environment, and marketing activities was evaluated a year later at the beginning of the 1970s; its objective was to include these new environmental terms developed previously to include them in your marketing processes. At the same time, the aim was to identify these responsible consumers who desire to buy ecological and environmentally friendly products. In 1980, with responsible consumption being a necessary behavior in business, changes in marketing trends began; it was identified that the problem with green products would be the volume of consumption rather than the quality of the products; this is how they began to develop strategies regarding their development, use, transport, and the valuable life cycle of these products [35].

However, there is a contradiction between consumer behavior and their attitude toward the environment. According to past research, consumers have a positive trend toward green consumption; however, this attitude does not necessarily translate into actual buying behavior, which means that although some consumers have attitudes that are favorable to more responsible and green consumption, this it rarely ends in an actual purchasing practice [6, 24, 46].

The two theories mentioned first are the most used [7, 29]. This incongruity has been registered in different countries [45], which makes the reasons for these incongruities highly requested to understand this behavior, and from these developed strategies, some of the conceptual theories to explain the green behavior are as follows: the theory of reasoned action of [2], according to this theory, the behavior of people is conditioned by their intention if they want to do something or not, but the intention is a balance of two things, our attitude and what we think we should do is, this theory takes into account social pressure. There is also the extension of this theory, which is planned behavior [1], among others. Despite this, the theory of planned behavior has limiters that make people who investigate them propose many variations. An example of this is the study by [10], where they expanded the theory of planned behavior to include environmental awareness, social impression, among others. Other investigations, such as [58], did the same, including long-term orientation factors and man-nature orientations in their framework.

This chapter talks about the importance of green marketing in the current market, its relationship with startups, and its success stories. In addition, the significant benefits and obstacles that arise in the application of green marketing are mentioned in the text.

2 Green Marketing in Today's Market

Green marketing results from the companies' efforts to follow and fulfill the consumer's needs and expectations regarding environmental awareness [11, 62]. Without a doubt, "green," "environmental," and "sustainable" marketing has grown significantly, with most of it centered on promoting "green" products, analyzing market niches and consumption patterns for eco-friendly items, and the importance of the environment in branding [15, 33, 38]. Green marketing encompasses various actions, including product modification, manufacturing process improvements, packaging adjustments, and advertising changes [11]. Some businesses are investing in green marketing strategies to be seen as environmentally and socially conscious [17] because of the idea that being environmentally friendly in the eyes of the consumer would increase their revenues [44, 47].

While environmental awareness and customer attitudes toward the environment are on the rise, empirical research suggests that attitudes toward green products rarely transfer into actual buying behavior as it has been mentioned before [27, 45, 61], but a favorable influence of buying intention on purchasing behavior for a greener product is amplified thanks to green marketing [61]. Environmental commitment presented in green marketing should be emphasized in advertising and promotional initiatives, but consumers do not buy simply because a company is environmentally friendly, products and services must fit the needs of customers in terms of quality, pricing, and usefulness [49, 61, 64].

It is necessary to consider the context in which we live where increasingly restrictive policies are developed to achieve a correct application of green marketing. In addition, it is necessary to consider the term green marketing mix and determine the 4P's before launching the green product on the market. For cases of this type, the first P for product must consider the impacts of its manufacture and use on the environment. Regarding the price, the development can be more expensive than a conventional product, but the consumer may be willing to pay an additional cost if he understands the added value of the green product and why it is expensive. On the other hand, for the promotion, this can be directed to the ecological product, promoting sustainable living, and to the company's environmental responsibility. Regarding the last P for point of sale, the channels must be by the proposed ecological strategy and enhance the business culture. The form of distribution must work as a team with the marketing area to optimize the chosen campaign and ensure a sustainable product distribution [12].

Parts of the population who care about the environment are more likely to use more concrete criteria to make purchasing decisions [60]. The presence of ecological seals and certificates, which can be found on product labels or via alternative distribution channels, is one manner that consumers may find appealing [19]. Based on "green certification" or labels, it can be used to promote environmental sustainability and the interest of consumers. Green certification is a licensed seal of approval that may be featured on a firm's Web site or in advertising to show customers that the company has been independently examined and certified as ecologically sustainable [11]. Environmental labels are given to customers to assist them in identifying between environmentally friendly and conventional products [9]. These assist in overcoming a lack of knowledge, environmental labeling would be superfluous if all important data about the product and the conditions of its manufacture were known [44]. Corporate clients usually demand that their suppliers show their adherence to environmental rules [22].

Businesses are expected to take the lead in terms of environmental sustainability because they are the main producers of unfriendly environmental activities and are projected to make a substantial impact [64]. Moreover, many companies have adopted green marketing in their operations processes [49, 53, 64, 66]. Green marketing provides financial incentives and opportunities for growth [44, 62]. It integrates the company's existing business, operations, and investment strategies, essentially assisting a company in making environmentally friendly decisions, considering that, while changes to corporate or manufacturing processes may incur initial expenses, they save money over time [36, 62]. These companies can promote financial progress, social prosperity, and environmental protection by following the green marketing philosophy. They help to resolve the tension between conflicting goals and the simultaneous focus on economic success, environmental quality, and social through green marketing, which significance in everyday life has made it unavoidable [11].

3 Opportunities and Challenges of Green Marketing for Entrepreneurship

In an era when consumers determine a company's fate, green marketing provides a strategic approach for these businesses to cater to the market by providing environmentally friendly products or services, or both, that eliminate or minimize any negative influence on the environment [11]. Nevertheless, as with any other strategy in the business world, green marketing poses some challenges or disadvantages in the present day that may make it challenging to apply for entrepreneurship in small companies [25].

3.1 Opportunities of Green Marketing

Marketers must recognize how green marketing is not solely altruistic; it can also be a successful venture for long-term development. Nowadays, green marketing has become a competitive advantage among companies in most industries due to the current environmental issues such as the inefficient use of natural resources and climate change [44]. As stated before, a green image would increase the chances of a product or company being preferred by consumers since they appreciate social and environmental responsibility. Thanks to customers' preference for these products, it would increase their value on the market and, therefore, their profits [44, 64]. However, it is not the only advantage produced by adopting a green marketing plan. Green marketing can save the company some money in the long term (e.g., investing in a renewable source of energy for the company's production line eventually would lead to savings on that matter) [62, 64], which could also provide long-term profitability through ensuring long-term growth with more disposition of investing [11, 64]. Another advantage of green marketing is that the relationship with the company's stakeholders improves due to the decreasing pressure put on the company's social responsibility and good reputation [44, 49, 64].

The practical implementation of green marketing can be noticed in some cases, such as The Body Shop and Levi´s; these are examples of big companies that have successfully applied green marketing in their business model [12].

A. The Body Shop

It has been one of the pioneer brands applying the green marketing mix, a brand present in several countries that sells food and beauty products inspired by nature and respect for the environment. They are also a brand that does not test its products on animals and actively promotes organic and vegan products. On the other hand, it accompanies its green products with a sustainable store design, making all its retail points of sale ecological by incorporating wood with the SSC certification of local origin in the construction, LED lighting, and non-toxic paints. An artificial grass space was also added inside the store precisely where payments are made to make consumers aware of their payment, helping to build in the long term to demonstrate these changes to their customers. This proposal to visually show users what is changing also translates into showing them the crops of the company's ingredients through videos. All the initiatives promoted by the company, such as Bio-Bridge, are transmitted to consumers through their marketing strategies. They are mainly exhibited in video format by social networks, and due to the age of regular consumers, the brand has dared to try channels such as Pinterest, Snapchat, and Instagram. The objective of the green marketing that this company uses is that consumers keep in mind that they are acquiring an eco-friendly product from a brand that is striving to be the most sustainable in the long term [26]. It also has several social initiatives in the process [12]. This brand has declared its desire to become the most truly sustainable ethic in the industry [26].

According to the study carried out by [8], ecological marketing positively influenced the loyalty of the customers of this brand, giving primary importance to the ecological product itself. In turn, green marketing proved to influence customer satisfaction positively. However, the combination of green marketing and customer loyalty through customer satisfaction proved the most influential (Fig. 1).

The success of The Body Shop brand focuses on the use of green marketing that focuses on keeping its customers satisfying their needs for quality green organic products that are constantly becoming more sustainable. In addition to giving importance to social issues that the brand strives to address through campaigns present on social networks.

Fig. 1 Relationship between green marketing, loyalty, and satisfaction of the body shop's customers. *Source* [8]

B. **Levi´s**

This clothing brand is recognized worldwide, and that has become very popular for using only denim or mezclilla for its garments, encourages ecological and sustainable practices such as the use of its primary material (denim) recycled or cotton that is of an ethical origin as well to apply innovative solutions to reduce the use of water during its manufacture. In addition, this brand pays equally to both its workers and community organizations [12].

In the beginning, the production of jeans was to make them resistant so that workers could use them in the mines; for this reason, materials such as anilines were used for the highly polluting dyeing process. On the other hand, natural resources such as water for the manufacturing process were abundant and could be wasted a lot. However, Levi's has been able to adapt over time both to the new needs of its customers and their change in mentality; this is how this brand plans to be more sustainable. In the first instance, the use of water used for washing the jeans has been replaced using stones that have been shown to have an effect like that of water. At the same time, the materials that are used to manufacture have changed, and, like many others, it has used recycling as part of its production process [40]

The brand's efforts to be greener have also been transferred to the company's marketing using campaigns such as "'Buy Better, Wear Longer', to promote the consumption of sustainable and responsible clothing. To this end,celebrities were used in its campaign, which is present on different platforms worldwide. The objective of this was to promote the use of clothing that lasts longer and raise awareness about the importance of the brand investing in being sustainable in the future and by using technology that improves the quality of its products to extend its cycle of life. In addition, the need for more environmentally friendly alternative materials such as cottonized hemp and organic cotton stands out a lot, and alternative plans to reduce the waste of natural resources in the production processes [31]. This brand has collaborated with others to make collections that use fewer polluting products and use fewer resources, creating the most sustainable jeans where its entire production process uses strategies that make it sustainable and environmentally friendly [48].

These are two examples of large companies with a long time in the market that have managed to change many aspects of their businesses to be more sustainable and transmit this message to their consumers through green marketing; however, these companies are not the only ones that have made use of this tool. There is also entrepreneurship that has managed to have a successful implementation of green marketing.

C. **"Too Good to Go"**

It connects consumers with stores or restaurants that report having surplus food that they have not sold; these are in good condition; however, over time, they would probably be thrown away for not having been sold; this is how this app connects you with users willing to buy this food that mostly has a much lower price. This venture mainly covers European cities and operates in North America [12]. In this business model, everyone benefits since one of the biggest problems in the food industry is partially solved, which is overproduction. The main objective of its creation was to reduce the exorbitant amount of waste that is generated, part of its marketing strategy is to make consumers and restaurants that are part of the app feel that both are benefiting from the transaction and that also are helping to reduce waste and prevent healthy food from going to waste [57]. On the other hand, this business culture is complemented by green marketing that presents a series of initiatives that help grow the company's image as a green one. He has developed companies that extend the idea of reducing waste to homes through tips and recipes to educate their customers to continue with the company's objective even in their homes. In addition, they also offer users and establishments projects that help them sell their products more safely within the app [41].

D. **WISErg**

It is the case of a sustainable enterprise that uses technology to convert food waste into fertilizers; this helps so that waste of this type does not end up in landfills or composters. The most popular product of this brand is called "Harvester," a machine that can be installed in any store or supermarket that has groceries. In this machine, people can deposit food, and it crushes them and capture their nutrients, and they are later converted into fertilizers that can be used for organic soils without any limitation. This machine also offers statistics of its use and a report of how much waste each establishment produces so that the owners can use this information to make better decisions [55].

For this undertaking, the brand focuses on offering clients the opportunity to turn their own businesses greener safely; this is reflected in its Web site that uses green colors to represent its image as a sustainable and environmentally friendly brand environment. It uses green marketing when offering its organic product, mainly highlighting the use of waste for other purposes such as fertilizer [65]. In addition, the brand has chosen to partner with Vivid Life Sciences to create 100% sustainable agricultural products.

E. **Resto-Zero**

It is the example of another venture that seeks to be sustainable by reducing waste and emissions generated by the production of Peruvian coffee. This brand seeks to propose a circular model for the coffee industry where coffee and its parts are used to the maximum [54]. It uses the coffee husks that were usually discarded and treated as waste and uses them to develop new products through innovation and technology. The green marketing of this brand is based on offering its consumers ecological products derived from what was usually considered waste and which are also nutritious products whose added value is that they contribute to the well-being of our planet by reducing waste.

3.2 *Challenges of Green Marketing*

As previously mentioned, the implementation of green marketing can sometimes be quite challenging, especially when compared to small companies or entrepreneurship [25]. Due to their severe financial constraints against them, small and medium-sized firms (SMEs) face a significant barrier in undertaking operations corresponding to the green marketing strategy compared with larger companies [25]. Green products necessitate the use of expensive renewable and recyclable materials. It also needs current technology, which necessitates a significant investment in research and development [64]. Moreover, since it is a more sustainable production strategy and a relatively new concept, it could take more time to obtain the expected revenues, making it imperative for marketers and stakeholders to be patient to appreciate their results in the long term [62]. SMEs and even entrepreneurship are severely behind in adopting innovative knowledge even though the application of creative knowledge has steadily transferred from academics to industry [34].

Green marketing myopia is defined as underestimating either or overemphasizing the former at the latter's expense [11, 62, 64]. Green marketing must concentrate on customer advantages or the major reason why people buy certain products, as previously mentioned. When done well, it encourages customers to switch products or even spend more for the sustainable option [44, 64]. According to [64], several green products have crashed due to marketers' myopic concentration on the "greenness" of their products rather than the broader expectations of others. It won't benefit if an eco-friendly product is developed that is entirely green in every way but fails to meet customer satisfaction criteria. Furthermore, if green products are too costly, they may lose market acceptance.

Finally, one of the most well-known problems for implementing green marketing in entrepreneurship and companies is "greenwashing." This term refers to the overlap of two company behavior patterns: harmful environmental performance and favorable advertising about said environmental performance [21]. Companies may be tempted to engage in greenwashing to gain legitimacy and improve their reputation for environmental stewardship [63]. Purchasers must frequently rely on their subjective assessments of suppliers' actual sustainability [56]. Due to greenwashing, those companies who apply the sustainable guidelines correctly are affected by a growing skepticism presented by customers [10, 17]. By wrongfully recognizing green characteristics in products for their benefit, companies set a precedent into customers' minds and affect the image of the unethical company [11] and of the companies that apply or are beginning to apply green marketing correctly. From this, those that could be most affected are those small companies and ventures with fewer monetary resources since a lousy reputation could cause their investments in sustainable practices not to be rewarded.

It is convenient to observe the cases of McDonald's and Coca-Cola and how they affected them to illustrate these challenges better.

A. **McDonald's**

In the case of this famous fast-food chain, in 2018, to be more sustainable, they decided to change the traditional straws that they delivered with their drinks to ones made of paper. At first, this change was very well received by its audience, especially among young people; however, shortly after, it was found that these new straws were even less recyclable than the previous ones. The brand's spokesperson admitted that although they could be recycled, the thickness of these straws prevents them from being processed and recycled. This act evidenced a greenwashing strategy where the company did not think about recycling these straws, but only pretended that they would do it so that consumers feel happy [42]. This event demonstrates the importance of investigating the components of the current materials of our business and the new materials that we want to use to ensure that there is a fundamental change and not fall into greenwashing.

B. **Coca-Cola**

The famous carbonated drink is a huge company that fell into greenwashing when, to attract more customers with eco-friendly beliefs, it launched a product that, instead of using traditional sugar, used stevia. This product was sold in green Coca-Cola bottles and was accompanied by green marketing with propaganda showing natural elements to show an eco-friendlier image to its consumers. However, the bottles in which they have only 30% vegetable fiber, which does not show enough commitment on the part of the brand for many [50]. This example demonstrates the importance of examining each part of our process and considering all the materials used in production. In the case of Coca-Cola, despite having changed one of the ingredients in its formula for a more natural one, the bottle material that was still contaminating took away much of the credibility of the brand's green image.

Recent Suggested Lectures About Green Approach

- Waste reduction and carbon footprint [28]
- Material selection for circularity and footprints (Morales-[43]
- 3D print, circularity, and footprints [16]

- Virtual tourism [16]
- Leadership for sustainability in crisis time [3]
- Virtual education and circularity [13]
- Circular economy for packaging [5]
- Students oriented to circular learning [18]
- Food and circular economy [14]
- Water footprint and food supply chain management [32]
- Waste footprint [4]
- Measuring circular economy [20]
- Carbon footprint [23].

4 Closing Remarks

It is undeniable that by this time, most companies are clear that they must include green marketing and have been doing it for all these years, some better than others. However, it is essential to understand the efforts of startups to apply green marketing and its effect on their consumers. As it has been mentioned previously, the use of this marketing has multiple benefits in all kinds of business; however, the brands that use it must not fall into using green marketing only for fashion or their benefit, since the use of it for the sole economic purpose will most likely end in negative consequences for the company. Entrepreneurs must be cautious not to fall into greenwashing, as this can seriously affect their reputation and affect how consumers perceive the added value of their products. Moreover, these must be aware of implementing it rightly and presenting it clearly since the reputation of other companies accused of greenwashing can affect them and, therefore, keep them away from success and the advantages of using green marketing.

In recent years, thanks to technology and innovation, many enterprises offer 100% ecological products; thus, their processes and marketing are 100% green. In this context, it is essential for consumers to feel encouraged to buy these products and understand the reasons behind and the fundamental changes your purchase makes. In addition, it is essential to consider purchase intentions and behaviors since, as explained above, these do not always go hand in hand and can be contradictory for entrepreneurs. In this way, strategies can be created, and products adapted so that the intention into a total purchase of the organic product.

References

1. Ajzen I (1991) The theory of planned behavior. Organ Behav Hum Decis Process 50(2):179–211
2. Ajzen I, Fishbein M (2005) The influence of attitudes on behavior
3. Alvarez-Risco A, Del-Aguila-Arcentales S, Villalobos-Alvarez D, Diaz-Risco S (2022) Leadership for sustainability in crisis time. In: Alvarez-Risco A, Muthu SS, Del-Aguila-Arcentales S

(eds) Circular economy: impact on carbon and water footprint. Springer, Singapore, pp 41–64. https://doi.org/10.1007/978-981-19-0549-0_3
4. Arias-Meza M, Alvarez-Risco A, Cuya-Velásquez BB, de las Mercedes Anderson-Seminario M, Del-Aguila-Arcentales S (2022) Fashion and textile circularity and waste footprint. In: Alvarez-Risco A, Muthu SS, Del-Aguila-Arcentales S (eds) Circular economy: impact on carbon and water footprint. Springer, Singapore, pp 181–204. https://doi.org/10.1007/978-981-19-0549-0_9
5. Castillo-Benancio S, Alvarez-Risco A, Esquerre-Botton S, Leclercq-Machado L, Calle-Nole M, Morales-Ríos F, Del-Aguila-Arcentales S et al (2022) Circular economy for packaging and carbon footprint. In: Alvarez-Risco A, Muthu SS, Del-Aguila-Arcentales S (eds) Circular economy: impact on carbon and water footprint. Springer, Singapore, pp 115–138. https://doi.org/10.1007/978-981-19-0549-0_6
6. CEAP (2007) China general public environmental awareness survey. Retrieved 01/01/2022 from http://www.chinaceap.org
7. Ceglia D, de Oliveira Lima SH, Leocádio ÁL (2015) An alternative theoretical discussion on cross-cultural sustainable consumption. Sustain Dev 23(6):414–424. https://doi.org/10.1002/sd.1600
8. Chairunnisa SS, Fahmi I, Jahroh S (2019) How important is green marketing mix for consumer? Lesson from the body shop. Jurnal Manajemen 23(2):321–337
9. Chemat F, Rombaut N, Sicaire A-G, Meullemiestre A, Fabiano-Tixier A-S, Abert-Vian M (2017) Ultrasound assisted extraction of food and natural products. mechanisms, techniques, combinations, protocols and applications a review. Ultrason Sonochem 34:540–560
10. Chen S-C, Hung C-W (2016) Elucidating the factors influencing the acceptance of green products: an extension of theory of planned behavior. Technol Forecast Soc Chang 112:155–163. https://doi.org/10.1016/j.techfore.2016.08.022
11. Choudhary A, Gokarn S (2013) Green marketing: a means for sustainable development. J Arts Sci Commerce 4(3):3
12. Colomo, A. (2021). Green marketing: buenas prácticas y ejemplos—Inlea. Inlea. Retrieved 01 Jan 2022 from https://inlea.com/es/geen-marketing-buenas-practicas-ejemplos
13. Contreras-Taica A, Alvarez-Risco A, Arias-Meza M, Campos-Dávalos N, Calle-Nole M, Almanza-Cruz C, Del-Aguila-Arcentales S et al (2022) Virtual education: carbon footprint and circularity. In: Alvarez-Risco A, Muthu SS, Del-Aguila-Arcentales S (eds) Circular economy: impact on carbon and water footprint. Springer, Singapore, pp 265–285. https://doi.org/10.1007/978-981-19-0549-0_13
14. Cuya-Velásquez BB, Alvarez-Risco A, Gomez-Prado R, Juarez-Rojas L, Contreras-Taica A, Ortiz-Guerra A, Del-Aguila-Arcentales S et al (2022) Circular economy for food loss reduction and water footprint. In: Alvarez-Risco A, Muthu SS, Del-Aguila-Arcentales S (eds) Circular economy: impact on carbon and water footprint. Springer, Singapore, pp 65–91. https://doi.org/10.1007/978-981-19-0549-0_4
15. Dangelico RM, Vocalelli D (2017) "Green Marketing": an analysis of definitions, strategy steps, and tools through a systematic review of the literature. J Clean Prod 165:1263–1279. https://doi.org/10.1016/j.jclepro.2017.07.184
16. De-la-Cruz-Diaz M, Alvarez-Risco A, Jaramillo-Arévalo M, de las Mercedes Anderson-Seminario M, Del-Aguila-Arcentales S (2022) 3D print, circularity, and footprints. In: Alvarez-Risco A, Muthu SS, Del-Aguila-Arcentales S (eds) Circular economy: impact on carbon and water footprint. Springer, Singapore, pp 93–112. https://doi.org/10.1007/978-981-19-0549-0_5
17. de Freitas Netto SV, Sobral MFF, Ribeiro ARB, da Luz Soares GR (2020) Concepts and forms of greenwashing: a systematic review. Environ Sci Eur 32(1):1–12. https://doi.org/10.1186/s12302-020-0300-3
18. de las Mercedes Anderson-Seminario M, Alvarez-Risco A (2022) Better students, better companies, better life: circular learning. In: Alvarez-Risco A, Muthu SS, Del-Aguila-Arcentales S (eds) Circular economy: impact on carbon and water footprint. Springer, Singapore, pp 19–40. https://doi.org/10.1007/978-981-19-0549-0_2

19. de Oliveira Júnior JC, da Silva AWP, Neto ARV, de Castro ABC, Lima DSVR (2020) Determining factors of environmental concern in purchasing decisions. REMark 19(4):888
20. Del-Aguila-Arcentales S, Alvarez-Risco A, Muthu SS (2022) Measuring circular economy. In: Alvarez-Risco A, Muthu SS, Del-Aguila-Arcentales S (eds) Circular economy: impact on carbon and water footprint. Springer, Singapore, pp 3–17. https://doi.org/10.1007/978-981-19-0549-0_1
21. Delmas MA, Burbano VC (2011) The drivers of greenwashing. Calif Manage Rev 54(1):64–87
22. Delmas MA, Montiel I (2007) The adoption of ISO 14001 within the supply chain: when are customer pressures effective?
23. Esquerre-Botton S, Alvarez-Risco A, Leclercq-Machado L, de las Mercedes Anderson-Seminario M, Del-Aguila-Arcentales S (2022) Food loss reduction and carbon footprint practices worldwide: a benchmarking approach of circular economy. In: Alvarez-Risco A, Muthu SS, Del-Aguila-Arcentales S (eds) Circular economy: impact on carbon and water footprint. Springer, Singapore, pp 161–179. https://doi.org/10.1007/978-981-19-0549-0_8
24. Eurobarometer (2011) Attitudes of European citizens towards the environment. Brussels, European Commission. Retrieved 01 Jan 2022 from https://ec.europa.eu/environment/eurobarometers_en.htm
25. Fang W, Wu T-H, Chang T-W, Hung C-Z (2021) What could entrepreneurial vision do for sustainable development? explore the cross-level impact of organizational members' green shared vision on green creativity. Sustainability 13(10):5364. https://doi.org/10.3390/su13105364
26. Faul J (2016) How marketing will help the body shop become the world's most ethical and sustainable brand. The Drum. Retrieved 01 Jan 2022 from https://www.thedrum.com/news/2016/02/22/how-marketing-will-help-body-shop-become-worlds-most-ethical-and-sustainable-brand
27. Gleim MR, Smith JS, Andrews D, Cronin JJ (2013) Against the green: a multi-method examination of the barriers to green consumption. J Retail 89(1):44–61. https://doi.org/10.1016/j.jretai.2012.10.001
28. Gómez-Prado R, Alvarez-Risco A, Sánchez-Palomino J, de las Mercedes Anderson-Seminario M, Del-Aguila-Arcentales S (2022) Circular economy for waste reduction and carbon footprint. In: Alvarez-Risco A, Muthu SS, Del-Aguila-Arcentales S (eds) Circular economy: impact on carbon and water footprint. Springer, Singapore, pp 139–159. https://doi.org/10.1007/978-981-19-0549-0_7
29. Hanss D, Böhm G, Doran R, Homburg A (2016) Sustainable consumption of groceries: the importance of believing that one can contribute to sustainable development. Sustain Dev 24(6):357–370. https://doi.org/10.1002/sd.1615
30. Haws KL, Winterich KP, Naylor RW (2014) Seeing the world through GREEN-tinted glasses: green consumption values and responses to environmentally friendly products. J Consum Psychol 24(3):336–354. https://doi.org/10.1016/j.jcps.2013.11.002
31. Joe T (2021) Levi's Unveils latest campaign "Buy Better, Wear Longer" to encourage sustainable fashion production practices. Green Queen. Retrieved 01 Jan 2022 from https://www.greenqueen.com.hk/levis-unveils-latest-campaign-buy-better-wear-longer-encourage-sustainable-fashion-production-practices
32. Juarez-Rojas L, Alvarez-Risco A, Campos-Dávalos N, de las Mercedes Anderson-Seminario M, Del-Aguila-Arcentales S (2022) Water footprint in the textile and food supply chain management: trends to become circular and sustainable. In: Alvarez-Risco A, Muthu SS, Del-Aguila-Arcentales S (eds) Circular economy: impact on carbon and water footprint. Springer, Singapore, pp 225–243. https://doi.org/10.1007/978-981-19-0549-0_11
33. Kemper JA, Hall CM, Ballantine PW (2019) Marketing and sustainability: business as usual or changing worldviews? Sustainability 11(3). https://doi.org/10.3390/su11030780
34. Kesting P, Günzel-Jensen F (2015) SMEs and new ventures need business model sophistication. Bus Horiz 58(3):285–293. https://doi.org/10.1016/j.bushor.2015.01.002
35. Kilbourne WE (1998) Green marketing: a theoretical perspective. J Mark Manag 14(6):641–655. https://doi.org/10.1362/026725798784867743

36. Kotler P (2011) Reinventing marketing to manage the environmental imperative. J Mark 75(4):132–135. https://doi.org/10.1509/jmkg.75.4.132
37. Kraus S, Palmer C, Kailer N, Kallinger FL, Spitzer J (2018) Digital entrepreneurship: a research agenda on new business models for the twenty-first century. Int J Entrep Behav Res. https://doi.org/10.1108/IJEBR-06-2018-0425
38. Kumar P (2016) State of green marketing research over 25 years (1990–2014): literature survey and classification. Mark Intell Plan. https://doi.org/10.1108/MIP-03-2015-0061
39. Leonidou CN, Katsikeas CS, Morgan NA (2013) "Greening" the marketing mix: do firms do it and does it pay off? J Acad Mark Sci 41(2):151–170. https://doi.org/10.1007/s11747-012-0317-2
40. Marketing Verde (2021) Levi's se vuelve verde gracias a una renovada fabricación ecológica—Marketing Verde: Por un mundo mejor. Retrieved 01 Jan 2022 from http://marketingverde.xyz/levis-se-vuelve-verde-gracias-a-una-renovada-fabricacion-ecologica
41. MarketingNews (2021) Too good to go: "Hemos conseguido crear una marca reconocida, pero sobre todo querida. Retrieved 01 Jan 2022 from https://www.marketingnews.es/marcas/noticia/1164602054305/too-good-to-go-hemos-conseguido-crear-marca-reconocida-querida.1.html
42. Martin A (2021) "Greenwashing": los trucos de las compañías para sacar tajada de la concienciación ambiental. Retrieved 01 Jan 2022 from https://hipertextual.com/2021/11/greenwashing-empresas
43. Morales-Ríos F, Alvarez-Risco A, Castillo-Benancio S, de las Mercedes Anderson-Seminario M, Del-Aguila-Arcentales S (2022) Material selection for circularity and footprints. In: Alvarez-Risco A, Muthu SS, Del-Aguila-Arcentales S (eds) Circular economy: impact on carbon and water footprint. Springer, Singapore, pp 205–221. https://doi.org/10.1007/978-981-19-0549-0_10
44. Moravcikova D, Krizanova A, Kliestikova J, Rypakova M (2017) Green marketing as the source of the competitive advantage of the business. Sustainability 9(12):2218. https://doi.org/10.3390/su9122218
45. Nguyen HV, Nguyen CH, Hoang TTB (2019) Green consumption: closing the intention-behavior gap. Sustain Dev 27(1):118–129. https://doi.org/10.1002/sd.1875
46. Nielsen (2014) Global consumers are willing to put their money where their heart is when it comes to goods and services from companies committed to social responsibility. New York, NY. Retrieved 01 Jan 2022 from https://www.nielsen.com/eu/en/press-releases/2014/global-consumers-are-willing-to-put-their-money-where-their-heart-is1
47. Nyilasy G, Gangadharbatla H, Paladino A (2014) Perceived greenwashing: the interactive effects of green advertising and corporate environmental performance on consumer reactions. J Bus Ethics 125(4):693–707. https://doi.org/10.1007/s10551-013-1944-3
48. Osborne H (2006) Levi's launches green jeans. The guardian; The Guardian. Retrieved 01 Jan 2022 from https://www.theguardian.com/environment/2006/nov/24/ethicalliving
49. Papadas K-K, Avlonitis GJ, Carrigan M, Piha L (2019) The interplay of strategic and internal green marketing orientation on competitive advantage. J Bus Res 104:632–643. https://doi.org/10.1016/j.jbusres.2018.07.009
50. Peiró E (2019) Greenwashing ¿Qué es y cómo funciona? Retrieved 01 Jan 2022 from https://bloo.media/blog/greenwashing
51. Peloza J, White K, Shang J (2013) Good and guilt-free: the role of self-accountability in influencing preferences for products with ethical attributes. J Mark 77(1):104–119. https://doi.org/10.1509/jm.11.0454
52. Rahman AS, Barua A, Hoque R, Zahir MR (2017) Influence of green marketing on consumer behavior: a realistic study on Bangladesh. Glob J Manage Bus Res
53. Roulet TJ, Touboul S (2015) The intentions with which the road is paved: attitudes to liberalism as determinants of greenwashing. J Bus Ethics 128(2):305–320. https://doi.org/10.1007/s10551-014-2097-8
54. RPP (2020) "Resto-Zero": Conoce a la marca que da una segunda vida a los residuos generados en la producción del café peruano. Retrieved 01 Jan 2022 from https://rpp.pe/campanas/contenido-patrocinado/resto-zero-conoce-a-la-marca-que-da-una-segunda-vida-a-los-residuos-generados-en-la-produccion-del-cafe-peruano-noticia-1286235?ref=rpp

55. Russo F (2021) 6 ideas de emprendimiento sustentable (green business) + bonus. Retrieved 01 Jan 2022 from https://tentulogo.com/6-ideas-emprendimiento-sustentable-green-business
56. Simula H, Lehtimäki T, Salo J (2009) Managing greenness in technology marketing. J Syst Inf Technol 11(4):331–346. https://doi.org/10.1108/13287260911002486
57. Southey F (2019) Too good to go turns food waste into business: "it really is a win-win-win concept. Retrieved 01 Jan 2022 from https://www.foodnavigator.com/Article/2019/09/04/Too-Good-To-Go-turns-food-waste-into-business-It-really-is-a-win-win-win-concept
58. Sreen N, Purbey S, Sadarangani P (2018) Impact of culture, behavior and gender on green purchase intention. J Retail Consumer Serv 41:177–189. https://doi.org/10.1016/j.jretconser.2017.12.002
59. Stone M (2007) Fundamentals of marketing. Routledge
60. Sun Y, Liu N, Zhao M (2019) Factors and mechanisms affecting green consumption in China: A multilevel analysis. J Cleaner Prod 209:481–493. https://doi.org/10.1016/j.jclepro.2018.10.241
61. Susanty A, Puspitasari NB, Prastawa H, Listyawardhani P, Tjahjono B (2021) Antecedent factors of green purchasing behavior: learning experiences, social cognitive factors, and green marketing. Front Psychol 12:777531. https://doi.org/10.3389/fpsyg.2021.777531
62. Thakkar R (2021) Green marketing and sustainable development challenges and opportunities. Int J Manage Public Policy Res 1(1):15–23
63. Torelli R, Balluchi F, Lazzini A (2020) Greenwashing and environmental communication: Effects on stakeholders' perceptions. Bus Strateg Environ 29(2):407–421. https://doi.org/10.1002/bse.2373
64. Wahab S (2018) Sustaining the environment through green marketing. Rev Integrative Bus Econ Res 7:71–77
65. WISErg (2018) About us. Retrieved 01 Jan 2022 from https://wiserg.com/about
66. Zhang L, Li D, Cao C, Huang S (2018) The influence of greenwashing perception on green purchasing intentions: the mediating role of green word-of-mouth and moderating role of green concern. J Clean Prod 187:740–750. https://doi.org/10.1016/j.jclepro.2018.03.201

Sustainable Entrepreneurship: How Create Firms and Support SDG

Romina Gómez-Prado, Aldo Alvarez-Risco, Jorge Sánchez-Palomino, María de las Mercedes Anderson-Seminario, and Shyla Del-Aguila-Arcentales

Abstract Over the years, the need to achieve sustainable development throughout the world shows a greater interest in society to preserve the state of the planet, a fact that influences the actions and decisions that companies take every day in their different functions. This situation makes the 17 Sustainable Development Goals (SDG), presented by the United Nations (UN), gain a position of great importance at the business level. These objectives are considered tools used in ventures and make them sustainable from birth. According to their classification, the SDGs were distributed in this framework's environmental, social, and economic dimensions. In addition, some cases of implementing strategies, measures, or action plans of sustainable enterprises that are developed to support the fulfillment of these goals are exposed.

Keywords Entrepreneurship · Sustainability · SDG · Social · Economic · Environmental · Entrepreneurs · Circularity · Circular · Sustainable

1 Introduction

Environmental degradation in the world takes on greater importance as the unsustainable use of scarce natural resources puts the well-being of humanity and the next generations at risk (Merino, 2018). Response measures must ensure both the care of the environment and the adaptability of current practices for human subsistence. A harmony between both needs is vital for developing more and better alternative solutions. Based on this, sustainable entrepreneurship is born from taking advantage of both social and ecological problems to build a business idea, ensuring care for the environment [10]. Among these problems are hunger, poverty, species extinction,

---

R. Gómez-Prado · A. Alvarez-Risco · J. Sánchez-Palomino ·
M. de las Mercedes Anderson-Seminario
Universidad de Lima, Lima, Peru

S. Del-Aguila-Arcentales (✉)
Escuela de Posgrado, Universidad San Ignacio de Loyola, Lima, Peru
e-mail: sdelaguila@usil.edu.pe

A. Alvarez-Risco et al. (eds.), *Footprint and Entrepreneurship*, Environmental Footprints and Eco-design of Products and Processes,
https://doi.org/10.1007/978-981-19-8895-0_10

global warming, among others [29]. This type of undertaking integrates the same objectives of sustainable development, which, according to the World Commission on Environment and Development, seek to satisfy human needs without damaging the resources of the future generation [90]. Implementing environmental guidelines to companies about carrying out their operations is an alternative commonly used to ensure sustainability from the beginning [53]. These strategies should improve social benefits while mitigating environmental damage [63]. International legal frameworks promote this type of practice, but it does not oblige them to take immediate action, which may not be fully executed [79].

Probably the best-known global environmental initiative today is the Sustainable Development Goals (SDGs). This UN proposal comprises 17 objectives and 169 targets segmented according to its approach, seeking to formulate regulatory and policy frameworks based on a more sustainable future [79, 90]. The goals of the SDGs are segmented into the 3 dimensions of sustainable development: social development, economic growth, and environmental protection [107]. The importance of the work of companies to promote sustainable development is also highlighted [62]. They also provide models that help other companies comply with the requirements of the SDG [62]. This chapter analyzes the importance of the SDGs in business, considering the role of companies through their activities related to the development and promotion of these global goals. The focus is on emerging companies, known as "start-ups". The SDGs are distributed in their corresponding dimension, either environmental, social, or economic, to develop or adopt a more optimal action plan based on the goals of the SDG [62], which allow a sustainable life with a better natural and social environment [88].

2 Origin of Sustainable Entrepreneurship

Today, emerging companies have an essential role in developing the world economy because they promote increased jobs and create valuable businesses that present a new and innovative vision [19, 103]. For this reason, start-ups are defined as organizations that have a business model that can be repeated and at the same time is scalable, which focuses on generating an innovative product for the market [67]. These characteristics of start-ups can be understood as a different entrepreneurial spirit [8].

Following this framework, [22] explain sustainable entrepreneurship as creating profitable businesses that positively impact the economic, social, and environmental situation. Its sustainable orientation demonstrates an entrepreneurial tendency that considers environmental care and the protection of society when establishing the objectives that he wants to achieve with his business, as cited in [13]. In the same way, the sustainable entrepreneur becomes a key agent in the transition to a sustainable economy by proposing unique solutions that have not only an economic value but also a social and ecological value [32]. The entrepreneurial spirit of a sustainable entrepreneur is seen as a solution to social inequality and environmental degradation that manifests itself through a business activity to design and launch sustainable products or services [49, 74, 101]. Entrepreneurship academics consider this as a driver and central source, quality of great importance in a market that demands new and better solutions to the challenges of sustainability [29, 75, 81].

Likewise, [13] point out that entrepreneurs who start a business are waiting for the emergence of market opportunities that generate benefits in various ways, bringing good results in the three dimensions of sustainability: economic and social, and environmental. On the other hand, although start-ups can generate wealth and help economic well-being, there must be a sound ecosystem around them that supports their formation and sustainable growth, since the issue of sustainable development is winning every time greater importance in the global system [101, 103]. In this sense, it is essential to understand better the meaning of a business ecosystem or, in this case, start-ups. This term refers to the collaboration between start-ups and their supporting elements in the form of a link to boost their growth and market share [64, 103]. On the other hand, an element that determines the continuity of said startup is the knowledge that the entrepreneur possesses [1]. The degree of experience can also be detrimental to the environmental reach of the start-up, being distorted to a more commercial approach [1, 2]. These scenarios should be considered to ensure the sustainable orientation of the start-up.

Hall et al. [49] highlight the importance of understanding the factors that drive entrepreneurs to achieve sustainable development goals and use them when forming a sustainable business, one of these factors is referred to in other studies as "sustainability intention" [37, 75, 93, 94]. However, before investigating why these entrepreneurs follow this sustainable approach, it is essential to examine their cognition and not just their intention [73].

In this way, [75] mention that researchers tend to distinguish values, motivations, and intentions when analyzing sustainable business behavior. A sustainable venture is born from its founder's motivation when using ecological and social issues and invents a promising business idea that ensures a more sustainable future [10]. It is essential to recognize the difference between a traditional entrepreneur and a sustainable one because different values and beliefs govern them. While the traditional one focuses more on generating economic benefits and gaining a better position in the market, the sustainable includes in its objectives both economic, environmental, and social values that promote the development of more sustainable enterprises [37, 109].

The procedure followed by a sustainable venture is described by Belz and Binder [10] in six critical stages, starting with recognizing an ecological or social problem, then proposing an innovative business idea. A solution is developed that has double results. The fourth stage includes a triple result and concludes with stages five and six, covering the financing plan and creating a sustainable company. This method connects the behavior of the sustainable entrepreneur and the starting point of sustainable start-ups based on the 17 SDGs [37].

Therefore, it is necessary to inquire about the approaches and practices that help evaluate the level of impact that sustainable start-ups comprise to serve investors and other interested parties as a guide in the decision-making of said companies to promote sustainable development [102]. In addition, efforts must also be made to counteract the effects of climate change and thus reduce inequality gaps to contribute to the implementation of the SDGs [6].

3 The SDGs

The SDGs are goals set out in the 2030 Agenda, a UN initiative that seeks to unify the efforts of humanity to act on the consequences of damage to the environment. Since its signing in September 2015, the UN has reached its proposals to the 193 member states, and, subsequently, they have implemented it in their respective regional agendas [90]. Many entities seek to achieve these objectives for peace and prosperity toward the current generation and the next and improve human life naturally and socially [88]. The proposal of the 2030 agenda is clear: the interconnection of the objectives to achieve progressive and joint progress [113]. The SDGs comprise 17 goals and 169 targets that integrate sustainable development through a triple-bottom-line approach: social, economic, and environmental dimension [89, 90, 107]. A graphical representation of the SDGs with a triple bottom line approach is shown in Fig. 1.

According to the [104], various authors classify them according to their focus or nature for a correct analysis of the SDGs. The most prominent indicators are demographics, income growth, culture, knowledge, technology, etc. However, the most used segmentation is the classification between social, economic, and environmental dimensions [107]. This classification allows measuring according to the consistency of each SDG, as cited in [113]. It originates from the worldwide concern for environmental deterioration and social tension to reverse this situation. In the last 30, there have been significant advances toward sustainable development [90]; however, limited access to resources, lack of support from capital markets [13], and indisputable global warming [113] put at risk both the fulfillment of several SDGs, as well as multilateral agreements such as the Paris Agreement on Climate Change [113].

As a result of the pandemic caused by COVID-19, many economies have been greatly affected. The top priority of such governments is focused on the rate of return and investment risk, regardless of the targets and indicators of the SDGs [113]. For this reason, different societies seek a shared approach between economic development, social inclusion, and environmental sustainability as a roadmap so that sustainable development is within reach of all countries in the world [89]. Along the same lines, the UN highlights the active participation of companies as key in problems that afflict the world, such as hunger, the protection of biodiversity, poverty, among others [105]. The SDGs cover different world problems, each one with its purposes and goals to be achieved, so it is vital to orient it toward investment projects or commercial opportunities, as well as a long-term strategy [82].

The COVID-19 pandemic is the causal economic recession and low investment rate in SDG projects [113]. For their part, several SDGs are also closely related to issues such as renewable energy, positive impacts on well-being, health, quality of

Fig. 1 SDG and its relationship with the 3 dimensions. *Source* Vinuesa et al. [107]

life, among others. However, the 2030 agenda emphasizes sustainability, designed at a regional level, and dedicated to the private sector [82]. On the other hand, an alternative solution to the lack of project financing for the SDGs is crowdfunding, a tool used to solve the capital necessary to start a project in a simple and automated way [68, 80]. This platform ensures the financing of social projects over and above commercial ones [13].

4 SDG of the Environmental Dimension

In addition to the SDGs, global environmental initiatives such as the New Urban Agenda (NAU) and the Paris Agreement result from the global trend to make the world more sustainable [79]. At present, the effects of environmental neglect can be appreciated, so the 17 SDGs seek to promote sustainable practices and solutions aimed at the main problems faced by the population [25, 69].

Taking actions in the face of environmental consequences has as its primary motivation the care of biodiversity, which is why it is not possible to speak of sustainability or the environment without involving biodiversity since living beings such as plants, animals, fungi, microorganisms, among others play a crucial role in human development [79]. The construction of transport, telecommunications, real estate, and other infrastructure makes civilization the main threat to biodiversity [78, 83], which translates into habitat loss, fragmentation, among others [5, 66].

Caring for biodiversity should not only be limited to protecting species and their habitats but rather in restoring and improving their living conditions and allowing positive human interaction with nature [79]. Pedersen Zari [83] affirms that civilization must contribute more to the care of biodiversity than it consumes it and, above all, to remedy any environmental damage caused. To achieve this ideal, all stakeholders must raise awareness about the consequences that biodiversity loss would entail at the cost of human well-being [79]. The UN proposes in its 2030 agenda to stimulate social, economic, and environmental action with a greater emphasis on environmental degradation [46, 114]. It also seeks to integrate policies toward a sustainable environment, such as, for example, proper management of green spaces [66]. In summary, environmental issues must be ensured from the beginning of the project to mitigate the negative impact of human actions toward nature [79].

Achieving the goals of the SDGs requires a more committed mindset toward the ecological and social mission of the goals. In this way, creative and innovative solutions have a more significant role and reduce global risks [69]. This initiative is reaching various actors of change such as politicians and NGOs to achieve social well-being [44], as well as being an information medium for those who wish to contribute to sustainability but have specific knowledge gaps [30, 99]. In addition, it also seeks to land specific objectives on the world agenda so that the goals are more realistic than idealistic [44]. At the insistence of different entities, the world is taking with more significant serenity the sustainable proposals to achieve the well-being of society. Added to this are various authors who propose models of greater understanding for the reach of both politicians and sectors of society itself [44], which is shown in many industries; for example, in the fashion sector, they emphasize less harmful inputs and assign a new function to waste. On the energy sector side, the increase in demand due to economic growth [48, 56] prompts to reduce greenhouse gas (GHG) emissions or to develop more conventional, efficient sources and renewable [12, 35, 72, 92] to achieve climate neutrality [48]. In agriculture, land conversion is avoided, and the restoration of damaged land is encouraged to reduce GHG emissions potentially [38, 85].

On the other hand, within the framework of the development of the SDGs with an environmental dimension is SDG 13, which establishes the taking of urgent measures in the fight against climate change and its effects [98]. SDGs 14 and 15 promote good resource management in caring for species and ecosystems. The focus of these SDGs stems from the damage caused by the excessive consumption of natural resources [79]. For this reason, this chapter segments these 3 SDGs according to the study by [108]. Figure 2 shows SDG 13–15, a classification that catalyzes actions on issues of

Fig. 2 SDG of the environmental dimension. *Source* Vinuesa et al. [107]

great importance to humanity and the planet and evaluates the progress and challenges involved [55].

Trautwein [102] mentions that sustainable entrepreneurship has become a valuable current research topic, highlighting its importance in the transition of society and the business world toward a more sustainable future. Some start-ups were the cause of the emergence of innovative sectors and markets such as renewable energy, organic food, or others in the agriculture sector, which are essential to achieve a transformation that follows the path of sustainability [36]. These emerging companies cement their strategies and make decisions based on the positive impact they may cause to promote and achieve sustainable development [110]. Therefore, they can create innovative market solutions to combat environmental and social problems such as damage to biodiversity, global warming, or inequality in education. In addition, these companies gain a better corporate image in the market and increase their attractiveness for investors who present a vision-oriented toward sustainability [102, 110].

5 SDG of the Social Dimension

Hibbert et al. [50] define the social entrepreneur as one who has a behavior attached to social purposes or seeks to create benefits for a population that needs support. He also has an entrepreneurial spirit that motivates him to carry out successful sustainable activities that later translate into social help [18, 115]. Therefore, social entrepreneurship is commonly considered a behavioral phenomenon that encompasses the factors of proactivity, innovation, and risk management [111], which has become one of the central themes of several economies [40], which shows the importance it has acquired today. However, it is necessary to complement social entrepreneurship to provide data to achieve an economically self-sustaining ecosystem [57].

This type of entrepreneurship plays a vital role in the framework of the SDGs [47], especially those that are linked to social equality about quality education (SDG 4) and equality gender (SDG 5) [20]. Xia et al. [112] point out that several SDGs directly relate to corporate social responsibility. The fact that the SDGs are considered when establishing their action plans and indicators means that these companies obtain more significant benefits and greater acceptance in the market [47].

A classification of the SDGs belonging to the social dimension was carried out according to the study by Vinuesa et al. [107] to have a better result in this chapter. Therefore, Fig. 3 shows SDGs 1–7, SDG 11, and SDG 16. Mugagga and Nabaasa [71] highlight the importance of SDG 6 focused on the sustainable management of water and its sanitation to achieve other goals such as SDG 1-3. This resource is an indispensable resource for the survival of all living beings on the planet; therefore, special attention should be paid to its progress and evolution as a goal set by the UN.

On the other hand, cities are responsible for 60% or 80% of GHG emissions worldwide, according to the evaluation of [41], so artificial intelligence (AI) can influence the creation of systems that reduce carbon emissions and, therefore, promote smart cities with circular economies that use efficient management of their resources. In the case of SDG 1, referring to the extermination of poverty, SDG 4 is related to the provision of quality education services. SDG 6 that pursues the sanitation of water and its supply as drinking water, SDG 7 on energy does not Pollutant and affordable and SDG 11 on sustainable communities, AI can facilitate and accelerate the achievement of these goals by supporting the provision of water, food, energy, and health services to the community [41].

Fig. 3 SDG of the social dimension. *Source* Vinuesa et al. [107]

It can be said that the SDGs comprise a large set of initiatives related to economic, social, and environmental development, which are linked and interconnected with each other. Let's take, for example, the progress of SDG 3 linked to health that depends on the development of SDG 1, that is, the reduction of existing poverty in countries around the world, bringing out intersectoral thinking [15]. The Social Enterprise UK Think Global Trade Social report published in 2015 highlights social enterprises' critical role in facing the challenges that arise and make it difficult to achieve the 17 SDGs proposed by the UN [62]. Therefore, considering that these companies can promote sustainable and inclusive development, it is essential to increase knowledge and understanding of how they can contribute to the progress of the SDGs [47, 62].

Once the need for a business commitment to achieve the SDGs has been recognized and its commitment to CSR [58, 61] and the entrepreneurial spirit that can contribute [18]. However, many SDGs do not provide sufficient recognition to such ventures, translating into a lack of initiative on the primary agents of change [62]. It is necessary to develop instruments to evaluate the impact of companies on the social SDGs and create various models of companies with a social purpose that can promote the SDGs in all parts of the world. Along the same lines, local and international policymakers must broaden their outlook beyond private, public, and nonprofit businesses and organizations to consider successful hybrid social enterprises in the SDGs, and the 2030 Agenda is fulfilled [62].

In the same way, new trends appear to combat emergency environmental and social situations related to the SDGs and propose efficient solutions recognized for their scalable and innovative nature. These trends focus on providing ideas that support business education and developing a social, ecological, and creative mindset and the skills necessary to achieve a green economy [69]. To illustrate the case, Sustainia100 presented 25 solutions incorporated to transform toward a more sustainable future [69]. Sustainia100 analyzes the latest global trends that mark the actions of many industries in the face of the social problems of a new world that seeks to be more sustainable. In addition to proposing innovative solutions, these align with the trend and respective sector. The main approaches are entrepreneurship education, social and ecological mindset, and green economy [69].

The trends that stood out the most in 2014 are concerns about water scarcity, better waste management in the supply chain, innovative business models, and alternative energies [69]. On the solutions side, it was found that the primary trend was the incorporation of the circular economy (CE) [69]. The solutions must be focused on creating value and climate resilience. As of 2018, priority was given to the progressive and even development of each SDGs, a cooperative business mentality, and the adoption of more circular consumption patterns [69]. As can be seen, the acceptance of the CE model in society has been widely accepted to achieve more environmental goals, implement more innovative and sustainable processes, and generate added value for the company. Its correct application depends on its adaptability in each sector, which has been seen satisfactorily in recent years.

6 SDG of the Economic Dimension

In the first instance, it is specified that, when mentioning economic activities, reference is made to all kinds of activity carried out by an organization that intends to develop, provide, sell, or buy any type of goods or services. Some of the most common activities are specified in the agriculture, manufacturing, or service sectors, causing international organizations such as the United Nations (UN) and the European Union to classify numerous individual economic activities, which can cause social, environmental, and economic impact. In this framework, it is necessary to know how economic activities affect the three dimensions of sustainable development to create an integrated monitoring and control procedure [106].

On the other hand, the participation of companies can be promoted through Responsible Innovation (RI); it can also be used as an evaluation method to identify the degree of the contribution they have in the implementation of the SDGs through 2030 Agenda. RI provides a framework focused on closing the gap between the orientation of innovation in business models toward society and the challenges companies face to achieve the SDGs globally. Therefore, RI generates an innovative sociotechnical integration procedure that includes diverse stakeholders [53]. The UN pointed out that the SDGs seek to achieve an economy that focuses on people and the planet [53], so it is necessary to have the support of companies, governments, and civil society in general [11]. This section specifically focuses on the SDGs addressed in the economic dimension of sustainability, as demonstrated in Fig. 4.

In the case of SDG 8, it promotes sustained economic growth inclusively and sustainably, in addition to entire, productive, and decent employment for all without distinction. Although this SDG defends labor rights, it is necessary to consider

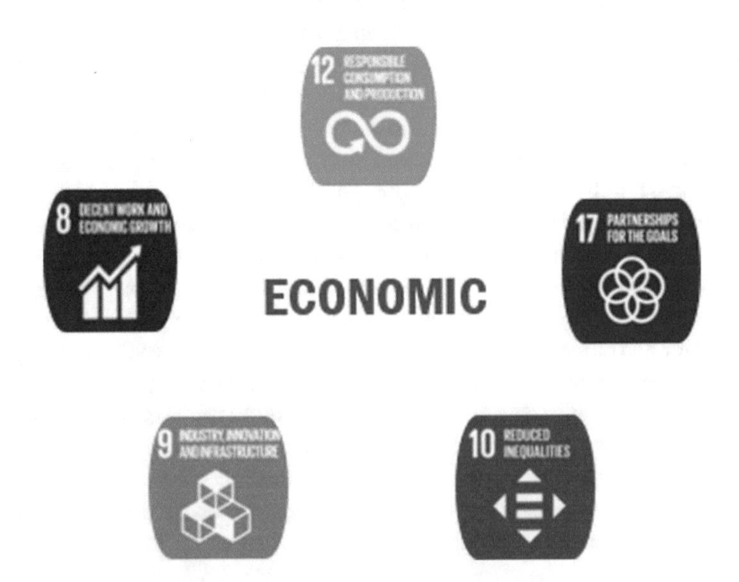

Fig. 4 SDG of the economic dimension. *Source* Vinuesa et al. [107]

specific economic growth measures such as per capita growth and GDP [86]. It is essential to mention gender equality concerning labor rights, offering equal work opportunities for men and women to do justice to their promise of inclusive and decent work [86]. However, the question arises as to whether the 2030 Agenda takes full and decent employment as an issue related to human rights or only as benefits generated by the growth of an economy. In addition to this, it is known that the Sustainable Development Agenda welcomes several institutional agreements that focus on the market, which would result in the creation of obstacles that prevent reaching the goals established by the UN [39].

If we look at the outlook under the COVID-19 pandemic, the economy of many countries suffered a negative impact, resulting in an economic crisis worse than the one at the end of the last decade. Some of the adverse effects were seen on employment and people's well-being. Likewise, the progress of the UN SDGs was delayed [42, 97]. Given this situation, the political leaders accepted specific measures that have the force to stop the economic decline and, on the contrary, promote the reactivation of the economy in the shortest possible time. However, many businesses did not have the strength to fight the containment measures and other government restrictions imposed to stop the contagions, which meant less demand. Thus, it is pertinent to evaluate and identify the factors that drive the creation of ventures during the COVID-19 situation and its connection with sustainable development since it would help businesses get closer to current consumers [42].

The adaptation of the CE model has been highlighted as crucial for sustainable urban transitions [96] due to high levels of consumption of global resources and human inaction in the face of pollution [76, 84]. Therefore, environmental policy instruments help developers, being the legal framework that allows eco-fallows. Some examples of these instruments are circular guidelines, taxes to avoid unsustainable resources, benefits for alternative resources, and public investment in R&D to reuse materials. Although governments must constantly intervene, the private sector is also part of this change to sustainability [4].

The global desire to achieve the SDGs is undeniable, and the CE is an option to achieve it [97] which requires social changes, new systems, and other innovations necessary to supply the current linear model and its inefficient management of resources [14, 59, 87, 91, 100]. This transition involves all participants in the supply chain, from suppliers to distributors, since it is defined as a critical element in the evolution of companies to be more circular [9, 23, 27, 34, 60]. One way to exemplify the need for sustainable supply chain improvements is through the ravages of COVID-19. Solid waste management during the pandemic is a sensitive issue given the large volume of waste generated. For example, India's amount of biomedical waste increased by 25% due to the pandemic [97]. The greater dependence on the delivery of food and supplies, the excessive control of plastic packaging, and single-use plastic waste are just some environmental consequences that have left the pandemic to the world.

Given those above, it is recommended to strengthen the prioritization for investment in green sectors, more sustainable supply chains, decentralization of waste management, among others [97]. Redesigning production chains and adopting the principles of the CE allows an ideal transition from the current model to one more adjusted to the needs of the planet [43, 52, 60, 65, 77, 95]. Finally, the economic viability of these circular models is a condition for their execution since, although better performance is sought without harming the planet's resources, said model must safeguard the interests of the companies that benefit from them, as well as the society [14].

Recent suggested lectures about green approach

- Waste reduction and carbon footprint [45]
- Material selection for circularity and footprints [70]
- 3D Print, circularity, and footprints [26]
- Virtual tourism [26]
- Leadership for sustainability in crisis time [3]
- Virtual education and circularity [21]
- Circular Economy for packaging [16]
- Students oriented to circular learning [28]
- Food and circular economy [24]
- Water footprint and food supply chain management [54]
- Waste footprint [7]
- Measuring circular economy [31]
- Carbon footprint [33]

7 Closing Remarks

Due to the increase in damage caused to the environment and ecological degradation, economic development has been damaged, in addition to threatening human lives and, therefore, their health [114]. Many times, commercial activities and their mismanagement of resources are responsible for the environmental consequences; However, they turn out to be the most efficient solution when they take a more sustainable approach, where green entrepreneurship obtains a vital role that must be considered in the business world [51]. Likewise, it should be noted that sustainable entrepreneurs present a strong feeling of responsibility for seeking solutions to ecological and social problems that arise every day, manifesting it through their prevention approach [10]. Therefore, they differ from traditional entrepreneurs by generating economic benefits as their primary objective.

The SDGs are a conglomerate of goals that ensure the care of the planet, which are divided into 17 approaches according to the area they involve [112]. This initiative seeks to combat the damage caused to the planet, which is currently perceived as the depletion of natural resources or global warming [13, 113]. The affiliation of companies to these sustainable guidelines is vital for achieving the objectives, otherwise; significant progress would not be achieved [53]. It is necessary to highlight the importance of all the SDGs being promoted and developed in parallel to move the global system toward fruitful and efficient results [17]. For this, it is essential to promote the growth of sustainable enterprises in the business world since, in this way, more sustainable practices are created that have a strategic base focused on obtaining a positive impact at an economic, social, and environmental level.

In the same way, both governments and organizations also seek to adapt their practices toward greater sustainability in their processes [62]; therefore, the goals set must be within the reach and understanding of all those who wish to join the environmental cause [89]. For their part, the SDGs provide a framework where affiliates can measure and evaluate their CSR performance [112]. For this reason, it is recommended that more research and theoretical and practical studies must be carried out on these topics to generate information that is useful for companies and at the same time supports the SDGs.

References

1. Abdelnaeim SM, El-Bassiouny N (2020) The relationship between entrepreneurial cognitions and sustainability orientation: the case of an emerging market. J Entrepreneurship Emerg Econ. https://doi.org/10.1108/JEEE-03-2020-0069
2. Adel HM, Mahrous AA, Hammad R (2020) Entrepreneurial marketing strategy, institutional environment, and business performance of SMEs in Egypt. J Entrepreneurship Emerg Econ. https://doi.org/10.1108/JEEE-11-2019-0171
3. Alvarez-Risco A, Del-Aguila-Arcentales S, Villalobos-Alvarez D, Diaz-Risco S (2022) Leadership for sustainability in crisis time. In: Alvarez-Risco A, Muthu SS, Del-Aguila-Arcentales S (eds) Circular economy: impact on carbon and water footprint. Springer Singapore, pp 41–64. https://doi.org/10.1007/978-981-19-0549-0_3
4. Ancapi FB (2021) Policy instruments for circular built environment implementation: a systematic literature review. IOP Conf Ser Earth Environ Sci 855(1):012019. https://doi.org/10.1088/1755-1315/855/1/012019
5. Antrop M (2000) Changing patterns in the urbanized countryside of Western Europe. Landscape Ecol 15(3):257–270. https://doi.org/10.1023/A:1008151109252
6. Apostolopoulos N, Al-Dajani H, Holt D, Jones P, Newbery R (2018) *Entrepreneurship and the sustainable development goals*. Emerald Publishing Limited. https://doi.org/10.1108/S2040-724620180000008005
7. Arias-Meza M, Alvarez-Risco A, Cuya-Velásquez BB, de las Mercedes Anderson-Seminario M, Del-Aguila-Arcentales S (2022) Fashion and textile circularity and waste footprint. In: Alvarez-Risco A, Muthu SS, Del-Aguila-Arcentales S (eds) Circular economy: impact on carbon and water footprint. Springer Singapore, pp 181–204. https://doi.org/10.1007/978-981-19-0549-0_9
8. Audretsch D, Colombelli A, Grilli L, Minola T, Rasmussen E (2020) Innovative start-ups and policy initiatives. Res Policy 49(10):104027. https://doi.org/10.1016/j.respol.2020.104027
9. Batista L, Bourlakis M, Liu Y, Smart P, Sohal A (2018) Supply chain operations for a circular economy. Prod Plann Control 29(6):419–424. https://doi.org/10.1080/09537287.2018.1449267
10. Belz FM, Binder JK (2017) Sustainable entrepreneurship: a convergent process model. Bus Strategy Environ 26(1):1–17. https://doi.org/10.1002/bse.1887
11. Buhmann K, Jonsson J, Fisker M (2019) Do no harm and do more good too: connecting the SDGs with business and human rights and political CSR theory. Corp Governance: Int J Bus Soc 19(3):389–403. https://doi.org/10.1108/CG-01-2018-0030

12. Bulut U, Inglesi-Lotz R (2019) Which type of energy drove industrial growth in the US from 2000 to 2018 ? Energy Rep 5:425–430. https://doi.org/10.1016/j.egyr.2019.04.005
13. Calic G, Mosakowski E (2016) Kicking off social entrepreneurship: how a sustainability orientation influences crowdfunding success. J Manage Stud 53(5):738–767. https://doi.org/10.1111/joms.12201
14. Calicchio Berardi P, Peregrino de Brito R (2021) Supply chain collaboration for a circular economy—from transition to continuous improvement. J Cleaner Prod 328:129511. https://doi.org/10.1016/j.jclepro.2021.129511
15. Carter DJ, Glaziou P, Lönnroth K, Siroka A, Floyd K, Weil D, … Boccia D (2018) The impact of social protection and poverty elimination on global tuberculosis incidence: a statistical modelling analysis of Sustainable Development Goal 1. Lancet Glob Health 6(5):e514–e522. https://doi.org/10.1016/S2214-109X(18)30195-5
16. Castillo-Benancio S, Alvarez-Risco A, Esquerre-Botton S, Leclercq-Machado L, Calle-Nole M, Morales-Ríos F, … Del-Aguila-Arcentales S (2022) Circular economy for packaging and carbon footprint. In: Alvarez-Risco A, Muthu SS, Del-Aguila-Arcentales S (eds) Circular economy: impact on carbon and water footprint. Springer Singapore, pp 115–138. https://doi.org/10.1007/978-981-19-0549-0_6
17. Cernev T, Fenner R (2020) The importance of achieving foundational sustainable development goals in reducing global risk. Futures 115:102492. https://doi.org/10.1016/j.futures.2019.102492
18. Chen R-L (2020) Trends in economic inequality and its impact on Chinese Nationalism. J Contemp China 29(121):75–91. https://doi.org/10.1080/10670564.2019.1621531
19. Cohen B (2006) Sustainable valley entrepreneurial ecosystems. Bus Strategy Environ 15(1):1–14. https://doi.org/10.1002/bse.428
20. Contreras-Barraza N, Espinosa-Cristia JF, Salazar-Sepulveda G, Vega-Muñoz A (2021) Entrepreneurial intention: a gender study in business and economics students from chile. Sustainability 13(9):4693. https://doi.org/10.3390/su13094693
21. Contreras-Taica A, Alvarez-Risco A, Arias-Meza M, Campos-Dávalos N, Calle-Nole M, Almanza-Cruz C, … Del-Aguila-Arcentales S (2022) Virtual education: carbon footprint and circularity. In: Alvarez-Risco A, Muthu SS, Del-Aguila-Arcentales S (eds) Circular economy: impact on carbon and water footprint. Springer Singapore, pp 265–285. https://doi.org/10.1007/978-981-19-0549-0_13
22. Crnogaj K, Rebernik M, Hojnik BB, Gomezelj DO (2014) Building a model of researching the sustainable entrepreneurship in the tourism sector. Kybernetes 43(3/4):377–393. https://doi.org/10.1108/K-07-2013-0155
23. Cruz-Sotelo SE, Ojeda-Benítez S, Jáuregui Sesma J, Velázquez-Victorica KI, Santillán-Soto N, García-Cueto OR, … Alcántara C (2017) E-waste supply chain in Mexico: challenges and opportunities for sustainable management. Sustainability 9(4):503. https://doi.org/10.3390/su9040503
24. Cuya-Velásquez BB, Alvarez-Risco A, Gomez-Prado R, Juarez-Rojas L, Contreras-Taica A, Ortiz-Guerra A, … Del-Aguila-Arcentales S (2022) Circular economy for food loss reduction and water footprint. In: Alvarez-Risco A, Muthu SS, Del-Aguila-Arcentales S (eds) Circular economy: impact on carbon and water footprint. Springer Singapore, pp 65–91. https://doi.org/10.1007/978-981-19-0549-0_4
25. Dantas TET, de-Souza ED, Destro IR, Hammes G, Rodriguez CMT, Soares SR (2021) How the combination of circular economy and industry 4.0 can contribute towards achieving the sustainable development goals. Sustain Prod Consumption 26:213–227. https://doi.org/10.1016/j.spc.2020.10.005
26. De-la-Cruz-Diaz M, Alvarez-Risco A, Jaramillo-Arévalo M, de las Mercedes Anderson-Seminario M, Del-Aguila-Arcentales S (2022) 3D print, circularity, and footprints. In: Alvarez-Risco A, Muthu SS, Del-Aguila-Arcentales S (eds) Circular economy: impact on carbon and water footprint. Springer Singapore, pp 93–112. https://doi.org/10.1007/978-981-19-0549-0_5

27. De Angelis R, Howard M, Miemczyk J (2018) Supply chain management and the circular economy: towards the circular supply chain. Prod Plann Control 29(6):425–437. https://doi.org/10.1080/09537287.2018.1449244
28. de las Mercedes Anderson-Seminario M, Alvarez-Risco A (2022) Better students, better companies, better life: circular learning. In: Alvarez-Risco A, Muthu SS, Del-Aguila-Arcentales S (eds) *Circular economy: impact on carbon and water footprint*. Springer Singapore, pp 19–40. https://doi.org/10.1007/978-981-19-0549-0_2
29. Dean TJ, McMullen JS (2007) Toward a theory of sustainable entrepreneurship: reducing environmental degradation through entrepreneurial action. J Bus Ventur 22(1):50–76. https://doi.org/10.1016/j.jbusvent.2005.09.003
30. Décamps A, Barbat G, Carteron J-C, Hands V, Parkes C (2017) Sulitest: a collaborative initiative to support and assess sustainability literacy in higher education. Int J Manage Educ 15(2, Part B):138–152. https://doi.org/10.1016/j.ijme.2017.02.006
31. Del-Aguila-Arcentales S, Alvarez-Risco A, Muthu SS (2022) Measuring circular economy. In: Alvarez-Risco A, Muthu SS, Del-Aguila-Arcentales S (eds) Circular economy: impact on carbon and water footprint. Springer Singapore, pp 3–17. https://doi.org/10.1007/978-981-19-0549-0_1
32. Diepolder CS, Weitzel H, Huwer J (2021) Competence frameworks of sustainable entrepreneurship: a systematic review. Sustainability 13(24):13734. https://doi.org/10.3390/su132413734
33. Esquerre-Botton S, Alvarez-Risco A, Leclercq-Machado L, de las Mercedes Anderson-Seminario M, Del-Aguila-Arcentales S (2022) Food loss reduction and carbon footprint practices worldwide: a benchmarking approach of circular economy. In: Alvarez-Risco A, Muthu SS, Del-Aguila-Arcentales S (eds) Circular economy: impact on carbon and water footprint. Springer Singapore, pp 161–179. https://doi.org/10.1007/978-981-19-0549-0_8
34. Farooque M, Zhang A, Thürer M, Qu T, Huisingh D (2019) Circular supply chain management: a definition and structured literature review. J Clean Prod 228:882–900. https://doi.org/10.1016/j.jclepro.2019.04.303
35. Fernández Fernández Y, Fernández López MA, Olmedillas Blanco B (2018) Innovation for sustainability: the impact of R&D spending on CO2 emissions. J Clean Prod 172:3459–3467. https://doi.org/10.1016/j.jclepro.2017.11.001
36. Fichter K, Tiemann I (2018) Factors influencing university support for sustainable entrepreneurship: insights from explorative case studies. J Clean Prod 175:512–524. https://doi.org/10.1016/j.jclepro.2017.12.031
37. Fischer D, Mauer R, Brettel M (2018) Regulatory focus theory and sustainable entrepreneurship. Int J Entrepreneurial Behav Res 24(2):408–428. https://doi.org/10.1108/IJEBR-12-2015-0269
38. Frank S, Havlík P, Soussana J.F, Levesque A, Valin H, Wollenberg E, … Obersteiner M (2017) Reducing greenhouse gas emissions in agriculture without compromising food security? Environ Res Lett 12(10):105004. https://doi.org/10.1088/1748-9326/aa8c83
39. Frey DF (2017) Economic growth, full employment and decent work: the means and ends in SDG 8. Int J Hum Rights 21(8):1164–1184. https://doi.org/10.1080/13642987.2017.1348709
40. Fritsch M, Wyrwich M (2017) The effect of entrepreneurship on economic development—an empirical analysis using regional entrepreneurship culture. J Econ Geogr 17(1):157–189. https://doi.org/10.1093/jeg/lbv049
41. Fuso Nerini F, Slob A, Ericsdotter Engström R, Trutnevyte E (2019) A research and innovation agenda for zero-emission European Cities. Sustainability 11(6):1692. https://doi.org/10.3390/su11061692
42. Galindo-Martín M-Á, Castaño-Martínez M-S, Méndez-Picazo M-T (2021) Effects of the pandemic crisis on entrepreneurship and sustainable development. J Bus Res 137:345–353. https://doi.org/10.1016/j.jbusres.2021.08.053
43. Ghisellini P, Ulgiati S (2020) Circular economy transition in Italy. Achievements, perspectives and constraints. J Cleaner Prod 243:118360. https://doi.org/10.1016/j.jclepro.2019.118360

44. Giannetti BF, Agostinho F, Eras JJC, Yang Z, Almeida CMVB (2020) Cleaner production for achieving the sustainable development goals. J Clean Prod 271:122127. https://doi.org/10.1016/j.jclepro.2020.122127
45. Gómez-Prado R, Alvarez-Risco A, Sánchez-Palomino J, de las Mercedes Anderson-Seminario M, Del-Aguila-Arcentales S (2022) Circular economy for waste reduction and carbon footprint. In: Alvarez-Risco A, Muthu SS, Del-Aguila-Arcentales S (eds) Circular economy: impact on carbon and water footprint. Springer Singapore, pp 139–159. https://doi.org/10.1007/978-981-19-0549-0_7
46. González Del Campo A, Gazzola P, Onyango V (2020) The mutualism of strategic environmental assessment and sustainable development goals. Environ Impact Assess Rev 82:106383. https://doi.org/10.1016/j.eiar.2020.106383
47. Günzel-Jensen F, Siebold N, Kroeger A, Korsgaard S (2020) Do the United Nations' sustainable development goals matter for social entrepreneurial ventures? A bottom-up perspective. J Bus Ventur Insights 13:e00162. https://doi.org/10.1016/j.jbvi.2020.e00162
48. Guzowska MK, Kryk B, Michalak D, Szyja P (2021) R&D spending in the energy sector and achieving the goal of climate neutrality. Energies 14(23):7875. https://doi.org/10.3390/en14237875
49. Hall JK, Daneke GA, Lenox MJ (2010) Sustainable development and entrepreneurship: past contributions and future directions. J Bus Ventur 25(5):439–448. https://doi.org/10.1016/j.jbusvent.2010.01.002
50. Hibbert SA, Hogg G, Quinn T (2005) Social entrepreneurship: understanding consumer motives for buying The Big Issue. J Consum Behav 4(3):159–172. https://doi.org/10.1002/cb.6
51. Hörisch J (2015) The role of sustainable entrepreneurship in sustainability transitions: a conceptual synthesis against the background of the multi-level perspective. Adm Sci 5(4). https://doi.org/10.3390/admsci5040286
52. Iacovidou E, Velenturf APM, Purnell P (2019) Quality of resources: a typology for supporting transitions towards resource efficiency using the single-use plastic bottle as an example. Sci Total Environ 647:441–448. https://doi.org/10.1016/j.scitotenv.2018.07.344
53. Imaz O, Eizagirre A (2020) Responsible innovation for sustainable development goals in business: an agenda for cooperative firms. Sustainability 12(17):6948. https://doi.org/10.3390/su12176948
54. Juarez-Rojas L, Alvarez-Risco A, Campos-Dávalos N, de las Mercedes Anderson-Seminario M, Del-Aguila-Arcentales S (2022) Water footprint in the textile and food supply chain management: trends to become circular and sustainable. In: Alvarez-Risco A, Muthu SS, Del-Aguila-Arcentales S (eds) Circular economy: impact on carbon and water footprint. Springer Singapore, pp 225–243. https://doi.org/10.1007/978-981-19-0549-0_11
55. Kakar N, Popovski V, Robinson NA (2021) Fulfilling the sustainable development goals. Routledge
56. Keong CY (2005) Energy demand, economic growth, and energy efficiency—the Bakun dam-induced sustainable energy policy revisited. Energy Policy 33(5):679–689. https://doi.org/10.1016/j.enpol.2003.09.017
57. Kitsios F, Kamariotou M, Grigoroudis E (2021) Digital entrepreneurship services evolution: analysis of quadruple and quintuple helix innovation models for open data ecosystems. Sustainability 13(21):12183. https://doi.org/10.3390/su132112183
58. Kolk A (2016) The social responsibility of international business: From ethics and the environment to CSR and sustainable development. J World Bus 51(1):23–34. https://doi.org/10.1016/j.jwb.2015.08.010
59. Korhonen J, Honkasalo A, Seppälä J (2018) Circular economy: the concept and its limitations. Ecol Econ 143:37–46. https://doi.org/10.1016/j.ecolecon.2017.06.041
60. Leising E, Quist J, Bocken N (2018) Circular economy in the building sector: three cases and a collaboration tool. J Clean Prod 176:976–989. https://doi.org/10.1016/j.jclepro.2017.12.010

61. Leisinger K (2015) Business needs to embrace sustainability targets. Nature 528(7581):165–165. https://doi.org/10.1038/528165a
62. Littlewood D, Holt D (2018) How social enterprises can contribute to the sustainable development goals (SDGs)–a conceptual framework. In: Entrepreneurship and the sustainable development goals. Emerald Publishing Limited, pp 33–46. https://doi.org/10.1108/S2040-724620180000008007
63. Liu Y, Samsami M, Meshreki H, Pereira F, Schøtt T (2021) Sustainable development goals in strategy and practice: businesses in Colombia and Egypt. Sustainability 13(22):12453. https://doi.org/10.3390/su132212453
64. Mäkinen SJ, Dedehayir O (2012) Business ecosystem evolution and strategic considerations: a literature review. In: 2012 18th international ICE conference on engineering, technology and innovation. IEEE, pp 1–10. https://doi.org/10.1109/ICE.2012.6297653
65. Masi D, Kumar V, Garza-Reyes JA, Godsell J (2018) Towards a more circular economy: exploring the awareness, practices, and barriers from a focal firm perspective. Prod Plann Control 29(6):539–550. https://doi.org/10.1080/09537287.2018.1449246
66. McKinney ML (2002) Urbanization, biodiversity, and conservation: the impacts of urbanization on native species are poorly studied, but educating a highly urbanized hsuman population about these impacts can greatly improve species conservation in all ecosystems. Bioscience 52(10):883–890. https://doi.org/10.1641/0006-3568(2002)052[0883:UBAC]2.0.CO;2
67. Melegati, J., Guerra, E., Wang X (2020) Understanding hypotheses engineering in software startups through a gray literature review. Inf Softw Technol 106465. https://doi.org/10.1016/j.infsof.2020.106465
68. Mollick E (2014) The dynamics of crowdfunding: an exploratory study. J Bus Ventur 29(1):1–16. https://doi.org/10.1016/j.jbusvent.2013.06.005
69. Moon CJ (2018) Contributions to the SDGs through social and eco entrepreneurship: new mindsets for sustainable solutions. In: Entrepreneurship and the sustainable development goals, vol 8. Emerald Publishing Limited, pp 47–68. https://doi.org/10.1108/S2040-724620180000008008
70. Morales-Ríos F, Alvarez-Risco A, Castillo-Benancio S, de las Mercedes Anderson-Seminario M, Del-Aguila-Arcentales S (2022) Material selection for circularity and footprints. In: Alvarez-Risco A, Muthu SS, Del-Aguila-Arcentales S (eds) Circular economy: impact on carbon and water footprint. Springer Singapore, pp 205–221. https://doi.org/10.1007/978-981-19-0549-0_10
71. Mugagga F, Nabaasa BB (2016) The centrality of water resources to the realization of Sustainable Development Goals (SDG). A review of potentials and constraints on the African continent. Int Soil Water Conserv Res 4(3):215–223. https://doi.org/10.1016/j.iswcr.2016.05.004
72. Munasinghe M (2002) The sustainomics trans-disciplinary meta-framework for making development more sustainable: applications to energy issues. Int J Sustain Dev 5(1–2):125–182. https://doi.org/10.1504/IJSD.2002.002563
73. Muñoz P (2018) A cognitive map of sustainable decision-making in entrepreneurship: a configurational approach. Int J Entrep Behav Res. https://doi.org/10.1108/IJEBR-03-2017-0110
74. Muñoz P, Cohen B (2018) Sustainable entrepreneurship research: taking stock and looking ahead [https://doi.org/10.1002/bse.2000]. Bus Strategy Environ 27(3):300–322. https://doi.org/10.1002/bse.2000
75. Muñoz P, Dimov D (2015) The call of the whole in understanding the development of sustainable ventures. J Bus Ventur 30(4):632–654. https://doi.org/10.1016/j.jbusvent.2014.07.012
76. Ness DA, Xing K (2017) Toward a resource-efficient built environment: a literature review and conceptual model. J Ind Ecol 21(3):572–592. https://doi.org/10.1111/jiec.12586
77. Nogueira A, Ashton WS, Teixeira C (2019) Expanding perceptions of the circular economy through design: eight capitals as innovation lenses. Resour Conserv Recycl 149:566–576. https://doi.org/10.1016/j.resconrec.2019.06.021

78. Nolan G, Hamilton M, Brown M (2009) Comparing the biodiversity impacts of building materials. Archit Sci Rev 52(4):261–269. https://doi.org/10.3763/asre.2009.0012
79. Opoku A (2019) Biodiversity and the built environment: Implications for the sustainable development goals (SDGs). Resour Conserv Recycl 141:1–7. https://doi.org/10.1016/j.resconrec.2018.10.011
80. Ordanini A, Miceli L, Pizzetti M, Parasuraman A (2011) Crowd-funding: transforming customers into investors through innovative service platforms. J Serv Manage 22(4):443–470. https://doi.org/10.1108/09564231111155079
81. Pacheco DF, Dean TJ, Payne DS (2010) Escaping the green prison: entrepreneurship and the creation of opportunities for sustainable development. J Bus Ventur 25(5):464–480. https://doi.org/10.1016/j.jbusvent.2009.07.006
82. Pedersen CS (2018) The UN sustainable development goals (SDGs) are a great gift to business! Procedia Cirp 69:21–24. https://doi.org/10.1016/j.procir.2018.01.003
83. Pedersen Zari M (2012) Ecosystem services analysis for the design of regenerative built environments. Build Res Inf 40(1):54–64. https://doi.org/10.1080/09613218.2011.628547
84. Pomponi F, Moncaster A (2017) Circular economy for the built environment: a research framework. J Clean Prod 143:710–718. https://doi.org/10.1016/j.jclepro.2016.12.055
85. Poore J, Nemecek T (2018) Reducing food's environmental impacts through producers and consumers. Science 360(6392):987–992. https://doi.org/10.1126/science.aaq0216
86. Rai SM, Brown BD, Ruwanpura KN (2019) SDG 8: Decent work and economic growth—a gendered analysis. World Dev 113:368–380. https://doi.org/10.1016/j.worlddev.2018.09.006
87. Ritzén S, Sandström GÖ (2017) Barriers to the circular economy—integration of perspectives and domains. Procedia Cirp 64:7–12. https://doi.org/10.1016/j.procir.2017.03.005
88. Rosati F, Faria LGD (2019) Business contribution to the sustainable development agenda: organizational factors related to early adoption of SDG reporting. Corp Soc Responsib Environ Manage 26(3):588–597. https://doi.org/10.1002/csr.1705
89. Sachs JD (2012) From millennium development goals to sustainable development goals. The lancet 379(9832):2206–2211. https://doi.org/10.1016/S0140-6736(12)60685-0
90. Salvia AL, Leal Filho W, Brandli LL, Griebeler JS (2019) Assessing research trends related to sustainable development goals: local and global issues. J Clean Prod 208:841–849. https://doi.org/10.1016/j.jclepro.2018.09.242
91. Sauvé S, Bernard S, Sloan P (2016) Environmental sciences, sustainable development and circular economy: alternative concepts for trans-disciplinary research. Environ Develop 17:48–56. https://doi.org/10.1016/j.envdev.2015.09.002
92. Saygin D, Rigter J, Caldecott B, Wagner N, Gielen D (2019) Power sector asset stranding effects of climate policies. Energy Sources Part B 14(4):99–124. https://doi.org/10.1080/15567249.2019.1618421
93. Schaltegger S (2002) A framework for ecopreneurship: leading bioneers and environmental managers to ecopreneurship. Greener Manage Int (38):45–58
94. Schaltegger S, Wagner M (2011) Sustainable entrepreneurship and sustainability innovation: categories and interactions. Bus Strategy Environ 20(4):222–237. https://doi.org/10.1002/bse.682
95. Schenkel M, Caniëls MCJ, Krikke H, van der Laan E (2015) Understanding value creation in closed loop supply chains—past findings and future directions. J Manuf Syst 37:729–745. https://doi.org/10.1016/j.jmsy.2015.04.009
96. Schröder P, Lemille A, Desmond P (2020) Making the circular economy work for human development. Resour Conserv Recycl 156:104686. https://doi.org/10.1016/j.resconrec.2020.104686
97. Sharma HB, Vanapalli KR, Samal B, Cheela VRS, Dubey BK, Bhattacharya J (2021) Circular economy approach in solid waste management system to achieve UN-SDGs: solutions for post-COVID recovery. Sci Total Environ 800:149605. https://doi.org/10.1016/j.scitotenv.2021.149605
98. Shaw K, Kennedy C, Dorea CC (2021) Non-sewered sanitation systems' global greenhouse gas emissions: balancing sustainable development goal tradeoffs to end open defecation. Sustainability 13(21):11884. https://doi.org/10.3390/su132111884

99. Soini K, Jurgilevich A, Pietikäinen J, Korhonen-Kurki K (2018) Universities responding to the call for sustainability: a typology of sustainability centres. J Clean Prod 170:1423–1432. https://doi.org/10.1016/j.jclepro.2017.08.228
100. Tebbatt Adams K, Osmani M, Thorpe T, Thornback J (2017) Circular economy in construction: current awareness, challenges and enablers. Proc Inst Civ Eng—Waste Resou Manage 170(1):15–24. https://doi.org/10.1680/jwarm.16.00011
101. Terán-Yépez E, Marín-Carrillo GM, del Pilar Casado-Belmonte M, de las Mercedes Capobianco-Uriarte M (2020) Sustainable entrepreneurship: review of its evolution and new trends. J Cleaner Prod 252:119742. https://doi.org/10.1016/j.jclepro.2019.119742
102. Trautwein C (2020) Sustainability impact assessment of start-ups–Key insights on relevant assessment challenges and approaches based on an inclusive, systematic literature review. J Cleaner Prod:125330. https://doi.org/10.1016/j.jclepro.2020.125330
103. Tripathi N, Seppänen P, Boominathan G, Oivo M, Liukkunen K (2019) Insights into startup ecosystems through exploration of multi-vocal literature. Inf Softw Technol 105:56–77. https://doi.org/10.1016/j.infsof.2018.08.005
104. United Nations Global Compact (2016) UN Global Compact Leaders Summit 2016. Retrieved 01/01/2022 from https://www.unglobalcompact.org/take-action/events/411-un-global-compact-leaders-summit-2016
105. Van der Waal JW, Thijssens T (2020) Corporate involvement in sustainable development goals: exploring the territory. J Clean Prod 252:119625. https://doi.org/10.1016/j.jclepro.2019.119625
106. van Zanten JA, van Tulder R (2021) Towards nexus-based governance: defining interactions between economic activities and sustainable development goals (SDGs). Int J Sust Dev World 28(3):210–226. https://doi.org/10.1080/13504509.2020.1768452
107. Vinuesa R, Azizpour H, Leite I, Balaam M, Dignum V, Domisch S, Felländer A, Langhans SD, Tegmark M, Fuso Nerini F (2020) The role of artificial intelligence in achieving the sustainable development goals. Nat Commun 11(1):233. https://doi.org/10.1038/s41467-019-14108-y
108. Vinuesa R, Azizpour H, Leite I, Balaam M, Dignum V, Domisch S, Fuso Nerini F (2020) The role of artificial intelligence in achieving the sustainable development goals. Nat Commun 11(1):233. https://doi.org/10.1038/s41467-019-14108-y
109. Vuorio AM, Puumalainen K, Fellnhofer K (2018) Drivers of entrepreneurial intentions in sustainable entrepreneurship. Int J Entrepreneurial Behav Res 24(2):359–381. https://doi.org/10.1108/IJEBR-03-2016-0097
110. Waas T, Hugé J, Block T, Wright T, Benitez-Capistros F, Verbruggen A (2014) Sustainability assessment and indicators: tools in a decision-making strategy for sustainable development. Sustainability 6(9):5512–5534. https://doi.org/10.3390/su6095512
111. Weerawardena J, Mort GS (2006) Investigating social entrepreneurship: a multidimensional model. J World Bus 41(1):21–35. https://doi.org/10.1016/j.jwb.2005.09.001
112. Xia B, Olanipekun A, Chen Q, Xie L, Liu Y (2018) Conceptualising the state of the art of corporate social responsibility (CSR) in the construction industry and its nexus to sustainable development. J Clean Prod 195:340–353. https://doi.org/10.1016/j.jclepro.2018.05.157
113. Yoshino N, Taghizadeh-Hesary F, Otsuka M (2021) Covid-19 and optimal portfolio selection for investment in sustainable development goals. Financ Res Lett 38:101695. https://doi.org/10.1016/j.frl.2020.101695
114. Zhang X-E, Li Q (2021) Does green proactiveness orientation improve the performance of agricultural new ventures in China? The mediating effect of sustainable opportunity recognition. SAGE Open 11(4):21582440211067224. https://doi.org/10.1177/21582440211067224
115. Zheng L, He X, Cao L, Xu H (2018) Making modernity in China: employment and entrepreneurship among the new generation of peasant workers. Int J Jpn Sociol 27(1):26–40. https://doi.org/10.1111/ijjs.12077

FinTech: An Innovative Green Entrepreneurship Model

Marco Calle-Nole, Aldo Alvarez-Risco, Anguie Contreras-Taica, María de las Mercedes Anderson-Seminario, and Shyla Del-Aguila-Arcentales

Abstract FinTech or financial technology has improved financial services through technology and innovation. These revolutionary financial companies are here to stay since they offer significant benefits to users, such as more incredible speed or lower commissions on their transactions. They are very competitive in the market, being more attractive than banks. Since the outbreak of the COVID-19 pandemic, many people are starting to start a business, and FinTechs represent an excellent opportunity to start a sustainable green business over time and fast-growing. Naturally, it is not an easy task to launch a new FinTech; it requires hard work, but it has great rewards. Due to those above, it is essential to analyze its implications deeply.

Keywords Financial services · Financial technology · FinTech · Entrepreneurship · Business model

1 Introduction

With the arrival of digitalization in the financial world, we have seen how everything evolved in a flurry of improvements and innovations for the financial world. With more interest in new technologies, companies were born that saw the great potential in optimizing some processes in an environmentally responsible way, linked to the products and services offered by banks. These companies decided to start offering specialized services according to what the market was looking for, a disruptive proposal that led to the generation of many investigations and successes, taking risks for a better future. Nowadays, in a different context, with greater social responsibility for the care of the environment, we see how everything is still in constant development and how these companies, far from giving up, continue to play at being a couple of big banks. Without a doubt, a challenging start full of ups and downs,

M. Calle-Nole · A. Alvarez-Risco · A. Contreras-Taica · M. de las Mercedes Anderson-Seminario
Universidad de Lima, Lima, Peru

S. Del-Aguila-Arcentales (✉)
Escuela de Posgrado, Universidad San Ignacio de Loyola, Lima, Peru
e-mail: sdelaguila@usil.edu.pe

A. Alvarez-Risco et al. (eds.), *Footprint and Entrepreneurship*, Environmental Footprints and Eco-design of Products and Processes,
https://doi.org/10.1007/978-981-19-8895-0_11

but thanks to the fact that they remained firm, we have the technology and infrastructure that we know today. In this research, point by point, how FinTech has been developing sustainability, their implications in the financial world, how they relate to banks, and the future innovations coming in a few years.

2 FinTech: Theoretical Framework

FinTech abbreviates "financial technology," which, as its name indicates, involves financial services and information technologies such as computer programs [47]. They offer innovative financial services and boost developments such as payment, wealth management, or trading [13, 24, 36]. Under the framework of the company's theory and the theory of strategy, it is consistent to analyze the industrial organization [49] the vision based on resources, organizational identity, and the transaction cost in a disaggregated way approach [39]. In the same way, the theories of transaction cost theory, resource dependency, organizational learning, strategic positioning, and institutional theory are included [37, 52]. By way of understanding, we can also find the integration with ideas of regional collaboration in the field of innovations, the implementation of clusters, the fundamental cooperation theory in the realistic theory, and the idealistic theory, which contrast individual interests and proven greater security.

In the same way, the globalist theory supports that the assimilation of different concepts and cultures is needed to create new processes and innovations in the industry, which leads to understanding the behavior of companies in the search for more sustainable solutions, capable of being replicated and adapted to improve improvements [43]. Fintech brings a new paradigm in which information technology is creating innovation in the financial industry [41]. Fintech is fast becoming a worldwide global phenomenon, led by innovators, researchers, and policy-makers [3, 45]. Currently, FinTech is seen as possessing the potential to provide the poor access to financial programs and support them to escape from poverty [40].

Three main characteristics of FinTech are recognized: technological innovation, process disruption, and service transformation [11, 25].

3 Types of FinTech

FinTech companies can be categorized into many different types [20]. In the following lines, some of the most representative categories are described.

3.1 Payments (PayTech)

First, it is possible to mention PayTech related to payments. It is a business model providing new and innovative payment solutions, such as mobile payment systems, e-wallets, billing, domestic transfers, and others [28, 29]. The payment industry is innovating thanks to financial technologies. Technology allows consumers to be better connected and protect their identity and money. As an example of PayTech companies, PayPal, WePay, Square, among others, can be highlighted [8].

3.2 Lending (LendTech)

FinTech companies sought the simplification of loans mainly strengthened by this digital medium quickly and efficiently; financial technology companies offer loans online, with approval in a matter of minutes, due to the automation that maintains the system and the quality of the digital infrastructure. A strategy is implemented to maintain people's privacy with software, and the borrower's solvency is calculated. Both the registration and the process are automated, gaining pace and time from traditional banks [8]. LendTech companies allow people to acquire goods reduce financial costs, personal credit, credit payroll, and working capital credit. It generates individual firms and start-ups to acquire financing [26]. A fascinating fact for this research is that it is estimated that there are currently 491 digital loan companies (Lendingtech companies) in the Latin American region [20]. As examples, we can mention start-ups that provide crowdfunding, crowdlending, microcredit, and factoring solutions [29].

3.3 Trading (TradeTech)

TradeTech is used to facilitate and optimize international trade by implementing new technologies in, for example, platforms for the distribution of assets, merchandise, and financing of the supply chain [8]. Naturally, this innovation allows companies in the industry to minimize costs and maximize profits in the context of international trade. As TradeTech companies' examples, it is possible to mention, for instance, Bitcointoyou, Jurus, Renda Fixa, among many others.

3.4 Insurance (InsurTech)

InsurTech refers to an emerging sector that applies technology to processes and products of the insurance activity. It is an abbreviation for Insurance Technology and is

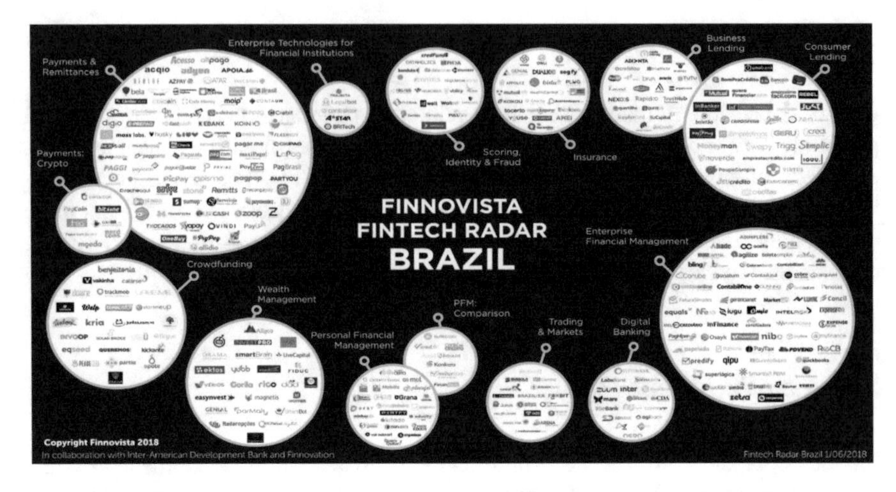

Fig. 1 Examples of Brazilian companies per type of FinTech. *Source* Finnovista [22]

defined as applying technological innovations in the insurance sector [20]. InsurTech is innovating insurance products and services offering outstanding benefits such as online marketplaces, more convenient and personalized approaches, customized profiting, and others [8].

3.5 Personal Finance (WealthTech)

WealthTech refers to technological and digital solutions to improve wealth management, administration, and investment [20]. WealthTech allows better management by those who invest their money. To do this, micro-investment platforms, automatic retirement, investment card management, among many others, are used [8] (Fig. 1).

4 FinTech as a Green Entrepreneurship

FinTech start-ups or financial technology start-ups revolutionize how customers experience financial services [42]. They offer innovative financial services and boost developments such as payment, wealth management, or trading [13, 36]. As emerging companies, the potential to develop processes that highlight the environmental factor is possible thanks to the infrastructure generated in creating new forms of innovation. Currently, there is evidence of an increase in the number of FinTech companies applying methodologies more in line with sustainable development objectives; while there are more cases of companies in line, more sustainable culture is generated.

The intense technological load that the software in charge of processing and securing each transaction tends to maintain is usually accompanied by increasingly practical routine processes; this means that so many viable ways of keeping track of processes remain less and less in such cases. As a result of the process, companies are reinventing themselves repeatedly to achieve a more environmentally conscious policy; with this in mind, many of them are incorporating highly rigorous process optimization processes; high productivity aligned to the low consumption of resources has become the key to have solidity in the market. These changes are mainly accompanied by the culture and relationship with other companies in the sector [7]. Many companies were born with that mentality of creating a solution capable of integrating many processes and ensuring users' confidence in the use of applications and Web platforms; it was a task without a doubt challenging for any company that wanted to start in this area in its beginnings. However, it was not until it was established that the need for people for faster processes from anywhere in the world set its sights on these companies capable of providing such improvement [1].

FinTech is being used in different sectors, such as agriculture [5, 30, 32, 33].

5 Evolution of FinTech in the Global Commerce

Sustainable finance and FinTech are issues of great political importance for most European governments. The European Commission evidence that the latter and some European Union countries have promoted various initiatives and encouraged complex research on sustainability and FinTech [7].

After forming regional clusters in the technological field, more alliances were developed by the entrepreneurs and the most significant financial entities. In this way, all collaborate to create new technological pieces capable of supporting and managing the stock market in an integrated manner, from the view to the face of the consumer. With an approximate sample of 100 French companies, it was shown that many of them are geographically grouped, making it difficult for FinTech companies to develop outside these spaces due to the sheer size of the associated groups and incubators. This integration reduces the risk of failures providing more safety to get involved; in the same way, larger clusters become effective to attract newer FinTech startups as the same incubators for the same reason. The alliances formed due to the growth and the good results that the cluster members demonstrated that with the right trust, sustainable growth could be achieved in collaboration. In the same way, the fact of collaborating more quickly integrated the organizational culture of each company that shared both knowledge and infrastructure to optimize processes in international trade [23].

Policies were developed following the environmental program of 2030 sustainable objectives; in that way, many companies created sustainable departments in their system to manage good practices within the workers, the mindset change in all the companies that integrated this department into their organizations, changing the usual behavior of the workers, change all the process more environmentally [23].

More regulation is needed to consider different sectors, global regulation, and new technologies [2, 9, 12, 18, 30, 35, 38, 51, 53–57].

6 FinTech in the COVID-19 Context

Due to the pandemic, a series of mechanisms were established to facilitate transactions online, which has become a standard today in the transmission of payment invoices and loans due to the growing demand for digital financial services worldwide [7] (Figs. 2 and 3).

During the pandemic, the digitalization of companies meant a significant change in the paradigm of people; there was confusion about what would happen in the

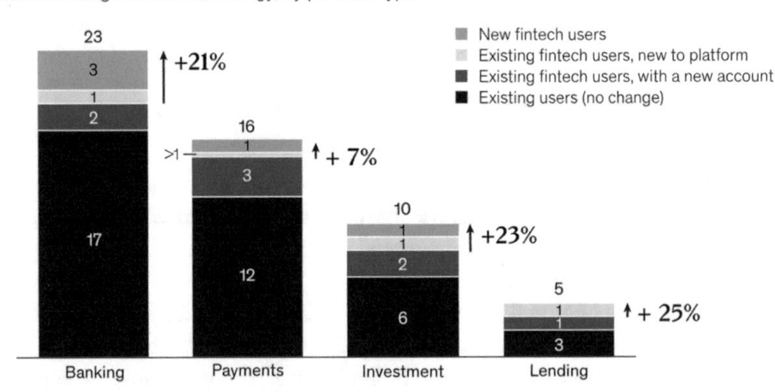

Fig. 2 Percentage of people who started using or opened a new account with each of the following types of financial technology companies since the start of COVID-19. *Source* Mckinsey and Company [44]

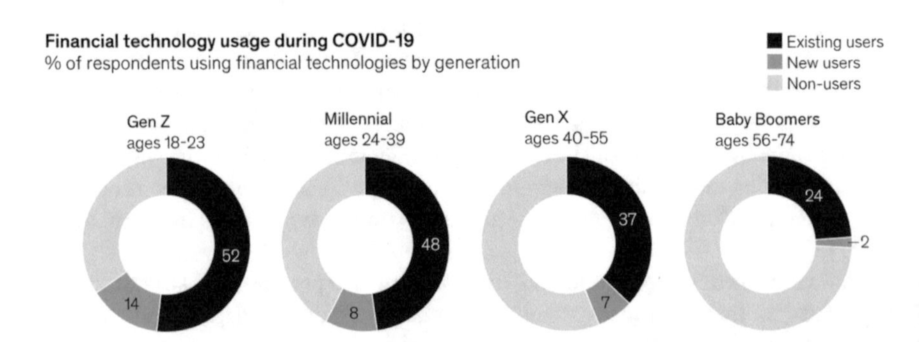

Fig. 3 Percentage of people using FinTech by generation (Gen Z, Millennial, Gen X, Baby Boomers). *Source* Mckinsey and Company [44]

coming times in the financial field that is where the companies that offered financial services managed to position themselves and find the best solution in which the banks take care of other matters. From then on, platforms capable of assisting in processes were developed, sometimes faster than the bank itself; due to this, the eyes of the world were placed on integrating these independent services into the banking system, providing them with more significant infrastructure and knowledge to develop further, more within the industry, which led to the competition in the fight to be the FinTech allied to a significant movement [31].

Below is a series of graphs developed by McKinsey & Company to illustrate how the FinTech sector has developed in the context of the COVID-19 pandemic (Figs. 4, 5 and 6).

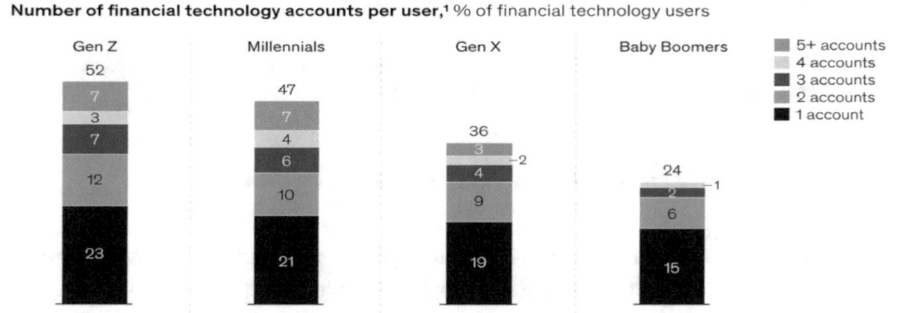

Fig. 4 Percentage of people generation according to the number of accounts per user. *Source* Mckinsey and Company [44]

COVID-19 crisis impact on current likelihood to use financial technology[1]
Percent of respondents "somewhat more likely" or "significantly more likely" by user type and generation

	Fintech users that did not open a new account since crisis	Fintech users that opened a new additional account since crisis	Fintech users who opened their first fintech account since the crisis	Non-user	Total
Gen Z	43	60	50	24	41
Millennial	26	59	47	15	29
Gen X	20	55	37	8	17
Baby Boomer	9	39	29	4	6
Total	19	57	41	7	16

Fig. 5 COVID-19 crisis impact by generation. *Source* Mckinsey and Company [44]

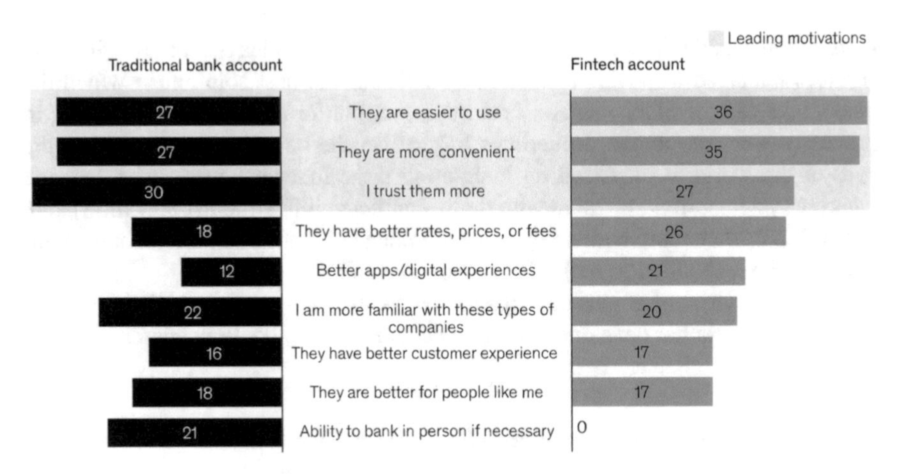

Fig. 6 Percentages of people opinions regarding a comparison between traditional bank account and FinTech account. *Source* Mckinsey and Company [44]

7 FinTech as a Solution on Actual Bank System

Since financial products and services were digitized, FinTech meant, among many other things, the beginning of the technological development of services; the understanding of such a changing market led to the need for a more specialized product that could meet the needs of the market over time. The change of time moved FinTech away from being a trend to become a standard; there are several cases of companies seeking the joint improvement of others in the undertaking of technological banking solutions; among them, we have TransUnion, which developed projects of innovation that facilitated process integration and global transfer. In this journey to FinTech, the client managed to take control of agile platforms, advanced data protection systems, and sophisticated user experience [48].

In this sense, we see an important correlation between the risk that the FinTech manages and the risk of the bank's operation. As we can see, the operators over time have adopted an "agency model" where the risk of the loan does not fall on the bank, in the company itself, something that differs a lot from the classic model of banks with a "wholesale model," where banks obtain income by buying funds from lenders and later reselling them to borrowers, keeping the credit risk in the accounting. This difference is also reflected in the income structure, which is likewise on the prices/interest rates paid to lenders, unlike digital platforms that instead earn money with fees, which reveals certain gaps in the scalability of these ventures and in the same way the obtaining of quality borrowers that can maintain the movements [50].

8 Future Research

It is vital to develop more effective ways to integrate the current practices of the financial industry with the development of strategies and sustainable innovations by the sources, which lead to increased care for the environment, and in the same way, the confidence and well-being of clients. It was seen that technological development had increased exponentially throughout these years, showing truly extraordinary innovations compared to what is expected by the industry. Being at the forefront of this technology allows these enterprises to have greater productivity and efficiency in their labor and further streamline the work of many other industries.

It is seen how these advancements, sometimes no matter how minimal they are, represent a considerable change in the experience and security of the buyers who look with good eyes that this upgrade is developed. In production, they streamline the processes and labor to achieve it as time goes by. In the same way, various tests may appear to create a more integrated platform; modern infrastructure research between companies may become the future that strengthens clusters, as two or more parties are integrated into a project that provides them with greater efficiency. The relationship of trust between the customer and the company is based on information, platform efficiency, active participation, and optimization of resources, which increases satisfaction in the medium term.

Undoubtedly, there is a great opportunity to generate more research on FinTech so that all its benefits can be known in detail, improve mechanisms, generate regulations, and evaluate its global expansion to benefit different sectors [40].

Recent suggested lectures about green approach

- Waste reduction and carbon footprint [27]
- Material selection for circularity and footprints [46]
- 3D print, circularity, and footprints [16]
- Virtual tourism [16]
- Leadership for sustainability in crisis time [4]
- Virtual education and circularity [14]
- Circular economy for packaging [10]
- Students oriented to circular learning [17]
- Food and circular economy [15]
- Water footprint and food supply chain management [34]
- Waste footprint [6]
- Measuring circular economy [19]
- Carbon footprint [21]

9 Closing Remarks

The relationship between FinTech and banks grows over time; as we have seen from a value perspective, cooperation between both entities only brings good results. That said, and based on practice, it is a reality that today, the specialization that most FinTech is developing brings essential innovations to the financial industry at levels not seen to date. In more than one sense, we see how day by day new infrastructures, new systems, new dynamic platforms, and applications are being developed that provide consumers with a unique experience in their transactions and transfers, in a way that has never been experienced before, and that in the future undoubtedly be optimized. Considering the system as it is today, we see that banks that reject and compete with FinTech are integrating more and more into their processes, either directly as part of their transactions or indirectly allowing access to part of its services for the good of the consumer. In some, there is even together advice on the creation and experimentation of new financial products or revealed tools in operations management.

References

1. Agarwal S, Chua YH (2020) FinTech and household finance: a review of the empirical literature. Chin Finance Rev Int. https://doi.org/10.1108/CFRI-03-2020-0024
2. Alam N, Gupta L, Zameni A (2019) Fintech Regulation. In: Alam N, Gupta L, Zameni A (eds) Fintech and Islamic finance: digitalization, development and disruption. Springer International Publishing, pp 137–158. https://doi.org/10.1007/978-3-030-24666-2_8
3. Alt R, Beck R, Smits MT (2018) FinTech and the transformation of the financial industry. Electron Mark 28(3):235–243. https://doi.org/10.1007/s12525-018-0310-9
4. Alvarez-Risco A, Del-Aguila-Arcentales S, Villalobos-Alvarez D, Diaz-Risco S (2022) Leadership for sustainability in crisis time. In: Alvarez-Risco A, Muthu SS, Del-Aguila-Arcentales S (eds) Circular economy: impact on carbon and water footprint. Springer Singapore, pp 41–64. https://doi.org/10.1007/978-981-19-0549-0_3
5. Anshari M, Almunawar MN, Masri M, Hamdan M (2019) Digital marketplace and FinTech to support agriculture sustainability. Energy Procedia 156:234–238. https://doi.org/10.1016/j.egypro.2018.11.134
6. Arias-Meza M, Alvarez-Risco A, Cuya-Velásquez BB, de las Mercedes Anderson-Seminario M, Del-Aguila-Arcentales S (2022) Fashion and textile circularity and waste footprint. In: Alvarez-Risco A, Muthu SS, Del-Aguila-Arcentales S (eds) Circular economy: impact on carbon and water footprint. Springer Singapore, pp 181–204. https://doi.org/10.1007/978-981-19-0549-0_9
7. Arner DW, Buckley RP, Zetzsche DA, Robin V (2020) Sustainability, FinTech and financial inclusion. Eur Bus Organ Law Rev 21(1):7–35
8. Bada A (2019) What are the different types of Fintech? Coinspeaker. Retrieved 01/01/2022 from https://www.coinspeaker.com/guides/different-types-of-fintech
9. Brown E, Piroska D (2022) Governing Fintech and Fintech as governance: the regulatory sandbox, risk washing, and disruptive social classification. New Political Economy 27(1):19–32. https://doi.org/10.1080/13563467.2021.1910645
10. Castillo-Benancio S, Alvarez-Risco A, Esquerre-Botton S, Leclercq-Machado L, Calle-Nole M, Morales-Ríos F, … Del-Aguila-Arcentales S (2022) Circular economy for packaging and

carbon footprint. In Alvarez-Risco A, Muthu SS, Del-Aguila-Arcentales S (eds) Circular economy: impact on carbon and water footprint. Springer Singapore, pp 115–138. https://doi.org/10.1007/978-981-19-0549-0_6
11. Chen MA, Wu Q, Yang B (2019) How Valuable Is FinTech Innovation? Rev Financ Stud 32(5):2062–2106. https://doi.org/10.1093/rfs/hhy130
12. Chiu IHY (2017) A new era in Fintech payment innovations? A perspective from the institutions and regulation of payment systems. Law Innov Technol 9(2):190–234. https://doi.org/10.1080/17579961.2017.1377912
13. Chuen DLK, Teo EG (2015) Emergence of FinTech and the LASIC principles. J Finan Perspect 3(3):24–36
14. Contreras-Taica A, Alvarez-Risco A, Arias-Meza M, Campos-Dávalos N, Calle-Nole M, Almanza-Cruz C, … Del-Aguila-Arcentales S (2022) Virtual education: carbon footprint and circularity. In: Alvarez-Risco A, Muthu SS, Del-Aguila-Arcentales S (eds) Circular economy: impact on carbon and water footprint. Springer Singapore, pp 265–285. https://doi.org/10.1007/978-981-19-0549-0_13
15. Cuya-Velásquez BB, Alvarez-Risco A, Gomez-Prado R, Juarez-Rojas L, Contreras-Taica A, Ortiz-Guerra A, … Del-Aguila-Arcentales S (2022) Circular economy for food loss reduction and water footprint. In: Alvarez-Risco A, Muthu SS, Del-Aguila-Arcentales S (eds) Circular economy: impact on carbon and water footprint. Springer Singapore, pp 65–91. https://doi.org/10.1007/978-981-19-0549-0_4
16. De-la-Cruz-Diaz M, Alvarez-Risco A, Jaramillo-Arévalo M, de las Mercedes Anderson-Seminario M, Del-Aguila-Arcentales S (2022) 3D print, circularity, and footprints. In: Alvarez-Risco A, Muthu SS, Del-Aguila-Arcentales S (eds) Circular economy: impact on carbon and water footprint. Springer Singapore, pp 93–112. https://doi.org/10.1007/978-981-19-0549-0_5
17. de las Mercedes Anderson-Seminario M, Alvarez-Risco A (2022) Better students, better companies, better life: circular learning. In: Alvarez-Risco A, Muthu SS, Del-Aguila-Arcentales S (eds) Circular economy: impact on carbon and water footprint. Springer Singapore, pp 19–40. https://doi.org/10.1007/978-981-19-0549-0_2
18. Degerli K (2019) Regulatory challenges and solutions for Fintech in Turkey. Procedia Comput Sci 158:929–937. https://doi.org/10.1016/j.procs.2019.09.133
19. Del-Aguila-Arcentales S, Alvarez-Risco A, Muthu SS (2022) Measuring circular economy. In: Alvarez-Risco A, Muthu SS, Del-Aguila-Arcentales S (eds) Circular economy: impact on carbon and water footprint. Springer Singapore, (pp 3–17. https://doi.org/10.1007/978-981-19-0549-0_1
20. Domínguez C (2021) Surgen nuevos modelos Fintech: Avanzan fuerte préstamos digitales o Lendingthec. Existen 491 empresas en AL que dan soluciones crediticias. Retrieved 01/01/2022 from http://fresno.ulima.edu.pe/ss_bd00102.nsf/RecursoReferido?OpenForm&id=PROQUEST-41716https://www.proquest.com/newspapers/surgen-nuevos-modelos-fintech/docview/2573327639/se-2?accountid=45277
21. Esquerre-Botton S, Alvarez-Risco A, Leclercq-Machado L, de las Mercedes Anderson-Seminario M, Del-Aguila-Arcentales S (2022) Food loss reduction and carbon footprint practices worldwide: a benchmarking approach of circular economy. In: Alvarez-Risco A, Muthu SS, Del-Aguila-Arcentales S (eds) Circular economy: impact on carbon and water footprint. Springer Singapore, pp 161–179. https://doi.org/10.1007/978-981-19-0549-0_8
22. Finnovista (2018) Brazil regains Fintech leadership in Latin America and surpasses the 370 startup barrier [Brasil recupera el liderazgo Fintech en América Latina y supera la barrera de las 370 startups]. Retrieved 01/01/2022 from https://www.finnovista.com/radar/brasil-recupera-el-liderazgo-fintech-en-america-latina-y-supera-la-barrera-de-las-370-startups
23. Gazel M, Schwienbacher A (2021) Entrepreneurial fintech clusters. Small Bus Econ 57(2):883–903
24. Goldstein I, Jiang W, Karolyi GA (2019) To FinTech and beyond. Rev Finan Stud 32(5):1647–1661. https://doi.org/10.1093/rfs/hhz025
25. Gomber P, Kauffman RJ, Parker C, Weber BW (2018) On the Fintech revolution: interpreting the forces of innovation, disruption, and transformation in financial services. J Manage Inf Syst 35(1):220–265. https://doi.org/10.1080/07421222.2018.1440766

26. Gomber P, Koch J-A, Siering M (2017) Digital finance and FinTech: current research and future research directions. J Bus Econ 87(5):537–580. https://doi.org/10.1007/s11573-017-0852-x
27. Gómez-Prado R, Alvarez-Risco A, Sánchez-Palomino J, de las Mercedes Anderson-Seminario M, Del-Aguila-Arcentales S (2022) Circular economy for waste reduction and carbon footprint. In: Alvarez-Risco A, Muthu SS, Del-Aguila-Arcentales S (eds) Circular Economy: Impact on Carbon and Water Footprint. Springer Singapore, pp 139–159. https://doi.org/10.1007/978-981-19-0549-0_7
28. Gromek M (2018) Clarifying the blurry lines of FinTech: Opening the Pandora's box of FinTech categorization 1. In: The rise and development of FinTech. Routledge, pp 168–189
29. Haddad C, Hornuf L (2019) The emergence of the global fintech market: economic and technological determinants. Small Bus Econ 53(1):81–105. https://doi.org/10.1007/s11187-018-9991-x
30. Hinson R, Lensink R, Mueller A (2019) Transforming agribusiness in developing countries: SDGs and the role of FinTech. Curr Opin Environ Sustain 41:1–9. https://doi.org/10.1016/j.cosust.2019.07.002
31. Hornuf L, Klus MF, Lohwasser TS, Schwienbacher A (2021) How do banks interact with fintech startups? Small Bus Econ 57(3):1505–1526
32. Hudaefi FA (2020) How does Islamic fintech promote the SDGs? Qualitative evidence from Indonesia. Qual Res Finan Markets 12(4):353–366. https://doi.org/10.1108/QRFM-05-2019-0058
33. Jiang S, Qiu S, Zhou H, Chen M (2019) Can FinTech development curb agricultural nonpoint source pollution? Int J Environ Res Public Health 16(22). https://doi.org/10.3390/ijerph16224340
34. Juarez-Rojas L, Alvarez-Risco A, Campos-Dávalos N, de las Mercedes Anderson-Seminario M, Del-Aguila-Arcentales S (2022) Water footprint in the textile and food supply chain management: trends to become circular and sustainable. In: Alvarez-Risco A, Muthu SS, Del-Aguila-Arcentales S. (eds) Circular economy: impact on carbon and water footprint. Springer Singapore, pp 225–243. https://doi.org/10.1007/978-981-19-0549-0_11
35. Khiaonarong T, Goh T (2020) FinTech and payments regulation: an analytical framework. J Payments Strategy Syst 14(2):157–171. https://www.ingentaconnect.com/content/hsp/jpss/2020/00000014/00000002/art00008
36. Kim Y, Choi J, Park Y-J, Yeon J (2016) The adoption of mobile payment services for "Fintech." Int J Appl Eng Res 11(2):1058–1061
37. Klus MF, Lohwasser TS, Holotiuk F, Moormann J (2019) Strategic alliances between banks and Fintechs for digital innovation: Motives to collaborate and types of interaction. J Entrepreneurial Financ 21(1):1
38. Knewtson HS, Rosenbaum ZA (2020) Toward understanding FinTech and its industry. Manage Financ 46(8):1043–1060. https://doi.org/10.1108/MF-01-2020-0024
39. Kohtamäki M, Parida V, Oghazi P, Gebauer H, Baines T (2019) Digital servitization business models in ecosystems: a theory of the firm. J Bus Res 104:380–392. https://doi.org/10.1016/j.jbusres.2019.06.027
40. Lagna A, Ravishankar MN (2022) Making the world a better place with Fintech research. Inf Syst J 32(1):61–102. https://doi.org/10.1111/isj.12333
41. Lee I, Shin YJ (2018) Fintech: ecosystem, business models, investment decisions, and challenges. Bus Horiz 61(1):35–46. https://doi.org/10.1016/j.bushor.2017.09.003
42. Mackenzie A (2015) The fintech revolution. Lond Bus Sch Rev 26(3):50–53. https://doi.org/10.1111/2057-1615.12059
43. Martinez APC (2007) The Theoretical Approach to International Cooperation in Education: Analysis of three International Cooperation Agencies: IMEXCI, Mexico; AECI, Spain and USAID, United States [El Enfoque Teorico de la Cooperacion Internacional en Educacion: Analisis de tres agencias de Cooperacion Internacional: IMEXCI, Mexico; AECI, Espana y USAID, Estados Unidos]. Retrieved 01/01/2022 from http://catarina.udlap.mx/u_dl_a/tales/documentos/lri/cid_m_ap/resumen.html

44. Mckinsey and Company (2020) How US customers' attitudes to Fintech are shifting during the pandemic. Retrieved 01/01/2022 from https://www.mckinsey.com/industries/financial-services/our-insights/how-us-customers-attitudes-to-fintech-are-shifting-during-the-pandemic
45. Mention A-L (2019) The future of Fintech. Res Technol Manage 62(4):59–63. https://doi.org/10.1080/08956308.2019.1613123
46. Morales-Ríos F, Alvarez-Risco A, Castillo-Benancio S, de las Mercedes Anderson-Seminario M, Del-Aguila-Arcentales S (2022) Material selection for circularity and footprints. In: Alvarez-Risco A, Muthu SS, Del-Aguila-Arcentales S (eds) Circular economy: impact on carbon and water footprint . Springer Singapore, pp 205–221. https://doi.org/10.1007/978-981-19-0549-0_10
47. Oxford English Dictionary. (n.d.). Definition of fintech. Retrieved 01/01/2022 from OxfordLearnersDictionaries.com
48. Portafolio (2019) Fintech is not a fad, but the new norm [Las fintech no son moda, sino la nueva norma]. Retrieved 01/01/2022 from https://www.portafolio.co/economia/finanzas/las-fintech-no-son-moda-sino-la-nueva-norma-534557
49. Porter M (1997) Competitive strategy, measuring business excellence
50. Pozzolo AF (2017) Fintech and banking. Friends or foes. European Economy–Banks, Regulation, and the Real Sector, Year, 3
51. Rupeika-Apoga R, Wendt S (2021) FinTech in Latvia: status quo, current developments, and challenges ahead. Risks 9(10). https://doi.org/10.3390/risks9100181
52. Thakor AV (2020) Fintech and banking: What do we know? J Financ Intermediation 41:100833. https://doi.org/10.1016/j.jfi.2019.100833
53. Tsai C-H, Peng K-J (2017) The FinTech revolution and financial regulation: the case of online supply-chain financing. Asian J Law Soc 4(1):109–132. https://doi.org/10.1017/als.2016.65
54. Wójcik D (2020) Financial geography II: the impacts of FinTech—financial sector and centres, regulation and stability, inclusion and governance. Prog Hum Geogr 45(4):878–889. https://doi.org/10.1177/0309132520959825
55. Xu Y, Bao H, Zhang W, Zhang S (2021) Which financial earmarking policy is more effective in promoting FinTech innovation and regulation? Ind Manage Data Syst 121(10):2181–2206. https://doi.org/10.1108/IMDS-11-2020-0656
56. You C (2018) Recent development of FinTech regulation in china: a focus on the new regulatory regime for the P2P Lending (loan-based crowdfunding) market. Capital Markets Law J 13(1):85–115. https://doi.org/10.1093/cmlj/kmx039
57. Zhou X, Chen S (2021) FinTech innovation regulation based on reputation theory with the participation of new media. Pac Basin Financ J 67:101565. https://doi.org/10.1016/j.pacfin.2021.101565

Intention of Green Entrepreneurship Among University Students in Colombia

Paula Viviana Robayo-Acuña, Gabriel-Mauricio Martinez-Toro, Aldo Alvarez-Risco, Sabina Mlodzianowska, Shyla Del-Aguila-Arcentales, and Mercedes Rojas-Osorio

Abstract This chapter shows the outcomes of field research that measured the effect of institutional support for developing entrepreneurship (ISDE) and country support for entrepreneurship (CSE) through entrepreneurial self-efficacy (ESE) on green entrepreneurial intention (GEI). 389 business students in Colombia completed the questionnaire using an online platform. Twenty-four questions measured the GEI. The study used SEM-PLS technical analysis. It was found that ISDE (0.224) and CSE (0.161) had a positive and significant effect on ESE, and ESE had a positive effect (0.705) on GEI. The model explained 49.7% of GEI. The information from the current study can be used to create new plans and strategies by university authorities.

Keywords Green entrepreneurship · Government policies · Sustainability · Sustainable development goals · Environment · Entrepreneurship · Students · Circularity · Entrepreneurs · Colombia

1 Introduction

Vocational education at university is composed of different contents and activities for learning. For this reason, knowing the factors behind certain interests is crucial for the planning of teaching since theoretical components must be combined with practical activities through business actors. The promotion of entrepreneurship by a country is associated with its level of competitiveness and is based on the training

P. V. Robayo-Acuña
Fundación Universitaria Konrad Lorenz, Bogota, Colombia

G.-M. Martinez-Toro
Universidad Autónoma de Bucaramanga, Bucaramanga, Colombia

A. Alvarez-Risco · S. Mlodzianowska
Universidad de Lima, Lima, Peru

S. Del-Aguila-Arcentales (✉) · M. Rojas-Osorio
Escuela de Posgrado, Universidad San Ignacio de Loyola, Lima, Peru
e-mail: sdelaguila@usil.edu.pe

A. Alvarez-Risco et al. (eds.), *Footprint and Entrepreneurship*, Environmental Footprints and Eco-design of Products and Processes,
https://doi.org/10.1007/978-981-19-8895-0_12

of new professionals, so it is critical that universities can play an active role in the generation of entrepreneurship-oriented professionals.

Green entrepreneurship refers to the practice of starting companies that focus on sustainability and social responsibility. Students who engage in green entrepreneurship may do so for personal reasons, such as wanting to make money or gain experience, or they may want to help others. In addition to these motivations, some students choose to pursue green entrepreneurship because they believe it is the right thing to do. Green entrepreneurship is not only a way to make money, but also a way to contribute to society. There are three ways that universities can support green entrepreneurship: (1) providing information about opportunities; (2) providing access to capital; and (3) offering mentorship. Universities should provide information about opportunities for students to participate in green entrepreneurship. Information includes details about the types of business models that exist, what type of funding is available, and how to apply for grants. Providing information about opportunities helps students decide if green entrepreneurship is something they would like to pursue. Therefore, the objective of this study is to measure variables that can explain the intention to implement a green venture. The variables evaluated are conceptual support, such as those developed in universities through promotion and training programs with complementary courses. Likewise, the support given by the country to generate ventures is evaluated, which is explained by the regulation to promote the creation and development of different types of ventures. This support is a central element that has been evaluated, since this implies that the contents of the courses include tools for the development of enterprises, which implies management, marketing, logistics, and local and international regulation. The mediating variable that is part of the study is entrepreneurial self-efficacy, i.e., how convinced the student is of being able to launch and grow a venture. Finally, the student's intention with respect to the generation and development of green businesses is measured.

2 Previous Studies

Table 1 shows relevant studies carried out in different regions on entrepreneurship. Table 2 shows important research developed about green entrepreneurship. Previous evidence is very valuable because it allows us to have the progress of each region or even in each country so that we can have a complete idea of the level of progress, the actions that are being developed mainly in the universities and the strategies that must be implemented to achieve greater implementation.

Table 1 Entrepreneurial education in universities

Authors	Country
Europe	
Frazier [21]	Poland
Johannisson [25]	Sweden
van der Sijde and van Alsté [60]	Netherlands
Tamkivi [59]	Estonia
Watkins and Stone [62]	UK
Raposo et al. [49]	Portugal
Asia	
Huu-Phuong and Soo-Jiuan [23]	Singapore
Menning [36]	India
Suzuki et al. [58]	Japan
Mok [41]	Hong Kong
Yu Cheng et al. [65]	Malaysia
Millman et al. [39]	China
Oceania	
Mitra [40]	Australia
Lee-Ross [29]	Australia
Maritz et al. [32]	Australia
Maritz et al. [33]	Australia
Luong and Lee [30]	New Zealand
Hardie et al. [22]	New Zealand
Africa	
Jesselyn Co and Mitchell [24]	South Africa
Owusu-Mintah [46]	Ghana
Olokundun et al. [45]	Nigeria
North America	
Chrisman et al. [13]	Canada
Dill [15]	United States
Cruz-Sandoval et al. [14]	Mexico
South America	
Bernasconi [6]	Chile
Postigo and Tamborini [47]	Argentina
Chafloque-Cespedes et al. [11]	Peru

Table 2 Green entrepreneurial education in universities

Authors	Country
Silajdžić et al. [55]	Bosnia and Herzegovina
Yan et al. [64]	China
Lynskey [31]	Japan
Bonnet et al. [8]	Netherlands
Soomro et al. [56]	Pakistan
Radović-Marković and Živanović [48]	Serbia
Alvarez-Risco et al. [3]	Peru

3 Theoretical Framework

3.1 Theory of Planned Behavior (TPB)

The theoretical basis for this research is Ajzen's TPB [1] is characterized by evaluating the factors that converge in people's final behaviors, with intention as the predictor variable of behavior. TPB infers that the intention of a behavior has a very close link with behavior; it is also mentioned that said intention is associated with two basic aspects: social influence and the nature of the person. The social factor is the perception of the social pressure of everyone to perform a specific action; this factor is called "subjective norm", which is related to the beliefs of significant people for the individual. Thus, when he/she feels that for other people the action is important then he/she will feel that he/she must do it. The personal factor is the evaluation about performing a behavior; this is called "attitude", which is linked to the consequences that an individual can expect from performing the specific behavior.

3.2 Social Cognitive Theory (SCT)

The SCT developed by Bandura [4] emphasizes that people's behavior is under their own control, and in this way, they can increase their self-efficacy. In SCT, a person has a self-system that allows them to evaluate control over their feelings and performance. This self-system is used as reference, regulation, and evaluation of behaviors. Thus, this serves as a self-regulatory function to convert individuals with the capacity to influence their own cognitive processes and actions and thus alter their environment. The model of the current study has been developed by the authors to evaluate how different types of support existing in the educational ecosystem, and at the governmental level, have an effect on students' entrepreneurial self-efficacy to increase green entrepreneurship intention. The combination of these factors is based on the planning that can be generated in universities to achieve an entrepreneurial ecosystem, involving students and professors, based on field results, with a focus on business, social and environmental sustainability.

4 Hypothesis

Green entrepreneurial intention (GEI)

Intention is the state of the individual that creates decisions, attention, and interest to carry out a specific action [7, 37, 51]. Several factors explain how a person plans a behavior [19]. The intention is the previous step of a behavior. The intention to perform a behavior have a positive effect on the development of entrepreneurship [9, 37, 43, 50]. Entrepreneurship can be developed by individuals or companies [38]. GEI is the implementation of innovative activities related to sustainability, offering products or services based in green process [16].

Country support for entrepreneurship (CSE)

This refers to regulation, in macro and micro level, in a country to promote the entrepreneurship [18]. The variable describes what the person thinks related to the country's support and promotion of entrepreneurial business initiatives. The perception of banking systems or social programs that can support the start-up of a business venture is also measured, this implies support in the initial process of the venture, bank loans, exemption from certain taxes, business-business-government linkages, among others. Thus, we have the following hypothesis:

H1. CSE has a positive and significant effect on ESE

Institutional support for entrepreneurship development (ISDE)

This variable involves the activities carried out by the educational institution to provide knowledge and promote competencies on business management, so that students are able to experience the theoretical components through involvement with successful entrepreneurs, start-up roundtables, seeking mainly that through non-curricular activities it is possible to generate the strength to create new businesses [17]. To this end, the following hypothesis is formulated.

H2. ISDE has a positive and significant effect on ESE

Entrepreneurial self-efficacy (ESE)

This variable is a person's belief in his or her ability to perform any activity successfully and incorporate specific behaviors into his or her daily schedule [5, 27]. Entrepreneurial self-efficacy means the confidence of one person to carry out entrepreneurial activities [44, 53]. Research shows the positive influence of entrepreneurial self-efficacy on entrepreneurial intention [12, 28, 34], which can

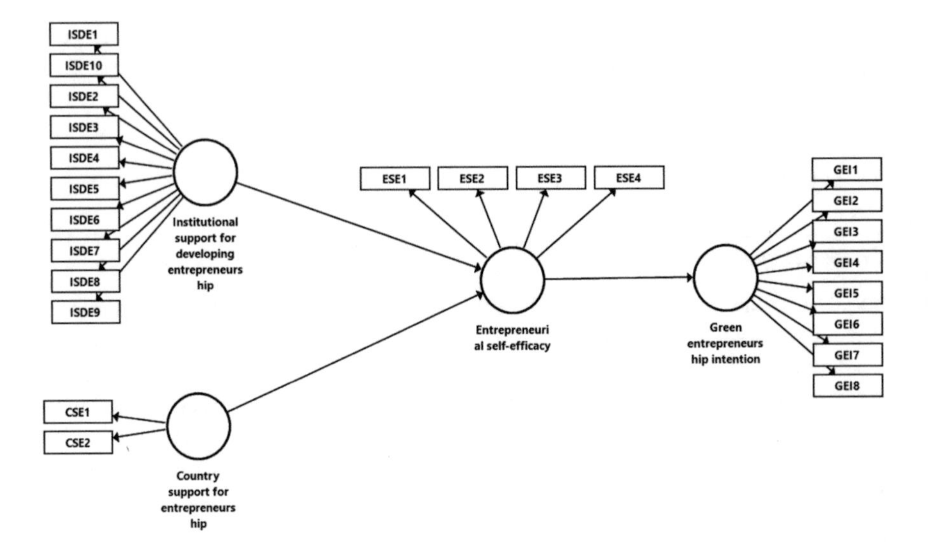

Fig. 1 Research model

be understood to mean that people who have a high level of self-efficacy are more likely to start businesses. Therefore, the following hypothesis is formulated.

H3. ESE has a positive and significant effect on GEI.

5 Model to Test

Figure 1 show the model to test. The proposed relationship between the variables of the study can be evidenced.

6 Methodology

6.1 Sample

The sample is based on university business students in Colombia, older than 18 years. A total of 389 students participated, 176 men (45.24%) and 213 women (54.76%), between 18 and 31 years old (mean = 22.34 years; SD = 5, 65 years).

6.2 Data Collection, Questionnaire and Analysis

To collect the data, it was used an online questionnaire, run online since March 16 to April 30, 2021. The questionnaire collects sociodemographic data and Likert scale data based on previous studies: ISDE [63], CSE (current authors), ESE [57], and GEI [63]. It was used a 5-point Likert-type scale (1 = completely disagree; 5 = completely agree. The analysis was using SmartPLS version 3.3.2.

7 Outcomes

7.1 Reliability

The analysis of internal consistency is showed in Table 3.

7.2 Validation with SEM-PLS

The Table 4 include the internal consistency of dimensions using composite reliability, average variance extracted, and discriminant validity.

7.3 Discriminant Validity Using SEM-PLS

It was calculated by the Fornell-Larcker criterion [20]. The fulfillment of this criterion, demonstrating the discriminant validity (Table 5).

Table 3 Reliability of scales

Variables	Items	Cronbach's alpha	Range of items scores	Composite reliability	Extracted variance
ISDE	9	0.937	0.721–0.850	0.946	0.639
CSE	2	0.933	0.830–0.949	0.885	0.794
ESE	4	0.857	0.803–0.875	0.903	0.699
GEI	9	0.884	0.640–0.911	0.934	0.643

Sample 389 business students

Table 4 Construct validity of the items

Scale–items	Factorial weight
Institutional support for developing entrepreneurship (My university……)	
…. Provides university ventures with financial and/or technical advice	0.810
…. Promotes contact networks between university entrepreneurs and investors	0.810
…. Promotes a favorable environment for the development of entrepreneurship among students and teachers	0.850
…. Promotes social entrepreneurial ideas through contests, fairs or contests	0.819
…. Has support programs for the creation of entrepreneurship (raising seed capital, incubators, etc.)	0.804
…. Supports the start-up of entrepreneurships through senior management and authorities	0.829
…. Offers subjects or courses related to entrepreneurship	0.758
…. Offers practical training in entrepreneurship (realization of projects, business plans, etc.)	0.807
…. Has specialized offices that provide advice for the development of entrepreneurship	0.780
…. Offers free virtual courses that train in entrepreneurship	0.721
Country support for entrepreneurship	
In my country, green entrepreneurs are encouraged by an institutional structure	0.949
The economy of my country offers many opportunities for entrepreneurs	0.830
Entrepreneurial self-efficacy	
Creating and maintaining an ecological enterprise is a task that I can carry out	0.803
I have the necessary knowledge to develop an ecological enterprise	0.810
I have sufficient skills to develop an ecological enterprise	0.875
I believe that in the future I will be able to develop a successful ecological enterprise	0.854
Green entrepreneurial intention	
I plan to develop an enterprise that addresses the ecological problems of my community	0.801
I recommend my colleagues to develop enterprises that solve ecological problems	0.640
My future initiatives will prioritize ecological benefits over financial ones	0.716
If I had the opportunity and the resources, I would definitely go green	0.692
I have seriously thought about becoming a green entrepreneur	0.849
I will do my best to start and run my own green business	0.911
I have the firm intention of starting an ecological enterprise one day	0.876

(continued)

Table 4 (continued)

Scale–items	Factorial weight
I intend to undertake and act in the management of my own ecological enterprise	0.882

Sample 389 business students

Table 5 Discriminant validity of sub-scales

Scales	CSE	ESE	GEI	ISDE
CSE	*0.891*			
ESE	0.254	*0.836*		
GEI	0.244	0.705	*0.802*	
ISDE	0.414	0.291	0.245	0.800

Sample 389 business students

Table 6 Significance of beta coefficients

Scales	Original sample	Mean sample	Standard deviation	t-statistic	P
CSE → ESE	0.161	0.163	0.053	3.036	0.002
ISDE → ESE	0.224	0.229	0.057	3.911	0.000
ESE → GEI	0.705	0.706	0.028	25.093	0.000

Sample 389 business students

7.4 Bootstrapping

The bootstrapping outcomes was based in resampling of 5000 times. Table 6 shows that the values are significant (p values < 0.01).

Figure 2 shows the research model tested. The results confirm that ESDE, CSDE, and CSE through ESE have an effect on GEI in business students.

7.5 Test of Hypothesis

Hypothesis 1 (H1): CSE has a positive and significant effect on ESE

CSE has a positive effect of 0.161 on ESE. The hypothesis was confirmed. Also, CSE and ESDE and CSDE explain 10.6% of ESE.

Hypothesis 2 (H2): ESDE has a positive and significant effect on ESE

ISDE has a positive influence of 0.224 on ESE. The hypothesis was confirmed.

Hypothesis 3 (H3): ESE have a positive and significant effect GEI

ESE has positive and significant of 0.705 on GEI. The hypothesis was confirmed.

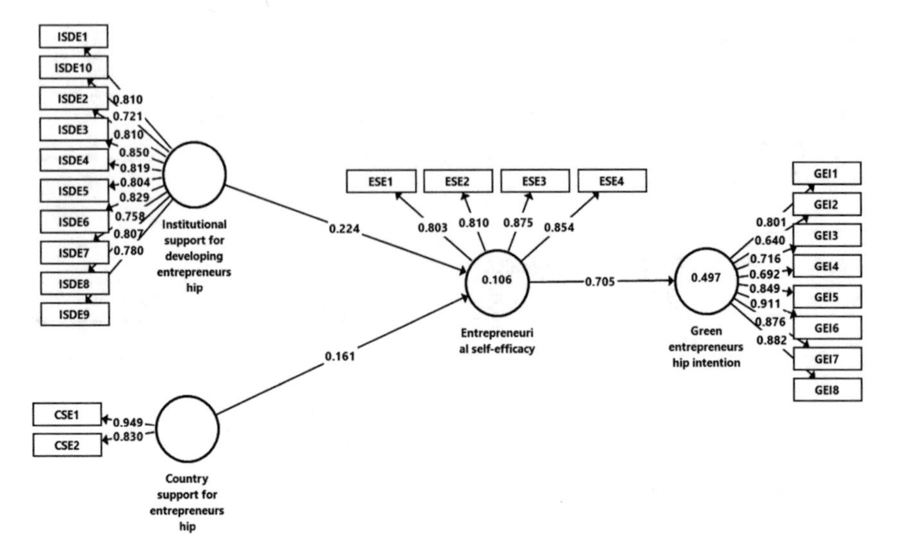

Fig. 2 Research model evaluated

The complete model explains the 49.7% of GEI.

8 Discussion

It has been possible to show that the questionnaire met the reliability criteria through internal consistency (Cronbach's alpha and composite reliability) and discriminant validity was met. Previously, [54] had confirmed by means of a study in 374 Chinese university students what has been corroborated in the present study: the effect of ESDE on ESE. This relationship was also previously verified in 560 university students in Lebanon by [42] and in 376 university students in Indonesia [61].

Through this study it is possible to demonstrate what is perceived from the student group regarding the entrepreneurial ecosystem in universities, that is, to be able to know if various contents are being reflected in the courses for the development of green enterprises. These efforts made by universities have already been reported by other institutions such as University of Melbourne [35] and Amsterdam Business School [52]. The fairs that allow showing the preliminary results of the undertakings, both of the students individually and organized with other students or even teachers, encourage the further development of green undertakings, especially when the issue of sustainability becomes an increasing need for business resilience and investments focused on the Sustainable Development Goals.

When correlational studies are developed, there is the great advantage of being able to show the prognoses of behavior that can occur in each population. Thus, when it is possible to demonstrate that the curricular components and, especially, the non-curricular ones, have a very relevant effect on self-efficacy and finally on GHG, investment plans can be generated to promote these activities that can be carried out independently different modes. Currently, the use of technologies to generate innovation in enterprises is being increasingly experimented with [2, 10, 26].

9 Conclusions

The study allows a contribution in the understanding of the influences of the variables in a university ecosystem that increasingly seeks to promote and carry out entrepreneurship. The positive and significant effects show that the model is valuable, on the one hand to understand reality and generate improvement plans in pursuit of more sustainable enterprises. On the other hand, it should be noted that the model may be used in future studies in other regions, which will serve to characterize these entrepreneurial ecosystems. Through the SEM-PLS analysis technique, it is possible to have a methodological strength; however, there are opportunities for future studies that can also use this methodology but are carried out through experimental design, which help to corroborate the relationships more reliably between variables and can describe reality more closely.

References

1. Ajzen I (1985) From intentions to actions: a theory of planned behavior. In: Kuhl J, Beckmann J (eds) Action control: from cognition to behavior (pp. 11–39). Springer Berlin Heidelberg. https://doi.org/10.1007/978-3-642-69746-3_2
2. Alvarez-Risco A, Del-Aguila-Arcentales S, Rosen MA, Yáñez JA (2022) Social cognitive theory to assess the intention to participate in the Facebook Metaverse by citizens in Peru during the COVID-19 Pandemic. J Open Innov Technol Market Complex 8(3):142. https://www.mdpi.com/2199-8531/8/3/142
3. Alvarez-Risco A, Mlodzianowska S, García-Ibarra V, Rosen MA, Del-Aguila-Arcentales S (2021) Factors affecting green entrepreneurship intentions in business university students in COVID-19 pandemic times: case of Ecuador. Sustainability 13(11):6447. https://www.mdpi.com/2071-1050/13/11/6447
4. Bandura A (1986) Social foundations of thought and action: a social cognitive theory. Prentice-Hall, Inc.
5. Bandura A (1992) Self-efficacy mechanism in psychobiologic functioning. In: Self-efficacy: thought control of action. Hemisphere Publishing Corp., pp 355–394
6. Bernasconi A (2005) University entrepreneurship in a developing country: the case of the P. Universidad Católica de Chile, 1985–2000. High Educ 50(2):247–274. https://doi.org/10.1007/s10734-004-6353-1
7. Bird B (1988) Implementing entrepreneurial ideas: the case for intention. Acad Manage Rev 13(3):442–453. https://doi.org/10.5465/amr.1988.4306970
8. Bonnet H, Quist J, Hoogwater D, Spaans J, Wehrmann C (2006) Teaching sustainable entrepreneurship to engineering students: the case of Delft University of Technology. Eur J Eng Educ 31(2):155–167. https://doi.org/10.1080/03043790600566979
9. Boubker O, Arroud M, Ouajdouni A (2021) Entrepreneurship education versus management students' entrepreneurial intentions. A PLS-SEM approach. Int J Manag Educ 19(1):100450. https://doi.org/10.1016/j.ijme.2020.100450

10. Buhalis D, Lin MS, Leung D (2022) Metaverse as a driver for customer experience and value co-creation: implications for hospitality and tourism management and marketing. Int J Contemp Hospitality Manage, ahead-of-print (ahead-of-print). https://doi.org/10.1108/IJCHM-05-2022-0631
11. Chafloque-Cespedes R, Alvarez-Risco A, Robayo-Acuña P-V, Gamarra-Chavez C-A, Martinez-Toro G-M, Vicente-Ramos W (2021) Effect of socio demographic factors in entrepreneurial orientation and entrepreneurial intention in university students of Latin American business schools. In: Jones P, Apostolopoulos N, Kakouris A, Moon C, Ratten V, Walmsley A (eds) Universities and entrepreneurship: meeting the educational and social challenges, vol 11. Emerald Publishing Limited, pp 151–165. https://doi.org/10.1108/S2040-724620210000011010
12. Chen X, Zhang SX, Jahanshahi AA, Alvarez-Risco A, Dai H, Li J, Ibarra VG (2020) Belief in a COVID-19 conspiracy theory as a predictor of mental health and well-being of health care workers in Ecuador: Cross-sectional survey study [Article]. JMIR Public Health Surveill 6(3), Article e20737. https://doi.org/10.2196/20737
13. Chrisman JJ, Hynes T, Fraser S (1995) Faculty entrepreneurship and economic development: the case of the University of Calgary. J Bus Ventur 10(4):267–281. https://doi.org/10.1016/0883-9026(95)00015-Z
14. Cruz-Sandoval M, Vázquez-Parra JC, Alonso-Galicia PE (2022) Student perception of competencies and skills for social entrepreneurship in complex environments: an approach with Mexican University students. Soc Sci 11(7):314
15. Dill DD (1995) University-industry entrepreneurship: The organization and management of American university technology transfer units. High Educ 29(4):369–384. https://doi.org/10.1007/BF01383958
16. Farinelli F, Bottini M, Akkoyunlu S, Aerni P (2011) Green entrepreneurship: the missing link towards a greener economy. ATDF Journal 8(3/4):42–48
17. Ferreira AdSM, Loiola E, Guedes Gondim SM (2017) Motivations, business planning, and risk management: entrepreneurship among university students. INMR—Innov Manage Rev 14(2):140–150. https://www.revistas.usp.br/rai/article/view/114344
18. Fichter K, Tiemann I (2018) Factors influencing university support for sustainable entrepreneurship: Insights from explorative case studies. J Cleaner Prod 175:512–524. https://doi.org/10.1016/j.jclepro.2017.12.031
19. Fishbein M, Ajzen I (1975) Belief, attitude, intention and behavior: an introduction to theory and research. Addison-Wesley, Reading, MA
20. Fornell C, Larcker DF (1981) Evaluating structural equation models with unobservable variables and measurement error. J Mark Res 18(1):39–50. https://doi.org/10.1177/002224378101800104
21. Frazier JW (1991) A partnership for environmentally-and educationally-based economic development in Poland, vol 14. Department of Geography, The University Center at Binghamton, State ….
22. Hardie B, Lee K, Highfield C (2022) Characteristics of effective entrepreneurship education post-COVID-19 in New Zealand primary and secondary schools: a Delphi study. Entrepreneurship Educ 5(2):199–218. https://doi.org/10.1007/s41959-022-00074-y
23. Huu-Phuong T, Soo-Jiuan T (1990) Export factoring: a strategic alternative for small exporters in Singapore. Int Small Bus J 8(3):49–57. https://doi.org/10.1177/026624269000800304
24. Jesselyn Co M, Mitchell B (2006) Entrepreneurship education in South Africa: a nationwide survey. Education + Training 48(5):348–359. https://doi.org/10.1108/00400910610677054
25. Johannisson B (1991) University training for entrepreneurship: Swedish approaches. Entrepreneurship Reg Dev 3(1):67–82. https://doi.org/10.1080/08985629100000005
26. Kraus S, Kanbach DK, Krysta PM, Steinhoff MM, Tomini N (2022) Facebook and the creation of the metaverse: radical business model innovation or incremental transformation? Int J Entrepreneurial Behav Res 28(9):52–77. https://doi.org/10.1108/IJEBR-12-2021-0984
27. Krueger NF, Reilly MD, Carsrud AL (2000) Competing models of entrepreneurial intentions. J Bus Ventur 15(5):411–432. https://doi.org/10.1016/S0883-9026(98)00033-0

28. Kumar R, Shukla S (2019) Creativity, proactive personality and entrepreneurial intentions: examining the mediating role of entrepreneurial self-efficacy. Glob Bus Revm, 0972150919844395. https://doi.org/10.1177/0972150919844395
29. Lee-Ross D (2017) An examination of the entrepreneurial intent of MBA students in Australia using the entrepreneurial intention questionnaire. J Manage Develop 36(9):1180–1190. https://doi.org/10.1108/JMD-10-2016-0200
30. Luong A, Lee C (2021) The influence of entrepreneurial desires and self-efficacy on the entrepreneurial intentions of New Zealand tourism and hospitality students. J Hospitality Tourism Educ 1–18. https://doi.org/10.1080/10963758.2021.1963751
31. Lynskey MJ (2004) Bioentrepreneurship in Japan: institutional transformation and the growth of bioventures. J Commer Biotechnol 11(1):9–37. https://doi.org/10.1057/palgrave.jcb.3040098
32. Maritz A, Nguyen Q, Bliemel M (2019) Boom or bust? Embedding entrepreneurship in education in Australia. Education + Training 61(6):737–755. https://doi.org/10.1108/ET-02-2019-0037
33. Maritz A, Nguyen QA, Shrivastava A, Ivanov S (2022) University accelerators and entrepreneurship education in Australia: substantive and symbolic motives. Education + Training, ahead-of-print (ahead-of-print). https://doi.org/10.1108/ET-08-2021-0325
34. Mei H, Ma Z, Jiao S, Chen X, Lv X, Zhan Z (2017) The sustainable personality in entrepreneurship: the relationship between big six personality, entrepreneurial self-efficacy, and entrepreneurial intention in the Chinese context. Sustainability 9(9):1649. https://www.mdpi.com/2071-1050/9/9/1649
35. Melbourne Uo (2021) Master of Entrepreneurship. https://study.unimelb.edu.au/find/courses/graduate/master-of-entrepreneurship/
36. Menning G (1997) Trust, entrepreneurship and development in Surat city India. Ethnos 62(1–2):59–90. https://doi.org/10.1080/00141844.1997.9981544
37. Meoli A, Fini R, Sobrero M, Wiklund J (2020) How entrepreneurial intentions influence entrepreneurial career choices: the moderating influence of social context. J Bus Ventur 35(3):105982. https://doi.org/10.1016/j.jbusvent.2019.105982
38. Miller D (1983) The correlates of entrepreneurship in three types of firms. Manage Sci 29(7):770–791. https://doi.org/10.1287/mnsc.29.7.770
39. Millman C, Matlay H, Liu F (2008) Entrepreneurship education in China: a case study approach. J Small Bus Enterp Dev 15(4):802–815. https://doi.org/10.1108/14626000810917870
40. Mitra J (2000) Nurturing and sustaining entrepreneurship: university, science park, business and government partnership in Australia. Ind High Educ 14(3):183–190. https://doi.org/10.5367/000000000101295039
41. Mok KH (2005) Fostering entrepreneurship: changing role of government and higher education governance in Hong Kong. Res Policy 34(4):537–554. https://doi.org/10.1016/j.respol.2005.03.003
42. Mozahem NA, Adlouni RO (2020) Using entrepreneurial self-efficacy as an indirect measure of entrepreneurial education. Int J Manage Educ 100385. https://doi.org/10.1016/j.ijme.2020.100385
43. Neneh BN (2019) From entrepreneurial intentions to behavior: the role of anticipated regret and proactive personality. J Vocat Behav 112:311–324. https://doi.org/10.1016/j.jvb.2019.04.005
44. Newman A, Obschonka M, Schwarz S, Cohen M, Nielsen I (2019) Entrepreneurial self-efficacy: a systematic review of the literature on its theoretical foundations, measurement, antecedents, and outcomes, and an agenda for future research. J Vocat Behav 110:403–419. https://doi.org/10.1016/j.jvb.2018.05.012
45. Olokundun M, Iyiola O, Ibidunni S, Ogbari M, Falola H, Salau O, … Borishade T (2018) Data article on the effectiveness of entrepreneurship curriculum contents on entrepreneurial interest and knowledge of Nigerian university students. Data Brief 18:60–65. https://doi.org/10.1016/j.dib.2018.03.011
46. Owusu-Mintah SB (2014) Entrepreneurship education and job creation for tourism graduates in Ghana. Education + Training 56(8/9):826–838. https://doi.org/10.1108/ET-01-2014-0001

47. Postigo S, Tamborini MF (2004) Entrepreneurship education in Argentina: the case of the San Andres University. In: Alon I, McLntyre JR (eds) Business education and emerging market economies: perspectives and best practices. Springer US, pp 267–282. https://doi.org/10.1007/1-4020-8072-9_17
48. Radović-Marković M, Živanović B (2019) Fostering green entrepreneurship and women's empowerment through education and banks' investments in tourism: evidence from Serbia. Sustainability 11(23):6826. https://www.mdpi.com/2071-1050/11/23/6826
49. Raposo M, do Paço A, Ferreira J (2008) Entrepreneur's profile: a taxonomy of attributes and motivations of university students. J Small Bus Enterp Develop 15(2):405–418. https://doi.org/10.1108/14626000810871763
50. Rauch A, Hulsink W (2015) Putting entrepreneurship education where the intention to act lies: an investigation into the impact of entrepreneurship education on entrepreneurial behavior. Acad Manage Learn Educ 14(2):187–204. https://doi.org/10.5465/amle.2012.0293
51. Santos Susana C, Liguori Eric W (2019) Entrepreneurial self-efficacy and intentions: outcome expectations as mediator and subjective norms as moderator. Int J Entrepreneurial Behav Res 26(3):400–415. https://doi.org/10.1108/IJEBR-07-2019-0436
52. School AB (2021) Entrepreneurship. https://abs.uva.nl/content/masters/entrepreneurship/entrepreneurship.html?cb
53. Shahab Y, Chengang Y, Arbizu Angel D, Haider Muhammad J (2019) Entrepreneurial self-efficacy and intention: do entrepreneurial creativity and education matter? Int J Entrep Behav Res 25(2):259–280. https://doi.org/10.1108/IJEBR-12-2017-0522
54. Shi L, Yao X, Wu W (2020) Perceived university support, entrepreneurial self-efficacy, heterogeneous entrepreneurial intentions in entrepreneurship education. J Entrepreneurship Emerg Econ 12(2):205–230. https://doi.org/10.1108/JEEE-04-2019-0040
55. Silajdžić I, Kurtagić SM, Vučijak B (2015) Green entrepreneurship in transition economies: a case study of Bosnia and Herzegovina. J Clean Prod 88:376–384. https://doi.org/10.1016/j.jclepro.2014.07.004
56. Soomro RB, Mirani IA, Sajid Ali M, Marvi S (2020) Exploring the green purchasing behavior of young generation in Pakistan: opportunities for green entrepreneurship. Asia Pac J Innov Entrepreneurship 14(3):289–302. https://doi.org/10.1108/APJIE-12-2019-0093
57. Soria-Barreto K, Zúñiga-Jara S, Ruiz Campo S (2016) Determinantes de la intención emprendedora: nueva evidencia. Interciencia 41(5):325–329
58. Suzuki K-I, Kim S-H, Bae Z-T (2002) Entrepreneurship in Japan and Silicon Valley: a comparative study. Technovation 22(10):595–606. https://doi.org/10.1016/S0166-4972(01)00099-2
59. Tamkivi R (1999) Support structures for innovation and research-based entrepreneurship in Estonia. Ind High Educ 13(1):46–53. https://doi.org/10.1177/095042229901300107
60. van der Sijde PC, van Alsté JA (1998) Support for entrepreneurship at the University of Twente. Ind High Educ 12(6):367–372. https://doi.org/10.1177/095042229801200607
61. Wardana LW, Narmaditya BS, Wibowo A, Mahendra AM, Wibowo NA, Harwida G, Rohman AN (2020) The impact of entrepreneurship education and students' entrepreneurial mindset: the mediating role of attitude and self-efficacy. Heliyon 6(9):e04922. https://doi.org/10.1016/j.heliyon.2020.e04922
62. Watkins D, Stone G (1999) Entrepreneurship education in UK HEIs: origins, development and trends. Ind High Educ 13(6):382–389. https://doi.org/10.5367/000000099101294726
63. Wegner D, Thomas E, Teixeira Eduardo K, Maehler Alisson E (2019) University entrepreneurial push strategy and students' entrepreneurial intention. Int J Entrep Behav Res 26(2):307–325. https://doi.org/10.1108/IJEBR-10-2018-0648
64. Yan X, Gu D, Liang C, Zhao S, Lu W (2018) Fostering sustainable entrepreneurs: evidence from china college students' "internet plus" innovation and entrepreneurship competition (CSIPC). Sustainability 10(9):3335. https://www.mdpi.com/2071-1050/10/9/3335
65. Yu Cheng M, Sei Chan W, Mahmood A (2009) The effectiveness of entrepreneurship education in Malaysia. Education + Training 51(7):555–566. https://doi.org/10.1108/00400910910992754

Printed in the United States
by Baker & Taylor Publisher Services